Molecular Evolution of Physiological Processes

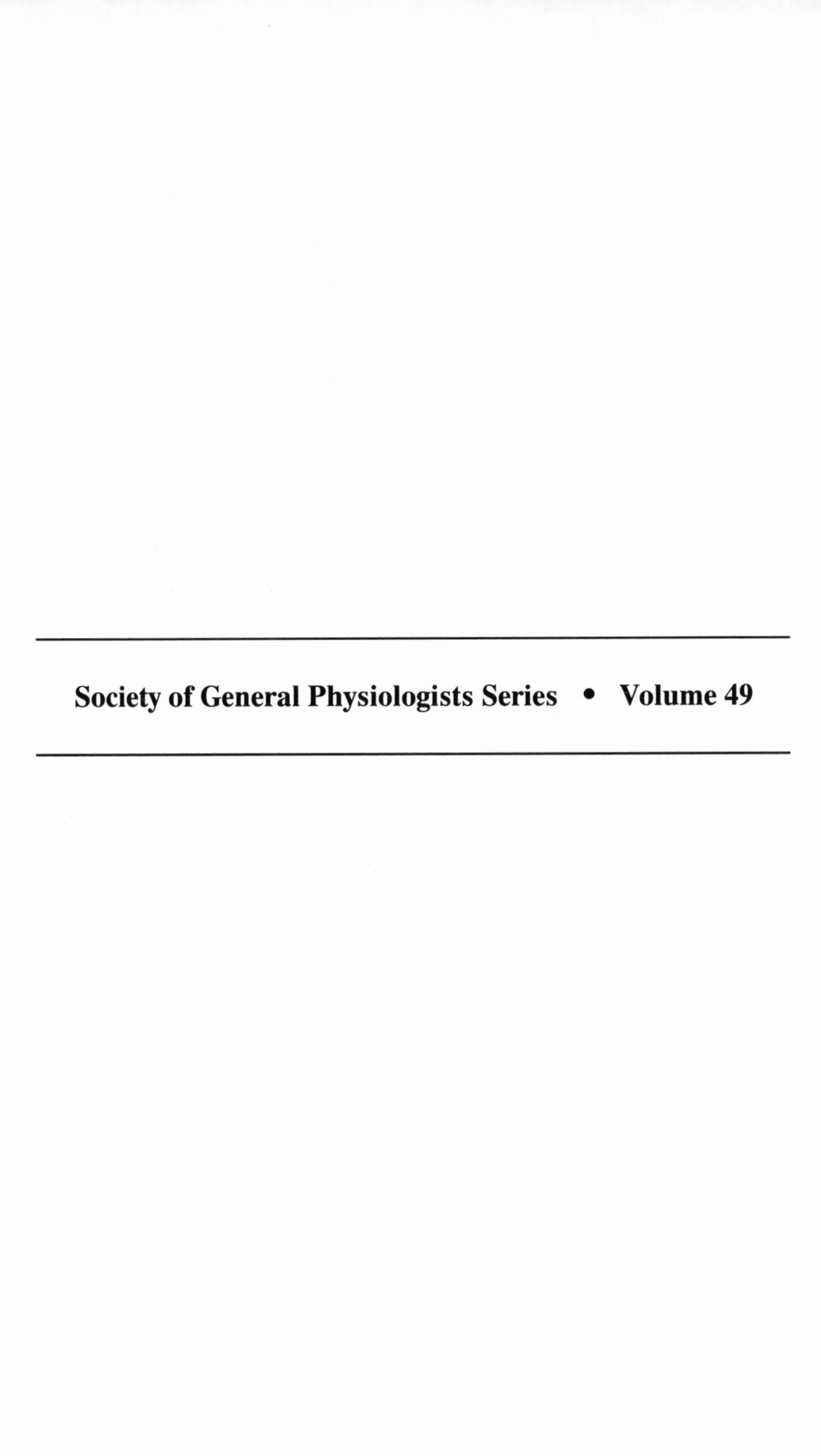

Society of General Physiologists Series • Volume 49

Molecular Evolution of Physiological Processes

Society of General Physiologists • 47th Annual Symposium

Edited by
Douglas M. Fambrough
The Johns Hopkins University, Baltimore, Maryland

Marine Biological Laboratory
Woods Hole, Massachusetts

8–11 September 1993

New York

Library of Congress Catalog Card Number 94-065642
ISBN 0-87470-055-8
Printed in the United States of America

Contents

Preface *ix*

Chapter 1 **Interpretation of Diversity: Homology Testing and Generation of Phylogenetic Trees**

Molecular Clocks Are Not as Bad as You Think
Walter M. Fitch and Francisco J. Ayala 3

Reconstructing the History of Evolutionary Processes Using Maximum Likelihood
Michael J. Sanderson 13

Introns, the Broken Transposons
Andrew J. Roger, Patrick J. Keeling, and W. Ford Doolittle 27

Monophyletic Origin of Animals: A Shared Ancestry with the Fungi
Patricia O. Wainright, David J. Patterson, and Mitchell L. Sogin 39

To Tree the Truth: Biological and Numerical Simulations of Phylogeny
David M. Hillis and John P. Huelsenbeck 55

Chapter 2 **Genome Organization and the Generation of Diversity**

Transposable Elements and the Evolution of Heterochromatin
Allan Spradling 69

Gene Conversion and the Evolution of Gene Families in Mammals
John C. Schimenti 85

Functional Polymorphism in Class I Major Histocompatibility Complex Genes
Peter Parham 93

The Generation of Protein Isoform Diversity by Alternative RNA Splicing
David M. Helfman 105

Chapter 3 **Evolution of the Actin and Myosin-related Families of Cytoskeletal Proteins**

Prolifins, Ancient Actin Binding Proteins with Highly Divergent Primary Structures
Thomas D. Pollard and Stephen Quirk 117

The Evolution of the Chicken Sarcomeric Myosin Heavy Chain Multigene Family
Everett Bandman, Laurie A. Moore, Maria Jesús Arrizubieta, William E. Tidyman, Lauri Herman, and Macdonald Wick 129

Molecular Evolution of the Myosin Superfamily: Application of Phylogenetic Techniques to Cell Biological Questions
Holly V. Goodson 141

Diversity of Myosin-based Motility: Multiple Genes and Functions
Alison Weiss, D. C. Ghislaine Mayer, and Leslie Leinwand 159

The Actin Protein Superfamily
Christine Fyrberg, Liza Ryan, Lisa McNally, Maura Kenton, and Eric Fyrberg 173

Chapter 4 **Structure, Function, Evolution of Plasma Membrane Proteins**

Ion Channels of Microbes
Yoshiro Saimi, Boris Martinac, Robin R. Preston, Xin-Liang Zhou, Sergei Sukharev, Paul Blount, and Ching Kung 179

Using Sequence Homology to Analyze the Structure and Function of Voltage-gated Ion Channels
H. Robert Guy and Stewart R. Durell 197

Molecular Evolution of K^+ Channels in Primative Eukaryotes
Timothy Jegla and Lawrence Salkoff 213

The Connexins and Their Family Tree
Michael V. L. Bennett, Xin Zheng, and Mitchell Sogin 223

Relationships of G-Protein–coupled Receptors. A Survey with the Photoreceptor Opsin Subfamily
Meredithe L. Applebury 235

Evolution of the G Protein Alpha Subunit Multigene Family
Thomas M. Wilke and Shozo Yokoyama 249

Molecular Evolution of the Calcium-transporting ATPases Analyzed by the Maximum Parsimony Method
Yan Song and Douglas M. Fambrough **271**

Appendix **285**

List of Contributors **287**

Subject Index **291**

Preface

This volume is a collection of contributions from the 47th Annual Symposium of the Society of General Physiologists, held September 8–11, 1993 at the Marine Biological Laboratory in Woods Hole, Massachusetts. The Symposium brought together scientists who study cytoskeletal and membrane functions with scientists whose research is focused upon molecular evolution or on mechanisms that generate genetic diversity. The theme for the 47th Symposium reflects a revolution in the study of cellular physiology. A decade ago most of the macromolecules that mediate cellular processes were known only indirectly through their functions in transport, motility, signal transduction, and so on. The development of molecular biological techniques made possible a definition of these macromolecules in terms of the nucleotide sequences of their encoding DNAs and RNAs. Among the surprises coming from the new molecular information was the discovery that nearly all cellular functions are performed by macromolecules that are members of gene families. Thus, commonalities in mechanism might be deduced from similarities in primary structure of macromolecules subserving related functions in different cell types or in different organisms. How best to evaluate these similarities and derive useful deductions from them? This is a problem currently being faced by many cell physiologists.

Nucleotide and amino acid sequences are character-rich and invite quantitative analyses. Such sequences have become major grist for studies of phylogenetic relationships, and the marriage of studies in molecular evolution with paleontological studies is among the truly exciting developments in modern biology. As contributions in this symposium volume attest, however, the deduction of phylogenetic and molecular relationships is fraught with complications, and quantitative approaches to such deduction have become exceedingly sophisticated. Some of the analytical problems that phylogeneticists face are also faced by cell physiologists interested in using comparative molecular structure to guide them in their studies of structure-function relationships. Given the progress molecular phylogeneticists have made in character analysis, the time seemed ripe for a meeting of experts in this discipline with representatives of the new molecular physiology. This volume well represents the status of the respective fields and indicates the extent to which cross-stimulation is necessary to resolve the major scientific issues.

We are grateful to Richard Harrison, Mitchell Sogin, and a large number of colleagues at the Johns Hopkins University for their advice during the planning stages of the meeting. Thanks also go to Jane Leighton and Sue Lahr for their continuous, tireless efforts, to the members of the Council of the Society of General Physiologists who were of enormous help in all phases, and to the director and staff of the Marine Biological Laboratory. We also thank all the participants in the meeting for providing the intellectual excitement that is reflected in this volume. Also, we wish to thank Carol Toscano at the Rockefeller University Press for her work in getting the proceedings published.

The symposium would not have been possible without the financial support of the Sloan Foundation and the National Science Foundation. An early commitment from the Sloan Foundation made the planning of the meeting a totally enjoyable task, free from financial worries. Support from the National Science Foundation allowed a number of young scientists to participate in the meeting.

Douglas M. Fambrough

Chapter 1

Interpretation of Diversity: Homology Testing and Generation of Phylogenetic Trees

Molecular Clocks Are Not as Bad as You Think

W. M. Fitch and F. J. Ayala

Department of Ecology and Evolutionary Biology, University of California at Irvine, Irvine, California 92717

In this paper we intend to show that whether a molecular clock seems to be accurate or not may well depend upon whether all the biological factors that might affect the clock have been taken into account. Such factors include the answer to questions like: (*a*) Can a site vary more than once? (*b*) How many ways can a site vary? (*c*) How many sites can vary? (*d*) Do variable sites sometimes become invariable and vice versa?

We will cover each of these questions in turn, ending with a study of an enzyme, the Cu-Zn superoxide dismutase (hereafter represented SOD), that is reputed to be a bad clock and show that, when all the biological factors are taken into account, SOD may be a good clock.

As this material is intended for an audience not particularly sophisticated with respect to molecular clocks, we shall start with definition. We shall assume knowledge of what DNA is, how mutations change DNA sequence, and how mutants may affect gene products.

A "clock" requires the periodic occurrence of observable events. A molecular clock will utilize molecular events that may be either "nucleotide substitutions" or "amino acid replacements". If these events occur with some regularity, there is a potential clock. While we may suppose that a clock exists in any particular case, it is a supposition subject to verification. This paper identifies factors that enter in that verification.

One should distinguish between replacements (substitutions) and mutations. A "mutation" is an alteration of the DNA in an individual, a "replacement" is an amino acid change (and a "substitution" a nucleotide change) that has spread throughout the population, the species. There are normally billions of mutations for every replacement (and for every substitution), and so they are definitely different phenomena.

Before we can test the regularity of replacements, we must be able to observe them. In fact, substitutions and replacements are unlikely to be observed. We can, however, observe "differences," and these imply that replacements (or substitutions) have occurred. If we line up the superoxide dismutase from humans with that from the rat, we will observe that the two sequences have different amino acids in 33 of the positions. These are differences and should be carefully distinguished from the replacements that we might hope are clocklike. True enough, each difference observed had at least one replacement in its history (as well as a substitution underlying that) but to treat them as identical is to assume that a replacement has never occurred twice at any given site. If one organism has an alanine at a position at which another organism has a leucine, they have one difference but their common ancestor might have had a valine at that position and, therefore, at least two replacements have occurred, one in each lineage. Even if the ancestor had one of the two amino acids (say, leucine), it is likely that there was an intermediate replacement (leucine by valine) before the replacement that yields alanine in one of the lineages. The point to keep in mind is that we observe differences whereas what we want to count are replacements.

There is a simple solution to this problem. If the next replacement occurs at a site that is essentially random with respect to previous replacements, then the distribution of replacements should be Poisson. Thus, if p_0 is the fraction of sites without differences between two sequences, then

$$r = \ln p_0, \qquad (1)$$

where r is the average number of replacements/site. If n is the number of sites, then the total number of replacements in both lineages since their common ancestor is rn.

Returning to the SOD example, there are 33 differences between human and rat out of 151 aligned positions. Thus, assuming all positions are variable, $p_0 = (151 - 33)/151 = 0.781$ and therefore, $r = 0.247$ replacements/site. The estimated total number of replacements is $151 \times 0.247 = 37.2$ which represents a 13% increase over the 33 differences observed.

We now move to question *b*. As we consider increasingly more distant relationships, there is an increasing probability that positions that do not differ have nevertheless been altered two or more times since their last common ancestor. For example, the common ancestor might be valine. After separation of the two lineages, a leucine replaced the valine at that position in each lineage. Thus, the sequences show zero differences in their character state (in the present amino acid at that site) but there have been two replacements in that character (site) since their last common ancestor.

This problem too can be simply handled, as shown by Jukes and Cantor (1969). For simplicity, we will first consider substitutions rather than replacements. We note that two random nucleotide sequences would be expected to match, if each nucleotide is equally probable, in 25% of their positions. This is simply because there is one chance in four that a randomly drawn nucleotide will match a given nucleotide, because only four are possible. Thus, divergence will approach 75% mismatch asymptotically as the maximum possible. Let δ = the fraction of sites that don't match $(1 - p_0)$. Then

$$r = -(3/4) \ln (1 - 4\delta/3), \qquad (2)$$

where δ, the fraction of sites that differ, equals d/n, the number of differences divided by the length of the sequence. Eq. 2 would be the same as Eq. 1 if the 3/4 and 4/3 were each 1 which would be approximately the case if, instead of only four kinds of nucleotides, there were very many. In Eq. 2, the units of r are substitutions/nucleotide site. For amino acid sequences, the fractions become 19/20 and 20/19 because there are 20 possible kinds of amino acids encoded:

$$r = -(19/20) \ln (1 - 20\delta/19), \qquad (3)$$

where r is now in replacements/codon site as in Eq. 1. For example, using the SOD data again, $\delta = d/n = 33/151$ and therefore, $r = (19/20) \ln (1 - 20\delta/19) = 0.248$ and the number of replacements becomes 37.5. This additional correction (compared to

TABLE I
Number of Codons That Change 0, 1, 2, 3 or More Times Among SOD Sequences

Number of substitutions	0	1	2	≥3
Expected_1	127.1	21.9	1.9	0.1
Observed	134	10	5	2
Expected_2	134.1	10.3	4.8	1.9

Expected_1 is the distribution of changes expected if all 151 codons are variable. Observed is the number of codons observed. Expected_2 is the distribution of changes expected if only 28 of the 151 are variable. In this case the zero class comprise 11.1 variable codons that by chance would not be expected to change plus 123 codons that were assumed to be invariable.

the correction in Eq. 2) is small, but becomes greater as δ increases and, therefore, has a large effect when the fraction of sites that are different is large.

Question *c* asks whether all sites are variable. Is it possible that some amino acids are so crucial that they cannot be changed without a debilitating loss of function sufficiently severe that selection will weed out the mutant? It is easy to imagine that the initiating methionine cannot be replaced without the complete loss of the protein, and additional examples are readily conjured up. Are such invariable positions present in significant numbers?

To answer this question we must first notice some features of the "parsimony method" of tree-making. The procedure asks, "what tree would permit relating these sequences to a common ancestor with the fewest number of substitutions (or the least amount of whichever kind of change is used, such as, for example, morphological)?" The procedure not only yields the relationship, but estimates how many substitutions are minimally required for each position of the sequence. These estimates allow one to obtain a picture of how the substitutions are spread over the gene and permits us to ask whether their distribution is random, i.e., Poisson.

SOD will serve as an example. In this case, the amino acid sequences are the data for the study, but we back-translate from the amino acids to their (ambiguous) codons. Thus, valine becomes the trinucleotide, GUN, as we switch from replacements to substitutions. "Replacement substitutions" is the name given by molecular biologists to the nucleotide substitutions that bring about amino acid replacements.

Consider a small portion of the SOD tree that involves only a few *Drosophila* species. Table I shows how many times each codon has changed one of its nucleotides. The first row shows the codon changes after 26 substitutions: 134 have not

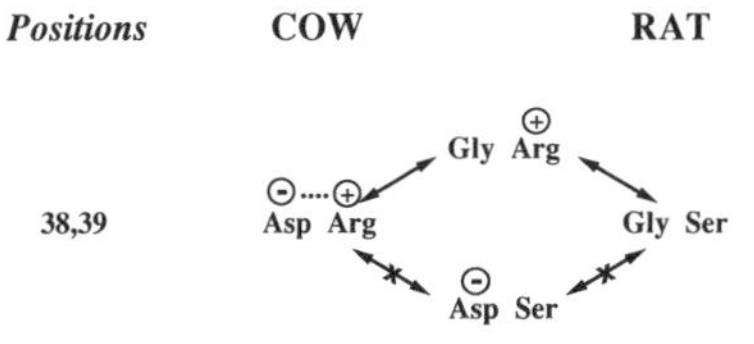

Figure 1. Amino acid site variability. Shown on the left and right are the amino acids present in positions 38 and 39 of the ribonuclease of cow and rat, respectively. In the middle are the two alternative intermediate stages required if replacements occur, generally one amino acid at a time. The lower intermediate would be ruled out if the positively charged arginine is required to protect the vital, positively charged lysine 41 from the deleterious effects of the negatively charged aspartate. See text for further details.

changed, 10 have changed only once, five have changed twice, and two have changed three times. What should we expect these numbers to be if the substitutions were randomly distributed? The expected numbers, shown in the second line, match very poorly and a test (Poisson dispersion statistic; see Markowitz, 1970) of the goodness of fit, indicates a probability of $< 10^{-6}$ that the fit would be worse than the one shown, if the distribution were random. The reason could be that some of the positions are

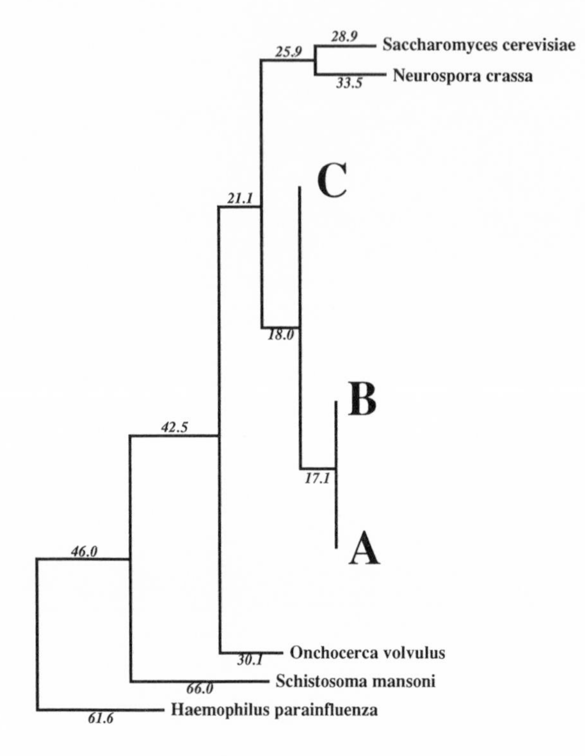

Figure 2. Superoxide dismutase tree. The amino acid sequences (back translated to ambiguous codons) were fit most parsimoniously to this tree (which is 1,300 nucleotide substitutions long) by the method of Fitch (1971) and Fitch and Farris (1974). It is not the most parsimonious tree which was not used for reasons discussed in the text. The complete tree is obtained by joining parts *A, B,* and *C* to the portion showing the outgroups. The GenBank numbers for these sequences are as follows: *Homo Sapiens* (human) [A00512]; *Oryctolagus cuniculus* (rabbit) [S01134]; *Rattus norvegicus* (rat) [P07632]; *Mus musculus* (mouse) [JQ0915]; *Sus scrofa* (pig) [A00514]; *Bos primigenius taurus* (bovine) [A00513]; *Ovis aries* (sheep) [P09670]; *Equus caballus* (horse) [A00515]; *Xenopus laevis* (African clawed toad-A) [S05021]; (African clawed toad-B) [S05022]; (African clawed toad) [S09568]; *Xiphias gladius* (swordfish) [P03946]; *Prionace glauca* (blue shark) [S04623]; *Pinus sylvestris* (Scots Pine-1) [S84896]; *Oryza sativa* (rice-1) [S00999]; (rice-2) [D01000]; *Zea mays* (maize) [A29077]; *Brassica oleracea* (white cabbage) [A25569]; *Lycopersicon esculentum* (tomato-1) [S08350]; *Spinacia oleracea* (spinach-1) [P22233]; *Pisum sativum* (garden pea-1) [M63003]; (Scots Pine-2) [S84902]; (spinach-2) [JS0011]; (garden pea-2) [S12313]; (tomato-2) [S08497]; *Petunia hybrida* (garden petunia) [P10792]; *Saccharomyces cerevisiae* (yeast) [A36171]; *Onchocerca volvulus* (nematode) [X57105]; *Schistosoma mansoni* (liver fluke) [A37029]; *Neurospora crassa* [M58687]; *Haemophilus parainfluenza* [M84013].

invariable. Indeed, if we assume that 123 positions are invariable and that the 26 substitutions are randomly distributed among the remaining 28 codons, one gets the distribution shown in the bottom row of the table. This is an excellent fit to the data, so good that 81% is the probability of getting a worse fit by chance.

The bad fit if we assume that all 151 codons are equally variable, and the excellent fit obtained by assuming that only 28 codons are variable (in this small group), demonstrates that some positions, perhaps many, may not tolerate replacing the encoded amino acid. What is the effect of this difference on computations of the number of replacements since the common ancestor of the two most divergent members of that group of *Drosophila?* They are *D. hydei* and *D. busckii* and they differ by nine amino acids. Treating differences as replacements, there were nine replacements. Using Eq. 3 to correct for multiple replacements/site we get:

$$r = -151(19/20) \ln [1 - (20/19)(9/151)] = 9.3 \text{ replacements.}$$

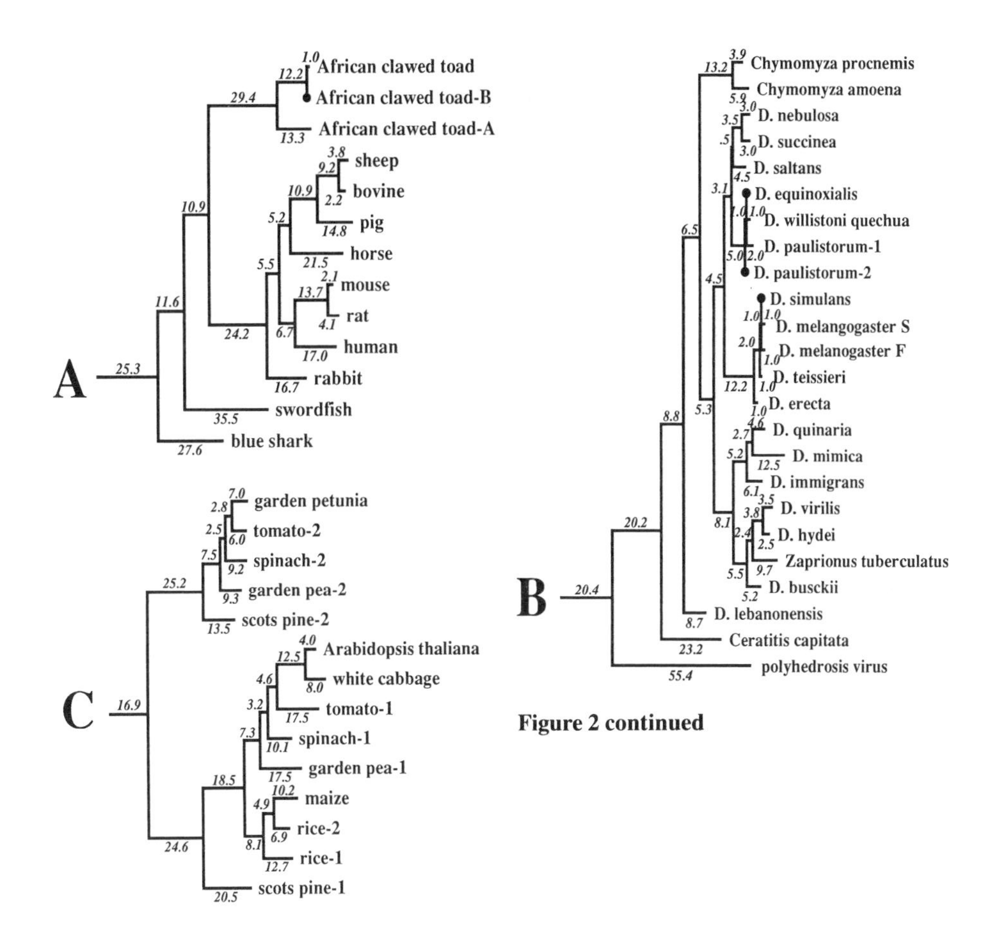

Figure 2 continued

If we now correct for the fact that only 28 of the sites appear to be variable, we get:

$$r = -28(19/20) \ln [1 - (20/19)(9/28)] = 11.6 \text{ replacements.}$$

This is an 8.7-fold increase in the amount of the correction. As δ increases, the magnitude of the correction is even greater.

Question *d* asks whether the codons that are invariable ever become variable (and vice versa) during the course of evolution. An affirmative answer is easily imagined. Consider the enzyme ribonuclease which has an essential amino acid, lysine at position 41, in the active site of the enzyme. Suppose that in the bovine ribonuclease (see Fig. 1, *left side*), the negatively-charged aspartate in position 38 would draw the positively-charged lysine 41 out of its normal position, were it not for the fact that the positively-charged arginine at position 39 makes an ionic bond with the aspartate (Wyckoff, 1968). Thus, the arginine is required for protection of the lysine and could thus be invariable. The aspartate, however, might be variable and change to glycine as shown in the upper middle part of Fig. 1. Once that happened, the arginine is no longer needed for protection, so that it could change, for example, to glutamine and produce the configuration found in rat ribonuclease (right side of Fig. 1). If this scenario is correct, the intermediate shown in the lower middle part of Fig. 1 would not be possible. After publication of Fig. 1 (Fitch and Markowitz, 1970) the sequence of the pig ribonuclease was found to have glycine and arginine at

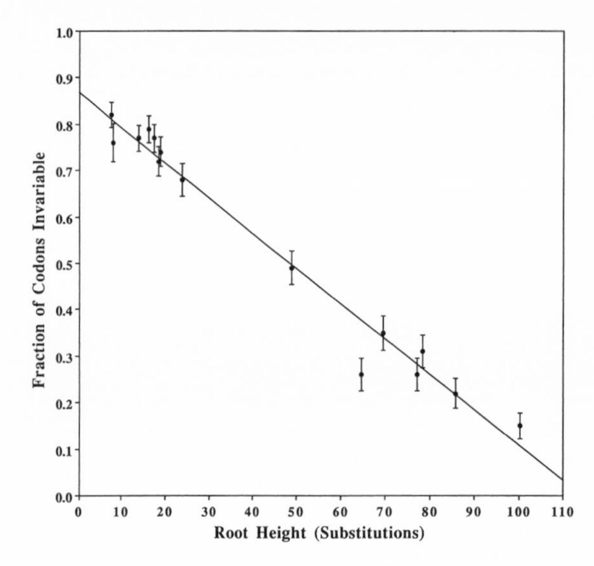

Figure 3. Estimation of number of covarions. For each clade in the tree in Fig. 2, we counted the number of codons subject to zero, one, two, . . . substitutions and fitted these data to a two-Poisson model by the method of Fitch and Markowitz (1970). One Poisson fits the varied codons and thus estimates a number of *variable but unvaried* codons. This number, subtracted from the number of codons with zero changes is the number of *invariable* codons which, as a fraction of the total, is plotted on the y-axis. On the x-axis is plotted the weighted average of the number of substitutions from the root of the clade to the descendants of that root. The y-intercept is the estimated fraction of codons that are not covarions. The vertical bars represent one standard deviation of the estimate of the fraction of invariable codons.

positions 38 and 39, as in the predicted intermediate, which confirms the plausibility of this example. Be that as it may, the point is that the evolutionary interchange of variability and invariability at a site is functionally quite reasonable.

We return now to the SOD data and ask whether the 28 variable codons estimated above for the subset of *Drosophila* species is typical for other regions of the tree. The answer is no. We have examined the tree shown in Fig. 2 and estimated the number of invariable codons for every clade in the tree. This number, expressed as a fraction of the total number of positions, is plotted against the range of species comprised by successively more inclusive clades; as the range widens, the node defining that clade gets closer to the root. The weighted average number of substitutions to the node defining each clade is used as the measure of the range (Fig. 3).

Fig. 3 shows that, as the range of species widens, the number of invariable sites decreases. We attribute this to the process just discussed by which positions that are not variable at one point in time may become variable later on, because replacements elsewhere alter what can or cannot occur at a given site.

Fig. 3 manifests a reasonably linear relation between the fraction of invariable sites and the range of species. There is no known reason why this should be so, but it makes it possible to extrapolate to the zero range of species and infer that 87% of the amino acid positions are invariable at any one point in time in any one species. This result in turn yields the estimate that only 21 codons are variable at any one time—a number that no doubt is not constant but represents a mean value. It may be noted that 21 does not compute as 13% of 151, because for the larger data set there are eight more positions present that do not occur in the animals and fungi. But if replacements can occur in only 21 codons (in any one species at any one point in time), then the occurrence of sequences that differ at, say, 69 positions, as human and yeast sequences do, implies that the set of variable sites changes over time. The set of codons that are variable at any one point in time in a particular species are called *co*comitantly *vari*able cod*ons,* or covarions for short.

We are now prepared to examine how bad is the SOD clock. Ayala (1986) examined SOD at a time when only eight sequences were known. He corrected the number of pairwise amino acid differences in order to account for superimposed replacements. The inferred rates of change implied by the data were 5.5 replacements/100 million years when animals and fungi were compared, but 27.8 replacements/100 million years when mammals were compared to each other, a greater than fivefold difference. The difference could in principle be the result of errors in setting the dates of common ancestry. But it is unreasonable to assume either that the set of mammals diverged more than 300 million years ago, or that the common ancestor of animals and fungi occurred as recently as 240 million years ago. The data would seem to justify the conclusion that the SOD clock is bad (Ayala, 1986). But, we shall argue here that the clock may be good and still yield pairwise differences apparently as discordant as those observed in the SOD data. The concept of covarions just presented will be central to the argument.

There is no simple formula to calculate how many changes have occurred between two sequences, based only on the number of differences, if the situation is as complicated as we have described above. We can, however, try to simulate the situation to ascertain whether we can create a model that gives differences comparable to those observed, which would lead us to conclude that our understanding of the process is reasonably good.

The SOD sequences in Fig. 2 require that the following features be present in the model. The sequence has 159 codons, 21 covarions, and incorporates changes at a clocklike rate. The model must yield differences comparable to those observed for SOD sequences, for any pair of species that diverged any particular number of years ago. Table II gives the observed differences, for different divergence times. The only variables in the model that we are free to determine are: (*a*) the rate of change in replacements/unit time; (*b*) the number of permanently invariable codons; (*c*) the rate at which codons comprising the covarions may interchange with the other presently invariable codons (i.e., with the other codons, except those permanently invariable); and (*d*) the number of alternative amino acids allowed at a variable site.

Table II shows (last column) the number of differences obtained in the model using six replacements/10 million years, 42 permanently invariable codons (giving 159 codons − 42 permanently invariable codons =117 codons, of which 21 are covarions and 96 are currently invariable codons), and a persistence (of covariability) of 0.01. A persistence of 0.01 means that the probability that the set of covarions will remain unaltered, after a replacement occurs in a lineage, is 1%. If the covarion set is to be altered, a single, randomly chosen covarion is switched with a single, randomly chosen codon, among the other currently invariable codons. The number of alternative amino acids at a site was set to average 2.5. The simulated differences obtained under this model should approximate those observed for SOD immediately to their left in Table II. To the degree that one feels that these pairs of numbers match, to

TABLE II
Amino Acid Replacements Between SOD Sequences from Different Species and Paleontological Dates Used

Sister Groups	Mya	Replacements	Observed SOD	Simulated
D. nebulosa–melanogaster	55 ± 5	33	18 ± 2	17 ± 3
D. hydei–melanogaster	60 ± ?	36	19 ± 3	17 ± 2
Chymomyza–melanogaster	65 ± ?	39	23 ± 2	19 ± 3
H. sapiens–B. taurus	70 ± 10	42	26 ± 3	20 ± 3
Ceratitis–D. melanogaster	100 ± ?	60	31 ± 2	24 ± 4
Monocot–Dicot	125 ± ?	74	29 ± 2	29 ± 3
Angiosperm–Gymnosperm	220 ± ?	132	33 ± 4	40 ± 4
Frog–Mammal	350 ± 1	210	48 ± 2	52 ± 4
Fish–Tetrapod	400 ± 2	240	45 ± 4	55 ± 5
Yeast–Neurospora	?		46	
Insect–Vertebrate	580 ± 2	348	56 ± 12	62 ± 5
Fungi–Metazoan	1,000 ± ?	600	67 ± 4	65 ± 6

Mya: millions of years ago. Replacements: numbers used in the simulation, which are equivalent to six replacements every 10 million years. Observed SOD: average number of observed differences between species from the two sister groups shown. The sister group names, e.g., *nebulosa-melanogaster*, should be understood as indicating the groups to which these species belong and not just two particular species. Simulated: estimated number of amino acid differences that would be observed by using the parameters of the model. The plus/minus values are crude estimates of error for Mya, but are standard deviations for observed and simulated differences. The simulated values are, in every case, the average for 40 simulated trials.

that degree one can accept that the SOD values might arise from a perfect stochastic clock because the simulated results do arise from such a clock.

Fig. 4 represents an alternative presentation of the same results. The filled circles represent the observed amino acid differences as a function of estimated paleontological times of divergence; they are clearly nonlinear. The open squares show the number of replacement differences obtained by simulation by means of the perfect clock.

We have not attempted to get an optimum fit of the simulated data to the observed SOD data, but rather were satisfied after trying several values of each of the four parameters and getting a reasonable fit. This, after all, is all that is required to show that the bad SOD clock might possibly be a good clock if we just make certain corrections that are biologically reasonable. A perfect match, one may note, is not really possible given that the model is monotonic (more years, more differences) whereas the observations are not. For example, there are fewer SOD differences at 400 Mya than at 350 Mya, although not significantly so. Thus, we do not wish to claim that the precise values used of a persistence of 0.01 and 42 permanently invariable codons, are definitive for the fungi-plant-metazoan SOD's.

We conclude with three broad observations. One is that the close correspondence between the observed and simulated differences demonstrates that pairwise differences that seem grossly nonclocklike may seem so because of a failure to account for relevant biological considerations.

The second is that the good fit of the simulated to the observed values lends further support, if more is needed, of the utility of the covarion concept.

The third is that the amount of divergence (number of replacements) estimated, even after the more customary corrections, may fall very short of the actual number

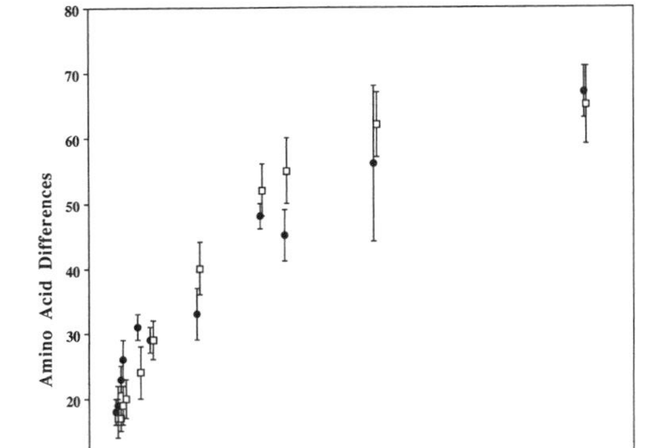

Figure 4. Comparison of observed SOD differences with those from a simulated perfect clock. The closed circles are observed differences in SOD, the open squares are simulated with a perfect clock (see text for clock parameters). The vertical bars show one standard deviation about the mean. Observed and simulated values are for the same date, although they are horizontally offset slightly for clarity.

whenever the number of covarions is small. For example, the 67 observed differences between metazoans and fungi becomes 89 replacements when corrected by Eq. 3. However, the actual number in the simulation was 600, a 6.8-fold difference.

An update of these analyses, employing more sequences is now in press (Fitch and Ayala, 1994).

Acknowledgements

We appreciate the computational assistance of Ms. Helene Van.

This work was supported by NIH grant GM42397 to F.J. Ayala and W.M. Fitch, and NSF grant DEB-9096152 to W.M. Fitch.

References

Ayala, F. J. 1986. On the virtues and pitfalls of the molecular evolutionary clock. *Journal of Heredity*. 77:226–235.

Fitch, W. M. 1971. Toward defining the course of evolution: minimum change for a specific tree topology. *Systematic Zoology.* 20:406–416.

Fitch, W. M., and F. J. Ayala. 1994. The superoxide dismutase molecular clock revisited. *Proceedings of the National Academy of Sciences, USA.* In press.

Fitch, W. M., and E. Markowitz. 1970. An improved method for determining codon variability in a gene and its application to the rate of fixations of mutations in evolution. *Biochemical Genetics.* 4:579–593.

Jukes, T. H., and C. R. Cantor. 1969. Evolution of protein molecules. *In* Mammalian Protein Metabolism. H. N. Munro, editor. Academic Press, New York. 21–132.

Markowitz, E. 1970. Estimation and testing goodness of fit for some models of codon variability. *Biochemical Genetics.* 4:595–601.

Wyckoff, H. W. 1968. Discussion. *Brookhaven Symposia in Biology.* 21:252.

Reconstructing the History of Evolutionary Processes Using Maximum Likelihood

Michael J. Sanderson

Department of Biology, University of Nevada at Reno, Reno, Nevada 89557

Introduction

The development of quantitative methods of phylogenetic analysis has revolutionized comparative biology (Harvey and Pagel, 1991) and provided powerful tools for molecular evolutionists (Felsenstein, 1982; Swofford and Olsen, 1990). Phylogenetic methods are aimed at estimating the branching pattern of relationships of taxa. The branching pattern typically reflects a historical process of speciation, gene duplication, or some other mechanism of lineage splitting that generates a nested set of groups related by common ancestry. "Taxa" may represent organisms, species, genes, or sets of these entities. Additional evolutionary information, such as the time of the splitting events and the duration of the branches of the tree is often of interest as well. However, divergence of taxa, which provides the signal to phylogenetic algorithms, is often obscured by noise due to evolutionary processes such as convergence, reversal, extinction, hybridization, recombination, lineage sorting, and concerted evolution, to name a few. In short, despite the methodological advances of the last 25 years, phylogeny reconstruction is still a complex and challenging enterprise.

The array of phylogenetic algorithms now readily available testifies to the diversity of opinion regarding the issues involved (Swofford and Olsen, 1990; for a compendium, see Felsenstein, 1993). Various classifications of these algorithms have been suggested, but for the present purpose, it suffices to distinguish between two kinds of algorithms: those that invoke probabilistic models of evolution and those that do not (or need not). The latter includes parsimony and distance methods, despite the latter's reliance on probability models in the calculation of the distance matrix. Distance methods can operate on any distance matrix, regardless of how the matrix is calculated and therefore, probabilistic models are not integral to the way they work.

Maximum likelihood (ML) approaches to phylogenetic inference, on the other hand, are obligately associated with probability models of evolution (Cavalli-Sforza and Edwards, 1967; Felsenstein, 1983). For computational reasons they are not widely used, although recent improvements in algorithms promise to change this (Olsen, Matsuda, Hagstrom, and Overbeek, 1993). Maximum likelihood is a method of estimation widely used in applied statistics (Lehman, 1983). It provides a prescription for how to estimate an unknown parameter of a probability distribution, given a set of observations on that distribution. The prescription relies on a quantity called likelihood, which is defined as follows:

$$L(P|D) = k \text{ Prob } (D|P). \tag{1}$$

The term on the left is the likelihood of the parameters, P, given the data, D, and it is proportional to the probability of observing the data, given the distribution, with its parameters (k is an arbitrary constant). Maximum likelihood estimation chooses that value of P, $\hat{P}$, such that Eq. 1 is a maximum. Thus, $\hat{P}$ is the maximum likelihood estimate, given the data. Fig. 1 illustrates the general scheme of maximum likelihood estimation.

In general, maximum likelihood possesses a number of desirable properties. Maximum likelihood estimates are statistically consistent and efficient. That is, as sample size increases, they converge to the true parameter values, and they have a minimum error variance among all possible estimators. They are also associated with hypothesis tests based on the likelihood ratio, which also has desirable properties. In the limit of large sample size, a simple function of the likelihood ratio converges to the χ^2 distribution, providing a readily calculated significance test.

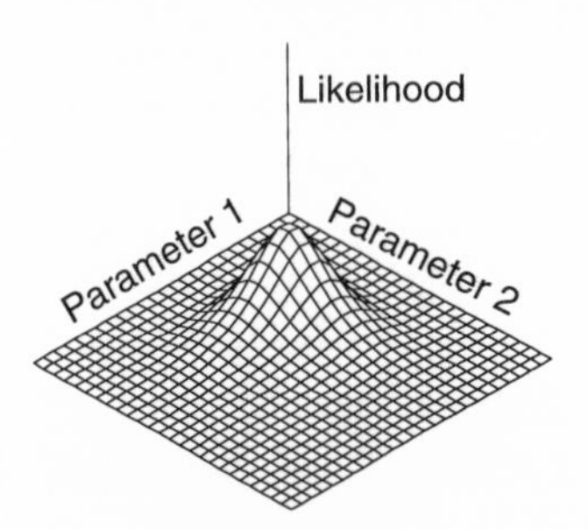

Figure 1. Illustration of maximum likelihood estimation. The surface is the likelihood function, which in this example depends on the data (which is given) and two unknown parameters. Maximum likelihood estimates of the parameters occur at those parameter values that correspond to the peak of the likelihood function. This likelihood function is actually the function used for the gene-duplication model, H^2, discussed later in the text, but the general principles of maximum likelihood estimation are the same regardless of the nature of the function.

In small samples, ML estimates and likelihood ratio tests can be biased or have high average error rates. However, even in small samples, maximum likelihood has an appealing philosophical interpretation. Likelihood is widely regarded as a measure of statistical weight of evidence. A ML estimate can always be regarded as the value of a parameter that corresponds to a maximization of the weight of evidence that a set of data confers on a hypothesis. This is true even if the estimate is biased. Bias is a concept that only applies to repeated trials of the estimation procedure, but for any one set of data, no repetition is feasible. The advocates of likelihood as a method of statistical inference reject the relevance of repeated sampling concepts, preferring the notion of weight of evidence of the data at hand (Edwards, 1972). Despite these statistical arguments over its ultimate justification, there is broad agreement about the power of likelihood methods, and it is not surprising that it was among the very first quantitative tools brought to bear on the issue of phylogenetic inference (Cavalli-Sforza and Edwards, 1967).

Background: Maximum Likelihood in Phylogenetic Inference

Application of maximum likelihood to phylogenetics proved to be a surprisingly difficult problem (see Goldman, 1990, for review). Despite the existence of an essentially complete formulation of the problem for over twenty years (Edwards, 1970; Thompson, 1975), only a subset of this complete problem has been in any sense solved, and even in that case the solutions are numerical rather than analytical. The complexity of the problem can be appreciated by considering the possibly relevant set of unknown parameters that must be estimated. These include: (*a*) a set of rate parameters associated with the stochastic model of character evolution (e.g., probabilities of amino acid or nucleotide substitution per unit time); (*b*) a set of distances between branching points in the tree; and (*c*) the character states in the common ancestor of all the taxa in the tree. A full likelihood solution might also include a set of rate parameters controlling branching and extinction of lineages (Edwards, 1970; Felsenstein, 1973; Thompson, 1975), and it might also include parameters associated with a model of how data are sampled by the investigator (Felsenstein, 1983). Recently, the uncertainties due to sequence alignment have been incorporated by including rate parameters for insertions and deletions in a sequence (Bishop and Thompson, 1986; Thorne, Kishino, and Felsenstein, 1991, 1992).

Given n sequences (or other kinds of data), there are $C_n = (2n-3)!/2^{n-2}(n-2)!$ phylogenies or cladograms (Cavalli-Sforza and Edwards, 1967; Edwards, 1970) and $H_n = (n-1)!n!/2^{n-1}$ possible dendrograms (sensu Page, 1991, or labeled histories sensu Edwards, 1970)—both large numbers whenever n is even moderately sized. A phylogeny is the branching diagram that indicates relative recency of common ancestry. A dendrogram is the set of branching relationships and the sequence of those branching events (see Page, 1991, for further discussion). In principle, the ML estimate of all the unknown parameters could be obtained (numerically at least) by reference to the following rather nasty likelihood equation:

$$L(\mathbf{x}_0, \Lambda, \Phi \mid \mathbf{x}, n) \propto \text{Prob}\,(\mathbf{x}, n \mid \mathbf{x}_0, \Lambda, \Phi)$$
$$= \sum_{H_n} \int_S p(\mathbf{x} \mid n, \mathbf{s}, F, \mathbf{x}_0, \Lambda, \Phi)\, p(n, \mathbf{s}, F \mid \mathbf{x}_0, \Lambda, \Phi)\, ds \qquad (2)$$

(Thompson, 1975; see Goldman, 1990, for a more elementary discussion). This equation requires some explanation. The data or observations consist of a matrix of characters (e.g., sequences) by taxa, $\mathbf{x}$, and the number of terminal taxa, n. The unknown parameters, those quantities to be estimated via maximum likelihood, consist of $\mathbf{x}_0$, the vector of character states of the most recent common ancestor of all n taxa; Λ, a set of branching rates (and possibly extinction rates as well) that governs the branching process; and Φ, a set of rates that governs character evolution, such as the rate of nucleotide substitution in a DNA sequence evolution model (see, e.g., Rodriguez, Oliver, Marin, and Medina, 1990, for discussion of various models, or Nei, 1987, for a more general discussion). Two other quantities in Eq. 2 are random variables: $\mathbf{s}$ is the vector of branch lengths, and F is the tree form or the phylogeny itself. The two terms inside the integral are probability densities. The first is the probability of the data matrix given all the unknown parameters and values of the two random variables. The second is the probability of observing a particular realization of the branching process given the branching rate parameters. The two probabilities in the second line arise from the single probability in the first line via the definition of conditional probability. The integral and sum are the result of considering the marginal probabilities involved in adding the variables F and $\mathbf{s}$, which do not appear in the probability on the first line, but are needed to actually calculate that probability under most models. The integral is over all possible branch lengths. The sum is over all H_n trees. The integral is difficult to evaluate and the sum extends over a potentially vast number of trees. These two factors contribute to the computational complexity of the present formulation. Moreover, the two quantities **s** and F are often the quantities of greatest interest, and yet it is clear under this formulation that they are not quantities that can be "estimated" in the likelihood sense, because they are random variables rather than parameters.

Below it will be seen that a slightly different formulation of the problem does away with this issue, but for now we could proceed to at least discuss the conditional probabilities of the tree shape and branch lengths, given the maximum likelihood estimates of the parameters. Given the various rate estimates, the most probable tree conditional on the data is obtained by substituting the ML estimates of $\mathbf{x}_0$, Λ, and Φ, into the appropriate conditional probability density obtained from Eq. 2 above (Edwards, 1970). However, this complete likelihood formulation has only been investigated for three taxa and one special model of phenotypic evolution (Thompson, 1975). Because of the difficult numerical integrations required, Eq. 2 is difficult to evaluate for one tree, much less for a large number, H_n, of trees, so simplifying assumptions are required.

Historically, the first and foremost simplifying assumption was the abandonment of the branching component in the model. Without some stochastic process that induces a set of prior probabilities on the branch times, **s**, and tree shapes, F, these former random variables become parameters. The parameter estimate for the topology is the tree that has highest likelihood among all trees (this is technically known as maximum relative likelihood and its merits have been debated elsewhere: Felsenstein and Sober, 1986; Sober, 1988). Estimation of the tree by this method is more straightforward than the full-blown method relying on Eq. 2 above, because no numerical integrations are required. Also, iterative methods that do not require evaluation of all possible internal branch times or the likelihood on all H_n trees (Thompson, 1975; Felsenstein, 1981) are available, although it is conceivable that they could also be brought to bear on the more difficult problem of solving Eq. 2. Such heuristic methods are not guaranteed to find the best tree under all conditions, but if the data are sufficiently clean they may provide rough estimates. Even with all these simplifications, however, ML phylogenetic inference is still computationally intense owing to the necessity of simultaneously estimating all the branch times (or in practice the branch lengths, see Felsenstein, 1983), which relies on the invocation of a multivariate numerical optimization algorithm for each tree examined. This is not necessary for parsimony searches because the calculation of branch lengths is neatly decomposed into contributions from each character and efficient algorithms for reconstructing parsimonious evolutionary histories of single characters exist (reviewed in Swofford and Maddison, 1987).

Other, noncomputational, issues can also pose problems. With sufficiently complicated models of character evolution, the desirable properties of maximum likelihood may no longer hold. For example, if rates are allowed to vary among branches and sites in a DNA sequence in an unconstrained way, there is simply not enough information to estimate all the unknown parameters, and the model is no longer guaranteed to be statistically consistent. Models that are too simple, on the other hand, may not provide an adequate goodness of fit to the data. For example, Goldman (1993) found that a molecular clock with equiprobable substitution rates between all nucleotides provided a very poor fit to data on primate ψη-globin pseudogenes. He also rejected even more complicated models that permitted different rates among branches for small subunit RNA genes in a phylogenetic analysis of basal relationships among known life forms. Various recent studies are slowly increasing model complexity by adding variation among sites, the possibility of invariant characters, and ambiguities due to alignment problems (Navidi, Churchill, and von Haeseler, 1991; Thorne et al., 1992). However, if the ultimate goal of a phylogenetic analysis is to address fairly complex evolutionary processes, such as rates of duplication in a multigene family (see below), a complete likelihood treatment would demand the addition of those potentially complex elements of the stochastic formulation on top of the already complex problem of reconstructing phylogenies.

While it is true that many of these problems can and will be solved eventually, it is also true that interesting evolutionary problems abound now, and interim solutions are both feasible and available. In particular, it is practical to use likelihood methods jointly with nonlikelihood phylogenetic algorithms in a kind of hybrid analysis of evolutionary processes.

Argument for a Hybrid Approach to Phylogenetic Studies of Evolution

Despite the raucous philosophical debates that have frequently occluded the agenda of phylogenetic biology, pragmatism has basically held the day. Discovery of the computational complexity of likelihood methods provided an impetus for the adoption of heuristic methods such as distance and parsimony (although these were also advocated on their own merits). However, the desirable properties of likelihood also provided a strong impetus to use it in molecular evolutionary studies, usually in combination with heuristic tree estimation algorithms. Thus, a kind of hybrid approach was born. One of its first incarnations was in the critical problem of rates of molecular evolution in relation to the molecular clock (Langley and Fitch, 1974). That study relied on a parsimony algorithm to reconstruct the tree and the number of substitutions per branch, and then maximum likelihood to estimate times of branching events and rates of evolution. A χ^2 test of goodness of fit of the observed to expected branch lengths permitted a test of the single-rate molecular clock model.

More recently, studies of a host of comparative evolutionary problems have been undertaken in the context of phylogenies generated by parsimony or other methods (Harvey and Pagel, 1991; Brooks and McClennan, 1991). Some of these have invoked maximum likelihood after the fact to address particular issues, including rates of branching (Hey, 1992; Sanderson and Bharathan, 1993; Nee, Mooers, and Harvey, 1992), character correlation (Harvey and Pagel, 1991), and character irreversibility (Sanderson, 1993).

The advantages of an approach that uses maximum likelihood inference on top of a previously derived phylogenetic tree include improved tractability. Moreover, it is possible to take advantage of mature software and volumes of published experience with heuristic algorithms, neither of which is likely to be matched by maximum likelihood algorithms for quite some time. Available ML programs (Felsenstein, 1993; Olsen et al., 1993) are neither as efficient nor as flexible and conducive to data exploration as current parsimony programs, for example (recently reviewed in Sanderson, 1990). The cumulative experience of users of parsimony methods is also orders of magnitude greater, which has meant that fundamental results have emerged regarding the properties of parsimony analyses that would not otherwise have been discovered. The existence and significance of multiple "islands" of equally parsimonious trees is a recent conspicuous example (Maddison, 1991).

On the down side, separating the inference problem into two components, one involving reconstruction of a phylogenetic tree, and one involving reconstruction of some other evolutionary pattern (rates, biases, correlations, constraints, et cetera) can adversely affect the power of the methods and presumably the accuracy of assessments of confidence (see also Maddison and Maddison, 1992, 65–66). Firm conclusions about this issue remain elusive because most hybrid investigations have either not considered the issue of power, error, and confidence at all, or have focused only on the error component associated with the second stage, ignoring the inevitable errors arising from inaccurate phylogenetic reconstruction. Numerical approaches to confidence estimation could be employed rather directly but have not been. For example, bootstrap estimates (Felsenstein, 1985) of sampling variance affecting the underlying phylogeny could be added in as an error term into the evolutionary model used in the second phase of analysis. This could readily be accomplished just by repeating the likelihood analysis on a sample of trees drawn from the bootstrap-resampled trees. The range of estimates that result would serve then as a robust indication of the phylogenetic component of error, and worst-case values of likelihood-based tests (i.e., those that are worst on some bootstrap tree) would serve as better estimates of true error of inference.

An Example: Rates of Gene Duplication

Here I present a worked example of the application of this hybrid approach to reconstruction of rates of diversification in multigene families. Maximum likelihood will be used to estimate parameters of a stochastic branching process model of gene duplication, using data that includes a given phylogeny derived presumably by nonlikelihood methods, and given reconstructions of the phylogenetic history of the duplication events, also probably obtained by other methods such as parsimony. It should be apparent that the actual data in this and similar examples consists only of observations on sequences (or other trait data) in taxa. It is a fiction to regard the tree reconstructed from these data and the ancestral states as actual observations, but a useful fiction that forms the basis of the hybrid approach.

Multigene families are conspicuous and ubiquitous components of eukaryotic genomes (Singer and Berg, 1991). They originate via a process of gene duplication, differentiate as a result of evolutionary forces of selection and drift, or they may converge or stay homogeneous via recombinational mechanisms involved in concerted evolution (Ohta, 1988, 1990). Individual gene copies may also go extinct if they become inactivated as pseudogenes. In many ways the differentiation of a gene family is analogous to the differentiation of a group of species, and presumably it can be modeled in analogous ways (Walsh, 1987). Splitting of lineages has been modeled by simple stochastic processes such as a Yule (1924) or pure birth model (Yule, 1924; Raup, 1985; Sanderson and Bharathan, 1993), or a Galton-Watson branching model (Gilinsky and Good, 1991; Guttorp, 1991). The Yule model is adopted in the following because a continuous-time model seems more natural than the discrete-time Galton-Watson model.

The Yule model is also a good choice for a model of splitting because it postulates that splitting times obey a Poisson process. Poisson processes are often appropriate because (*a*) empirically, they provide excellent fits to many kinds of data (Feller, 1968) and, (*b*) more fundamentally, Poisson processes can be derived from a bare minimum of assumptions—specifically, a constant probability of splitting in any interval, and independence of those intervals. It is difficult to imagine a one-parameter model on the open interval $[0, \infty)$ that could be any simpler than the Poisson. Postulating a Yule model therefore does not make reconstructing branching rates an overly theory-laden procedure.

To avoid some tedious but tractable mathematical complications, it will also be assumed that the approximate times of the duplication events are known. For some well-studied gene families in which the phylogeny is known with some confidence and a fairly rich fossil record is also available, reasonable constraints on the duplication times are available, although concerted evolution may confound estimation of actual duplication times (Goodman, Czelusniak, Moore, Romero-Herrera, and Matsuda,

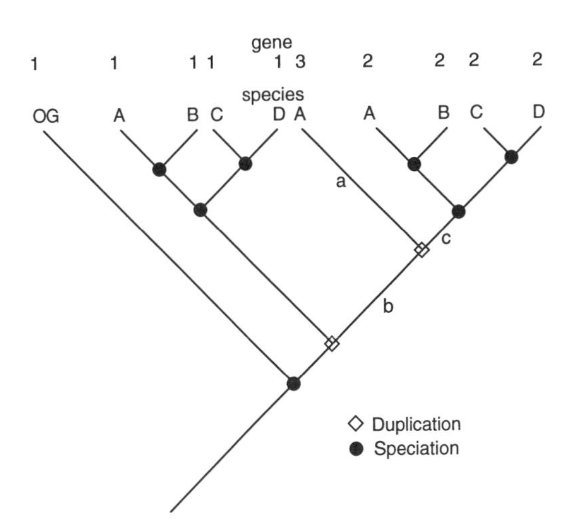

Figure 2. Phylogeny of a hypothetical multigene family with four species (*A–D*), an outgroup species (*OG*), and three genes (*1, 2, 3*). The outgroup has a single gene in the gene family; species *B*, *C*, and *D* have two genes, and species *A* has three genes, entailing two duplication events. Examples of the three kinds of branches discussed in the text are indicated by lower case letters: (*a*) branches ending in a terminal taxon, (*b*) branches ending in a duplication event, and (*c*) branches that end in a speciation event. Distinguishing among these classes of branches is necessary for derivation of the likelihood function for gene duplication.

1979). Elsewhere, methods that do not rely on this information have been discussed (Sanderson and Bharathan, 1993; Sanderson and Donoghue, submitted for publication). These calculate integrated likelihoods using the predicted times of splitting based on the Yule model itself, constrained but not fixed by evidence about the time of occurrence of fossil taxa at the tips of the tree (Novacek and Norell, 1982). This more careful method is preferable but does not add anything to the present discussion.

The data for the present problem consist of a tree, F, a set of times of branching, t_i, associated with node i, branch lengths, d_i, derived from $d_i = t_i - t_{ancestor(i)}$, and the inferred number of duplications, j, on the tree. One complication is that branching arises both from speciation and from gene duplication (Fig. 2). The only times relevant to reconstructing duplication rates are the times between duplication events. This can be demonstrated by the following argument. There are three kinds of branches on a phylogeny of a multigene family (Fig. 2): branches that end in a terminal taxon, branches that end in a duplication event, and branches that end in a speciation event. Under a Poisson model the probability of observing a branch ending in a duplication is $\lambda \exp(-\lambda d)$; the probability of observing a branch ending in a terminal taxon is $\exp(-\lambda d)$, which can be obtained by integrating the first expression from d to infinity (i.e., we have no way of knowing when after the present the next duplication event will occur). Finally, the probability of observing an internal branch that ends in a speciation event but not a duplication event is $\exp(-\lambda d)$, because with respect to the duplication process this case is the same as if the speciation event were a terminal tip. The overall likelihood is then obtained by multiplying these contributions for each branch, which is warranted by virtue of the stochastic independence of different branches in a Yule model. The likelihood reduces to

$$L = \lambda^{j} \exp(-\lambda D) \tag{3}$$

where D is the sum of all branch lengths, Σd_i. The maximum likelihood estimate of rate is obtained by solving

$$\frac{d}{d\lambda}[\lambda^{j} \exp(-\lambda D)] = 0 \tag{4}$$

which leads to

$$\hat{\lambda} = j/D. \tag{5}$$

This matches the intuition that the rate of duplication should be proportional to the number of duplications, and inversely proportional to the path length over which those duplications could have occurred.

Generally it is desirable not merely to make estimates but also to test hypotheses about those estimates. A natural hypothesis to test for this kind of branching rate model is a contrast between the simplest one-rate model and a slightly more complicated two-rate model in which one part of the tree has one rate and the rest of the tree is allowed to have a different rate. In a sense this hypothesis tests the goodness of fit of the one-parameter model to the data. Rejection of the one rate model then might lead the investigator to seek explanations for the differences in rate entailed by the two rate-model. The one-rate model will be termed H^1 and the two-rate model H^2. Maximum likelihood methods are associated with a natural test known as the likelihood ratio, which is the ratio of the likelihoods for the two hypotheses, evaluated at their respective (usually different) maximum likelihood points. Thus,

$$LR = \frac{\max\{L[H^1(\hat{\lambda})]\}}{\max\{L[H^2(\hat{\lambda}_1, \hat{\lambda}_2)]\}}. \tag{6}$$

The maximum likelihood of H^2 will generally be greater than that of H^1 (because it is easier to fit a more complicated model to any data set), and hence the LR will vary between zero and one. If LR is sufficiently small one can be confident in the statistical sense that H^1 should be rejected in favor of H^2. The confidence level can be assessed by a variety of means. In the limit of large samples, for example, $G = -2 \ln(LR)$

tends to a χ^2 distribution with one degree of freedom (in this case). Thus, if $G > \chi^2(0.05)$, model H^1 can be rejected with a probability of type I error of 0.05. For smaller sample sizes, Monte Carlo simulation provides a more accurate assessment of the properties of the likelihood ratio test (Goldman, 1993).

For the present model the likelihood ratio of Eq. 6 reduces to a simplified form. For the one rate model, the ML estimated rate is just j/D, and the likelihood at that point is (from Eq. 3) $(j/D)^j \exp(-j)$. If the tree is subdivided into two parts, with j_1 and j_2 duplications and length totals of D_1 and D_2 respectively, then the one rate model is clearly maximized at $\lambda = (j_1 + j_2)/(D_1 + D_2)$, and the corresponding likelihood is just $[(j_1 + j_2)/(D_1 + D_2)]^{j_1+j_2} \exp[-(j_1 + j_2)]$. The maximum value of the likelihood function for the two-rate model is obtained by noting that the overall likelihood equation will be just the product of two terms like Eq. 3 for each subtree: $L = \lambda_1^{j_1} \exp(-\lambda_1 D_1)\, \lambda_2^{j_2} \exp(-\lambda_2 D_2)$, which is maximized at $(\hat{\lambda}_1, \hat{\lambda}_2) = (j_1/D_1, j_2/D_2)$. The likelihood ratio of Eq. 6 is then

$$LR = \frac{\left(\dfrac{j_1 + j_2}{D_1 + D_2}\right)^{j_1+j_2}}{\left(\dfrac{j_1}{D_1}\right)^{j_1}\left(\dfrac{j_2}{D_2}\right)^{j_2}}. \tag{7}$$

An Application to Globin Gene Evolution

Sample size is likely to be a problem in small gene families, but many gene families are large, diverse, and some are even imbedded in superfamilies, as in the globins and immunoglobulins. Discovery of differences in rates of duplication in a phylogeny of some gene family might bear on hypotheses of structural, functional, or other evolutionary constraints. Nei (1987, 134), for example, speculated that the number of genes in the α-globin gene cluster of vertebrates is evolutionarily more canalized (and hence its rate of duplication restricted) in comparison to the β-δ globin cluster and other globins.

The globin family is one of the best studied both from the perspective of its molecular biology and evolutionary history (Hardison, 1991). Phylogenetic analysis using a diverse set of algorithms generally agree on the basic outline of the history of the family, and approximate dates of duplications events can be inferred (caveat emptor; see Gillespie, 1991) from the rich vertebrate fossil record and from molecular clocks. Cursory examination of the number of duplications in various parts of the globin phylogeny (Fig. 3) suggests that rates of duplication have been highest in the β-globin lineage of higher vertebrates, relative to either the α-globins or the globins found in lower vertebrates such as agnathans (jawless fishes).

This crude numerical perception is lent credence by a more detailed likelihood analysis along the lines suggested above. Two separate sets of hypotheses associated with different hierarchical levels in the gene phylogeny were tested. In each, the null hypothesis was that the rate of duplication was the same throughout the phylogeny; the alternative hypothesis was that the rates differed in each of the sister groups descended from the most recent common ancestor of the group. The first test focused on node A in Fig. 3, which is the group of genes descended from the duplication that lead to the α-β globin clade. The second test backed off to node B and contrasted the agnathan globins to the other vertebrate globins. Maximum

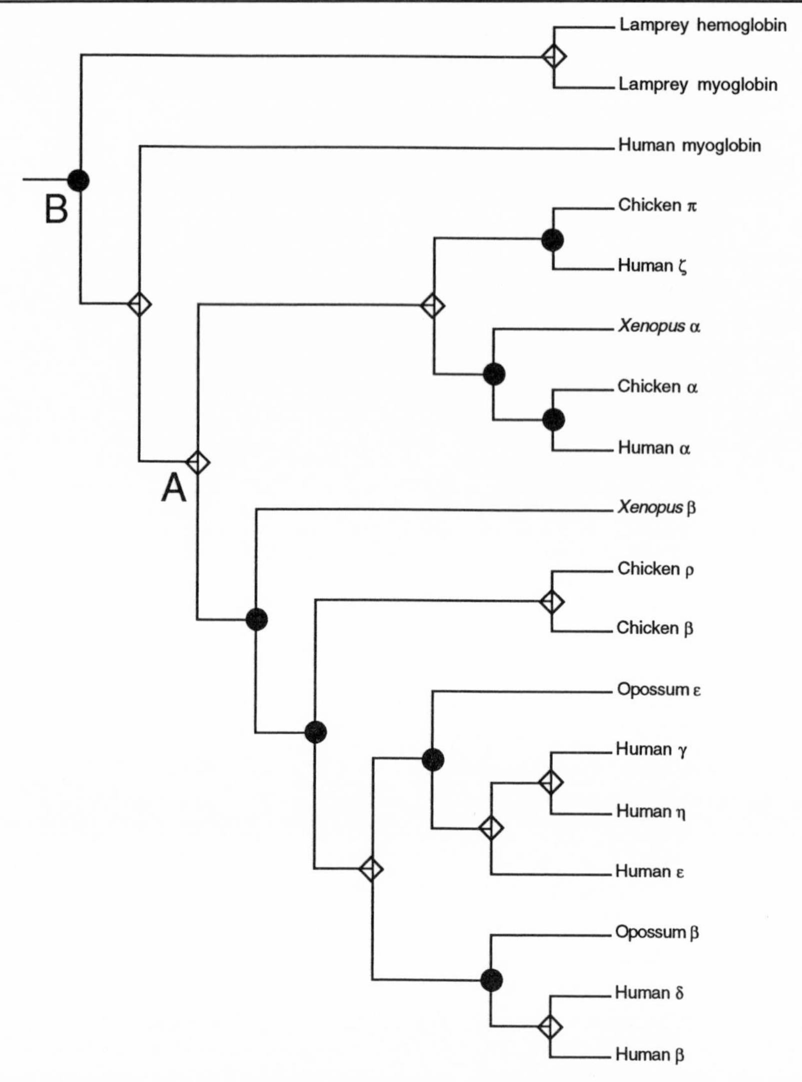

Figure 3. Phylogeny of major components of the vertebrate globin multigene family (after Hardison, 1991). Branch lengths in this figure are not colinear with time; for exact branch lengths, see Hardison (1991). Nodes labeled *A* and *B* denote the position in the phylogeny that divides the tree into subtrees for the two hypotheses discussed in the text and Table I.

likelihood estimates were obtained according to Eq. 5 and the likelihood ratios for the null versus alternative hypotheses were calculated using Eq. 7. Results (Table I) indicate that it is impossible to reject the single rate model in either case (possibly due to limited sample size), but that of the two contrasts, the contrast between the α and β globins is closer to significance than is the contrast between the agnathan and gnathastome (jawed fishes and tetrapod vertebrates) globins. This provides weak support for Nei's (1987: see above) ideas that quantitatively different duplication rates are acting in the α- versus β-globins.

Conclusions

This example serves to illustrate the dilemma confronted by an investigator interested in a particular evolutionary pattern or process, such as the evolution of a multigene family. The phylogenetic history of a gene family obviously bears on inferences that can be made about evolutionary processes such as duplication, gene conversion, et cetera. At the same time, those processes are likely to affect any kind of inference procedure used to reconstruct the phylogeny itself (Sanderson and Doyle, 1992). A full-blown likelihood assault on the problem would require (*a*) a model of gene duplication and extinction, (*b*) a model of sequence evolution, including (*c*) a model of recombination (concerted evolution), (*d*) a model of insertion/deletion to permit alignment estimation, and possibly (*e*) a model of speciation and extinction. Currently available likelihood methods incorporate only one or (very recently) two of these elements, and they are computationally difficult as is. Needless to say the results obtained by the hybrid analysis performed in this paper would be impossible to obtain with currently available maximum likelihood techniques.

TABLE I
Likelihood Ratio Tests of One-Versus Two-Parameter Gene Duplication Models for Phylogenetic Contrasts in Vertebrate Globin Gene Family

Node/ contrast	j_1, j_2	D_1, D_2*	$\lambda, \lambda_1, \lambda_2$‡	*LR*	*G*§	*P*‖
A	1, 5	1,775, 2,075	1.56, 0.56, 2.40	0.31	2.32	0.13
B	1, 8	950, 4,375	1.69, 1.05, 1.82	0.86	0.31	0.52

See Fig. 3 for node/contrasts.
*In units of millions of years, estimated on the basis of approximate dates of duplication events (Hardison, 1991).
‡$\times 10^{-3}$/million years.
§$G = -2 \ln (LR)$.
‖Significance levels calculated based on assumed convergence of G to χ^2 with one d.f.

That an answer (of uncertain reliability) can be obtained by some method is not by itself a strong argument in favor of that method. But additional arguments for a hybrid approach can be made. The primary one is that nonlikelihood methods, especially parsimony, appear to provide a reasonable reconstruction of evolutionary history. Parsimony is the most widely used phylogenetic algorithm (Sanderson et al., 1993), and its accuracy is supported by a growing body of simulation evidence that suggests that it does as well as or better than other nonlikelihood methods, and significantly better than some nonlikelihood methods. Maximum likelihood can out perform parsimony, because conditions under which parsimony will fail but likelihood will succeed are well-known (Felsenstein, 1978)—but only if the maximum likelihood model is correct. There are always conditions under which maximum likelihood will fail to find the right tree (or fail to be consistent) given deviations from the assumed model (J. Kim, personal communication). Thus, for workers interested in studying interesting evolutionary processes now, a not unreasonable compromise is to begin with a phylogeny derived by parsimony and overlay it later with maximum likelihood (or other statistical) methods.

Acknowledgments

The author is grateful to Doug Fambrough for extending an invitation to speak at the 1993 SGP meeting and to the Society for supporting his travel and accomodations.

References

Bishop, M. J., and E. A. Thompson. 1986. Maximum likelihood alignment of DNA sequences. *Journal of Molecular Biology.* 190:159–165.

Brooks, D. R., and D. A. McLennan. 1991. Phylogeny, Ecology, and Behavior: A Research Program in Comparative Biology. University Chicago Press, Chicago. 434 pp.

Cavalli-Sforza, L. L., and A. W. F. Edwards. 1967. Phylogenetic analysis: models and estimation procedures. *Evolution.* 21:550–570.

Edwards, A. W. F. 1970. Estimation of branching points of a branching diffusion process. *Journal of the Royal Statistical Society, B.* 32:155–174.

Edwards, A. W. F. 1972. Likelihood. Cambridge University Press, Cambridge, UK. 275 pp.

Feller, W. 1968. An Introduction to Probability Theory and its Applications. 3rd edition. John Wiley and Sons, Inc., New York. 509 pp.

Felsenstein, J. 1973. Maximum-likelihood estimation of evolutionary trees from continuous characters. *American Journal of Human Genetics.* 25:471–492.

Felsenstein, J. 1978. Cases in which parsimony and compatibility will be positively misleading. *Systematic Zoology.* 27:401–410.

Felsenstein, J. 1981. Evolutionary trees from DNA sequences: a maximum likelihood approach. *Journal of Molecular Evolution.* 17:368–376.

Felsenstein, J. 1982. Numerical methods for inferring evolutionary trees. *Quarterly Review of Biology.* 57:127–141.

Felsenstein, J. 1983. Statistical inference of phylogenies. *Journal of the Royal Statistical Association, A.* 146:246–272.

Felsenstein, J. 1985. Confidence limits on phylogenies: an approach using the bootstrap. *Evolution.* 39:783–791.

Felsenstein, J. 1993. PHYLIP (Phylogeny Inference Package) Version 3.5c. Computer program distributed by the author.

Felsenstein, J., and E. Sober. 1986. Parsimony and likelihood: an exchange. *Systematic Zoology.* 35:617–626.

Gilinsky, N. L., and I. J. Good. 1991. Probabilities of origination, persistence, and extinction of families of marine invertebrate life. *Paleobiology.* 17:145–166.

Gillespie, J. H. 1991. The Causes of Molecular Evolution. Oxford University Press, New York. 336 pp.

Goldman, N. 1990. Maximum likelihood inference of phylogenetic trees, with special reference to a poisson process model of DNA substitution and to parsimony analyses. *Systematic Zoology.* 39:345–361.

Goldman, N. 1993. Statistical tests of models of DNA substitution. *Journal of Molecular Evolution.* 36:182–198.

Goodman, M., J. Czelusniak, G. W. Moore, A. E. Romero-Herrera, and G. Matsuda. 1979. Fitting the gene lineage into its species lineage: a parsimony strategy illustrated by cladograms constructed from globin sequences. *Systematic Zoology.* 28:132–168.

Guttorp, P. 1991. Statistical Inference for Branching Processes. John Wiley and Sons, Inc., New York. 211 pp.

Hardison, R. C. 1991. Evolution of globin gene families. *In* Evolution at the Molecular Level. R. K. Selander, A. G. Clark, and T. S. Whittam, editors. Sinauer Associates, Inc., Sunderland, MA. 272–289.

Harvey, P. H., and M. D. Pagel. 1991. The Comparative Method in Evolutionary Biology. Oxford University Press, Oxford, UK.

Hey, J. 1992. Using phylogenetic trees to study speciation and extinction. *Evolution.* 46:627–640.

Huelsenbeck, J. P., and D. M. Hillis. 1993. Success of phylogenetic methods in the four-taxon case. *Systematic Biology.* 42:247–264.

Langley, C. H., and W. M. Fitch. 1974. An examination of the constancy of the rate of molecular evolution. *Journal of Molecular Evolution.* 3:161–177.

Lehman, E. L. 1983. Theory of Point Estimation. John Wiley and Sons, Inc., New York. 516 pp.

Maddison, D. 1991. The discovery and importance of multiple islands of most-parsimonious trees. *Systematic Zoology.* 40:315–328.

Maddison, W. P., and D. R. Maddison. 1992. MacClade: Analysis of Phylogeny and Character Evolution. Sinauer Associates, Inc., Sunderland, MA. 398 pp.

Navidi, W. C., G. A. Churchill, and A. von Haeseler. 1991. Methods for inferring phylogenies from nucleic acid sequence data using maximum likelihood and linear invariants. *Molecular Biology and Evolution.* 8:128–143.

Nee, S., A. O. Mooers, and P. H. Harvey. 1992. Proceedings of the National Academy of Sciences, USA. 89:8322–8326.

Nei, M. 1987. Molecular Evolutionary Genetics. Columbia University Press, New York.

Novacek, M. J., and M. A. Norell. 1982. Fossils, phylogenies, and taxonomic rates of evolution. *Systematic Zoology.* 31:366–375.

Ohta, T. 1988. Multigene and supergene families. *In* Oxford Surveys in Evolutionary Biology. V. P. H. Harvey, and L. Partridge, editors. Oxford University Press, Oxford, UK.

Ohta, T. 1990. How gene families evolve. *Theoretical Population Biology.* 37:213–219.

Olsen, G. J., H. Matsuda, R. Hagstrom, and R. Overbeek. 1993. FastDNAml 1.0. Computer program distributed by the authors.

Page, R. D. M. 1991. Random dendrograms and null hypotheses in cladistic biogeography. *Systematic Zoology.* 40:54–62.

Raup, D. M. 1985. Mathematical models of cladogenesis. *Palaeobiology.* 11:42–52.

Rodriguez, F., J. L. Oliver, A. Marin, and J. R. Medina. 1990. The general stochastic model of nucleotide substitution. *Journal of Theoretical Biology.* 142:485–501.

Sanderson, M. J. 1990. Flexible phylogeny reconstruction: a review of phylogenetic inference packages using parsimony. *Systematic Zoology.* 39:414–420.

Sanderson, M. J. 1993. Reversibility in evolution: a maximum likelihood approach to character gain/loss bias in phylogenies. *Evolution.* 47:236–252.

Sanderson, M. J., B. G. Baldwin, G. Bharathan, C. S. Campbell, C. von Dohlen, D. Ferguson, J. M. Porter, M. F. Wojciechowski, and M. J. Donoghue. 1993. The growth of phylogenetic information, and the need for a phylogenetic data base. *Systematic Biology.* 42:562–568.

Sanderson, M. J., and G. Bharathan. 1993. Does cladistic information affect inferences about branching rates? *Systematic Biology.* 42:1–17.

Sanderson, M. J., and J. J. Doyle. 1992. Reconstruction of organismal and gene phylogenies from data on multigene families: concerted evolution, homoplasy, and confidence. *Systematic Biology.* 41:4–17.

Singer, M., and P. Berg. 1991. Genes and Genomes. University Science Books, Mill Valley, CA. 929 pp.

Sober, E. 1988. Reconstructing the Past. Massachusetts Institute of Technology Press, Cambridge, MA. 265 pp.

Swofford, D., and W. Maddison. 1987. Reconstructing ancestral character states under Wagner parsimony. *Mathematical Bioscience.* 87:199–229.

Swofford, D., and G. J. Olsen. 1990. Phylogeny reconstruction. *In* Molecular Systematics. D. M. Hillis and C. Moritz, editors. Sinauer Associates, Inc., Sunderland, MA. 411–501.

Thompson, E. A. 1975. Human Evolutionary Trees. Cambridge University Press, Cambridge, UK.

Thorne, J. L., H. Kishino, and J. Felsenstein. 1991. An evolutionary model for maximum likelihood alignment of DNA sequences. *Journal of Molecular Evolution.* 33:114–124.

Thorne, J. L., H. Kishino, and J. Felsenstein. 1992. Inching toward reality: an improved likelihood model of sequence evolution. *Journal of Molecular Evolution.* 34:3–16.

Walsh, J. B. 1987. Sequence-dependent gene conversion: can duplicated genes diverge fast enough to escape conversion? *Genetics.* 117:543–557.

Yule, G. U. 1924. A mathematical theory of evolution, based on the conclusions of Dr. J. C. Willis, F. R. S. *Proceedings of the Royal Society of London.* 213:21–87.

Introns, the Broken Transposons

Andrew J. Roger, Patrick J. Keeling, and W. Ford Doolittle

Canadian Institute for Advanced Research, Department of Biochemistry, Dalhousie University, Halifax, Nova Scotia, B3H 4H7

The origin of spliceosomal introns is a classic molecular evolutionary puzzle that still inspires debate 16 years after their first discovery in eukaryotic genes, debate which still centers on the initial polarization into two general theories on intron origins. The "introns-early" theory arose by combining Gilbert's ideas about how the exon/intron organization of eukaryotic genes could speed evolution by the process of "exon-shuffling" (Gilbert, 1978) with Darnell's and Doolittle's (1986) assertions that introns were relics of the assembly of genes in a primitive ancestor of all living cells. According to this theory, modern prokaryotes lost their introns through genomic streamlining, while eukaryotes retained the primitive introneousness of their genes. By contrast, proponents of the "introns-late" models argued that most introns are likely the husks of mobile genetic elements that had the special property of splicing out of genes on the RNA-level. Genes were split, the argument goes, by these mobile introns relatively recently after the origin of the eukaryotic nucleus and therefore had nothing to do with the origin of genes.

Our position is that current data best support a view where introns were inserted into pre-existing full-length genes. In this paper we will briefly discuss the data which has led us to reject the introns-early theory in favor of insertional models. However, there are many reviews which give a more complete treatment of the data relevant to this issue (Rogers, 1990; Palmer and Logsdon, 1991; Patthy, 1991). We instead wish to concentrate on elaborating new ideas with testable implications regarding the origin and spread of spliceosomal introns, so that experiments can be done to illuminate the remaining dark corners of intron evolution.

What Kinds of Introns Are Relevant to the Debate?

Since the first proposals of the theories explaining intron origins much has been learned about the diversity of intron splicing mechanisms and the genomes that introns inhabit. It is now clear that there are many different intron types, distinguishable by splicing mechanisms and phylogenetic distribution (reviewed in Lambowitz and Belfort, 1993):

Of these types, only two are relevant to the original theories: group II self-splicing introns (and their degenerate group III form) commonly found in eukaryotic organellar genomes, and spliceosomal introns which are abundant in nuclear genes and are spliced by a multimolecular RNA/protein complex called the spliceosome. These demonstrate similarity in structure and splicing mechanism of the sort which would suggest convincingly that they shared a common ancestor (Jacquier, 1990; Weiner, 1993; Lambowitz and Belfort, 1993; and Lamond, 1993). Thus, the general term "intron" probably does not refer to any evolutionarily coherent group of elements, perhaps only to a molecular niche. We feel that attempts to derive unitary scenarios to explain the origin and evolution of all introns are, therefore, misguided.

Introns Early: A Theory in Search of Evidence

Introns-early, a theory which addresses spliceosomal introns specifically, holds that they are evolutionary leftovers of the processes that produced the first full-length protein coding genes by "exon-shuffling." Proponents of this theory rely on evidence falling into two general categories: (*a*) Exon/exon boundaries of ancient genes found in eukaryotes are thus claimed to delineate units of protein structure and/or function as a result of their assembly by exon-shuffling (Blake, 1983); and (*b*) the conservation of intron positions between phylogenetically distant organisms indicates that they were ancestrally present in the gene (Gilbert, Marchionni, and McKnight, 1986).

Of these two claims, the first has never been supported convincingly and the second, in the light of recent data, increasingly argues for a late rather than an early origin of spliceosomal introns.

Exon-shuffling, Intron-positions, and Protein Structure

As a mode of gene evolution, exon-shuffling has been confirmed for many genes encoding extracellular proteins in vertebrates (Patthy, 1991). The origins of these genes, however, occurred relatively recently in evolution, perhaps just before the evolution of the chordate phylum 500 million years ago. The central claim of introns-early is not that such shuffling has occurred sometime in evolution, but that it was the dominant mode of evolution of the first protein-coding genes before the divergence of all extant cellular life. To prove such a claim, therefore, one must show that some of the genes common to all living organisms are chimaeric and that their recombinant junctions correspond to known intron positions. Examples of ancient exon-shuffling have been suggested for proteins such as alcohol dehydrogenase (ADH), glyceraldehyde-3-phosphate dehydrogenase (GAPDH), lactate dehydrogenase (LDH), pyruvate kinase (PK) (Duester, Jornvall, and Hatfield, 1986) and triose isomerase (TPI) (Gilbert et al., 1986). However, as Patthy (1987) points out, such proposals seem dubious as these proteins do not show the hallmarks of exon-shuffling discerned from known recent vertebrate examples. For instance, one would expect that exon-shuffling would produce genes sharing homologous exons bounded by introns of the same phase; a pattern not observed in these genes. In addition, the second claim of correlation between exon/exon junctions and protein structural boundaries has been suggested qualitatively in all of these cases. An analysis carried out in our lab (A. Stoltzfus, personal communication) testing this correlation in some ancient genes (e.g., TPI, globins, PK, and ADH) found that random intron placement cannot be statistically excluded. To further compromise the case for early exon shuffling, an attempt by Gilbert's group intended to estimate the size of the underlying "exon universe" (Dorit, Schoenbach, and Gilbert, 1990) failed to detect any examples of shared exons between ancient genes.

Until protein-structure/exon structure correlations are rigourously established for ancient genes and a single case of bona fide ancient exon-shuffling is identified, the first claim of introns-early is without any empirical evidence.

The Phylogenetic Distribution of Spliceosomal Introns within Eukaryotic Genes

Several arguments have been made that ancient intron positions shared between genes found in eukaryotic groups as diverse as plants, animals and fungi (Gilbert et al., 1986) are evidence for an early origin of introns. However, these examples can be as easily interpreted as inherited insertions of introns which occurred before the plant, animal, and fungal divergence, a relatively recent event. A better test makes use of the fact that the protists are a multiply paraphyletic group from which multicellular eukaryotes evolved (Fig. 1 and Cavalier-Smith, 1993), allowing us to repeatedly test how parsimonious an early origin for introns is. In globins, for example, two intron positions are shared between plants and animals out of a total 11 known, but neither is found in the only known protist homolog, possessed by *Paramecium* (Stoltzfus and Doolittle, 1993). Similarly, in the case of TPI, out of 14 known intron positions, only five are shared between plants and animals and none of these are found in this gene from representatives of three of the early diverging

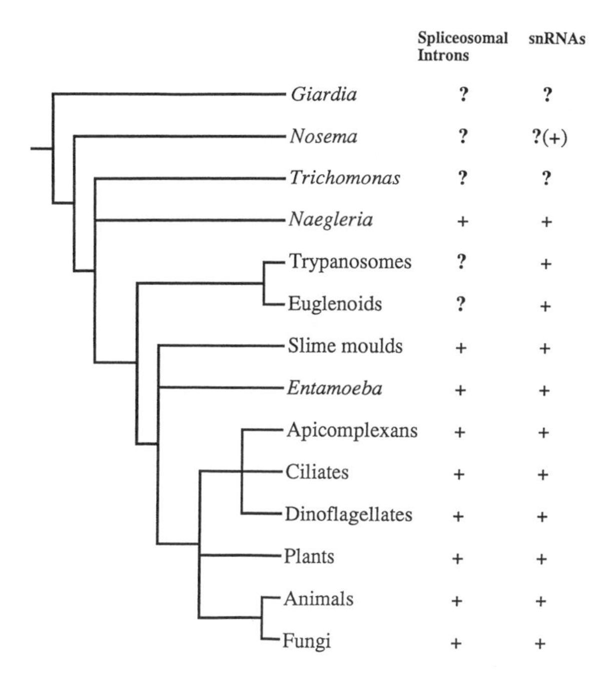

Figure 1. Our current knowledge of the phylogenetic distribution of spliceosomal introns and snRNAs within eukaryotes. This provisional phylogenetic tree is based on a synthesis of small-subunit rRNA trees with ultrastructural data (Leipe et al., 1993; and Cavalier-Smith, 1993) and is rooted with archaebacterial and eubacterial outgroups. In many cases, published tree topologies conflict. We have, in these cases, either made a decision on what is the most likely phylogeny of the organisms or collapsed the equivocal node. The finding of at least one spliceosomal intron or snRNA in a taxon is indicated by a "+" and, conversely, the lack of either, based on current data, is indicated by a question mark.

protist lineages, *Giardia lamblia* (M. Mowatt, Weinbach, Howard, and Nash, 1994), *Trichomonas vaginalis* and *Entamoeba hystolytica* (A. Roger, unpublished data).

These and other examples (see Dibb and Newman, 1989; Palmer and Logsdon, 1991; and Kwiatowski, Skareky, and Ayala, 1992) clearly indicate that the small number of intron positions that are conserved between kingdoms of eukaryotes are phylogenetically restricted to plants, fungi and animals. Furthermore, the majority of intron positions appear unique to lineages within these kingdoms. To reconcile this with introns-early, one must postulate vast numbers of parallel events of intron loss from a hypothetical intron-riddled ancestral gene. Instead, we suggest a more parsimonious interpretation of the data; that the current intron distribution reflects processes of independent intron gain and loss that have operated within these lineages during the course of eukaryote evolution (Palmer and Logsdon, 1991).

Spliceosomal Intron Positions Conserved Between Prokaryotic and Eukaryotic Genes

Recently, several instances of introns that are shared between nuclearly encoded cytoplasmic targetted genes and organellar homologs of eubacterial origin have been suggested (Shih, Heinrich, and Goodman, 1988, Iwabe, Kuma, Kishino, Hasegawa, and Miyata, 1990, and Juretic, Mattes, Ziak, Christen, and Jaussi, 1990). Although in some of these cases the introns are not actually in identical places (Palmer and Logsdon, 1991), other examples are more convincing (Kersanach, Brinkman, Liaud, Zhang, Martin, and Cerff, 1994). Such instances have been interpreted, in support of introns-early, as evidence that the common ancestor of the eubacterial endosymbiont and the nuclear genome had these introns. However, if one accepts that prokaryotic genomes presently lack introns due to streamlining selection, then the assertion that these intron positions are ancestral is very unparsimonious. Previous "streamlining" theories have had to invoke the global loss of introns twice; once in the ancestor of all eubacteria and a second time in the lineage leading to archaebacteria (Doolittle, 1991). To account for the presence of shared intron positions between the nuclear and organellar genes, one must posit that introns have been independently lost in archaebacteria as well as nine major eubacterial groups (Woese, 1987). Moreover, they must have been maintained in the cyanobacterial and purple bacterial lineages for half of the history of life, only to be cataclysmically lost after the origin of organelle-bearing eukaryotes (Palmer and Logsdon, 1991).

The alternative, that introns have been inserted into the same positions in these homologs independently, is quite reasonable if one assumes that introns insert into sites with a specific target sequence. If sites compatible for insertion (for instance the "proto-splice sites" of Dibb and Newman, 1989) are limited to a small fraction of all possible insertion sites in a gene, then, provided that the gene is highly enough conserved in sequence, multiple events of intron insertion at any particular site are expected.

Insertional Models of Spliceosomal Intron Origin

While the accumulating evidence on intron distribution may argue against the primitive presence of spliceosomal introns, it remains necessary to develop models that provide a plausible mechanism for the spread of introns in the eukaryotic genome. These must address both the means by which introns are gained and lost, as well as how the splicing machinery evolved in the first place.

Spliceosomes May Have Evolved from Retroposing Group II Introns

The initial characterization of self-splicing group II introns revealed that they shared similar splice-sites and branchpoint sequences as well as a lariat splicing mechanism with spliceosomal introns, hinting at a possible relationship between the two types. While these features are somewhat superficial, better support for their homology has been provided by recently discovered structural similarities between the intermolecular secondary structure formed by small nuclear RNAs in the spliceosome and the intramolecular elements of group II introns (Jacquier, 1990; Newman and Norman, 1992; Madhani and Guthrie, 1992; Lamond, 1993; and Weiner, 1993). The idea that these self-splicing introns gave rise to spliceosomal introns by fragmentation (Hickey et al., 1989) rests partly on the intuitive appeal of a transition from the relatively simple unimolecular self-splicing introns to a complex multimolecular RNA/protein spliceosome. Better evidence is provided by the observation that group II introns can be fragmented into pieces in vitro without losing their spliceability (Suchy and Schmelzer, 1991) and by the precedent of such intron fragmentation having occurred in several independent organelle lineages (Bonen, 1993).

In addition, the finding that many group II introns contain reverse transcriptase-like open reading frames (Michel and Lang, 1985) together with the subsequent demonstration of reverse splicing in vitro (Augustin, Mueller, and Schweyen, 1990) provided preliminary evidence that they may be mobile elements. These suggested a mechanism of spreading to new genes by reverse splicing into heterologous RNAs, reverse transcription and recombination of the complementary DNA with the intronless, genomic copy. The case for group II intron mobility now seems sewn up by two reports of their actual transposition into new sites in the mitochondrial genomes of *Saccharomyces* (Mueller, Allmaier, Eskes, and Schweyen, 1993) and *Podospora* (Sellem, Lecellier, and Belcour, 1993) where both events appear to have occurred on the RNA level. A retrotransposition model for group II introns is further supported by the finding that one of the intron-encoded ORFs in the yeast *cox1* gene is an intron- and exon-specific reverse transcriptase (Kennel, Moran, Perlman, Butow, and Lambowitz, 1993).

When Did Spliceosomal Introns Evolve from Group II Introns?

In 1991, Cavalier-Smith (1991) proposed that spliceosomal introns evolved from group II introns which first entered eukaryotes in the α-purple bacterial endosymbiont which gave rise to the mitochondrion. Group II introns, he argued, invaded nuclear protein coding genes by retrotransposition, and then one subsequently fragmented, producing the genes for small nuclear RNAs which could splice all of the group II introns present in the nuclear genome in *trans,* thereby making autocatalysis redundant.

This theory, henceforth referred to as the mitochondrial origin theory, predicts that spliceosomes and the introns they splice will not be found in prokaryotes or the earliest diverging eukaryotic lineages, the Archezoa, which are thought to lack mitochondria primitively (Cavalier-Smith, 1991). We suggest two alternatives to this theory: a nuclear origin theory where group II introns were converted into the spliceosome during the evolution of the nucleus and a prokaryotic origin theory where group II introns fragmented to form the spliceosome before the origin of the nucleus in its prokaryotic ancestors (Fig. 2). To decide between these alternatives, the phylogenetic distribution of group II and spliceosomal introns must be considered.

The Phylogenetic Distribution of Group II and Spliceosomal Introns

Until recently, group II introns had only been found in mitochondrial genomes of fungi, and the mitochondrial and chloroplast genomes of various algal groups and land plants. However, a survey using PCR has identified group II introns in a cyanobacterium and a γ-purple bacterium, representatives of eubacterial groups which gave rise to plastids and mitochondria respectively (Ferat and Michel, 1993). This seems, at first glance, strong evidence for Cavalier-Smith's (1991) hypothesis. However, the true diversity of organisms possessing group II introns is unknown as this PCR search was restricted to only species within the eubacterial groups which gave rise to organelles. A comprehensive survey for these introns in diverse groups of eubacteria, archaebacteria and early eukaryotes is necessary to pin down the phylogenetic distribution of this intron type. If group II introns are found in both prokaryotic groups (the Eubacteria and the Archaebacteria), then such introns may have been ancestral to eukaryotic nucleii.

Fig. 1 shows a hypothetical phylogeny of eukaryotes with the presence and absence of spliceosomal introns highlighted. In support of Cavalier-Smith's (1991) contention, it can be seen that no spliceosomal introns have yet been identified in the

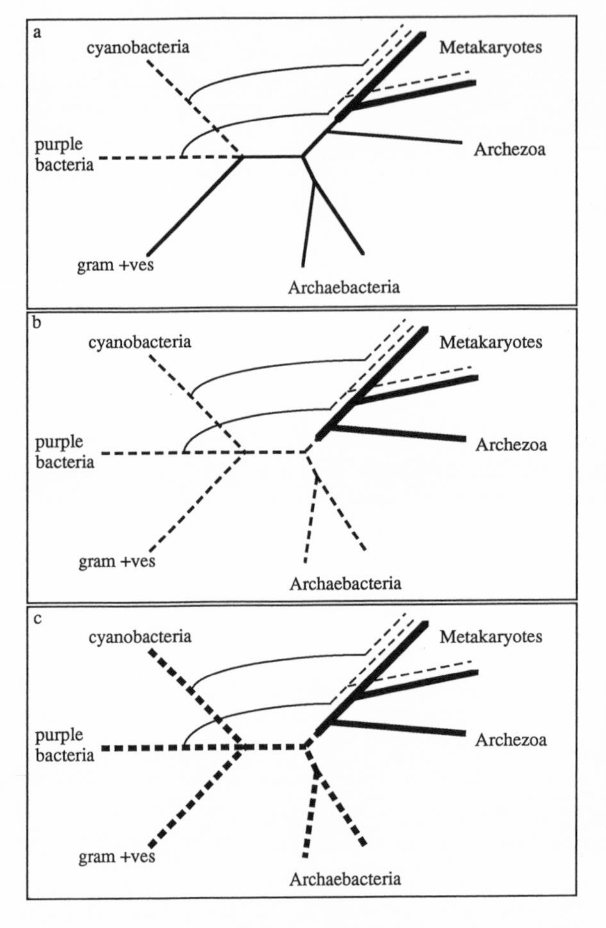

Figure 2. Schematic diagram outlining three possible scenarios of intron evolution. A unrooted tree of major extant cellular lineages is shown with the presence of group II introns (*hatched lines*) and spliceosomal introns (*heavy lines*) indicated. (*a*) The organelle origin of introns; here group II introns from the mitochondrial endosymbiont spread to the nucleus and degenerate into spliceosomal introns. (*b*) In the nuclear origin, group II introns are inherited by the nucleus and degenerate into spliceosomal introns. (*c*) This prokaryotic origin model has the degeneration of group II introns in a prokaryote before the divergance of the eukaryotes. It implies the coexistance of group II and spliceosomal introns in prokaryotes (*heavy hatched lines*).

earliest amitochondrial eukaryotic lineages. The lack of introns found in the Archezoa comes almost exclusively from only a dozen or so genes sequenced in *Giardia.* Clearly more data from this and other archezoans are needed. Recent work in our lab suggests that one such organism, the microsporidian *Nosema locustae,* where no introns have been discovered, appears to possess the highly conserved gene encoding the spliceosomal U6 small nuclear RNA (A. Roger, unpublished data). If microsporidia are truly primitively amitochondrial (a point over which reasonable people could disagree), then these data suggest that at least some of the spliceosomal machinery evolved before the endosymbiotic origin of mitochondria in contrast to Cavalier-Smith's theory and in favour of a nuclear or prokaryotic origin of the spliceosome. Unfortunately, this kind of analysis depends critically on our confidence in the phylogeny of the organisms in question, and at best can only exclude the organellar origin of spliceosomal introns; to understand where the transition between group II and spliceosomal introns took place it is necessary to also consider prokaryotes.

A Prokaryotic Origin for the Spliceosome?

Because spliceosomal splicing is notoriously slow compared with self-splicing, it is often suggested that transcriptional/translational coupling in prokaryotes would prevent the origin of spliceosomal splicing; the frequent translation of unspliced messages would be too heavy a selective burden (Cavalier-Smith, 1991; and Doolittle, 1991). The origin of the nucleus, however, would have freed self-splicing group II introns from this selection and allowed them to fragment and spread. Such an argument favors, a priori, a nuclear origin over a prokaryotic origin for spliceosomes. However, the presence of group II introns in the eubacteria raises the possibility that they were also present in the common ancestor of prokaryotes and eukaryotes. The fragmentation of one of these introns to form a spliceosome may therefore have occurred before the origin of the nucleus. The discovery in *Mycoplasma capricolum* of an abundant small RNA species which is $>50\%$ identical to most U6 snRNA genes in eukaryotes (Ushida and Muto, 1993) may be evidence for bacterial spliceosomes. If other eubacteria as well as archaebacteria turn out to possess this RNA, and it is found to be involved in splicing, then the origin of at least some parts of the spliceosome may harken back to the common ancestor of all living cells. Why such introns would fail to spread through bacterial genomes is not certain, but may be limited by transcriptional/translational coupling. More plausibly, the *Mycoplasma* U6-like RNA could be the result of an independent fragmentation of a group II intron in this lineage, quite separate from the origin of the spliceosome in eukaryotes. The fact that this RNA is more similar to U6 snRNAs than any characterized group II intron sequence may therefore betray the existance of an undetected family of group II introns which share sequence similarity to both the *Mycoplasma* U6-like RNA and the eukaryotic spliceosomal snRNAs, and are ancestral to both.

However, such speculation is premature until more data about the phylogenetic distribution of snRNAs, spliceosomal introns and group II introns within the Archezoa, Archaebacteria, and Eubacteria become available.

How Are Spliceosomal Introns Invading New Sites Now?

A single transposition burst of group II introns in the early eukaryote nuclear genome and their subsequent slow loss (by the recombination of reverse-transcribed, spliced mRNAs with their parent alleles) could possibly explain the current intron distribution in eukaryotic genes. However, many of the phylogenetic difficulties that introns-early is prone to also apply to a view where all introns were inserted at a single time. Because most intron positions are restricted to the most recently evolved eukaryotic lineages, it is most parsimonious to assume a model where introns have been spreading since their initial insertion as group II introns.

There are a handful of examples of DNA transposons which generate spliceosomal introns by transposition, but none generate precise splice-site boundaries. The widespread insertion of introns requires that transposition leads to splice-sites which do not alter the coding sequence; the most tenable mechanism for this is by an RNA based transposition using reverse-splicing.

In this vein, Cavalier-Smith (1991) suggests that the conversion of all of the primary invading group II introns to spliceosomal introns by their gradual dependence on the snRNP-mediated splicing may have taken hundreds of millions of years to complete (Cavalier-Smith, 1991). In this lingering group II introns theory, remaining group II introns are suggested to have been retroposing to new sites relatively recently in evolution. Others have suggested that these introns could be continually introduced into nuclear genes from organellar genomes containing them (Rogers, 1989; Palmer and Logsdon, 1991). The finding of a mitochondrial group II intron remnant in the nuclear genome of *Dolichos biflorus* suggests that these introns are transferred to the nucleus on occasion (Knoop and Brennicke, 1991). The subsequent degeneration of group II introns into spliceosomal introns would be relatively simple requiring only a single nucleotide replacement to convert the consensus sequence for the former into the latter (Rogers, 1989). Such lateral transfer, however, could only be responsible for recent intron gain in plants, fungi and protists, because group II introns appear to be absent from animal mitochondrial genomes (Gray, 1989).

Whereas both of these ideas may have merit, they also predict the existence of a so far undetected, intact, group II intron in nuclear genes. A simpler explanation for recent intron spread, first offered by Sharp (1985), is that the spliceosomal machinery, itself, may introduce introns into RNAs by catalyzing the reverse-splicing reaction. In this way, introns could be introduced into sites compatible with splicing. If the spread of introns into new sites in RNAs depends on accidental rare reverse-splicing events, a reverse transcriptase activity could be supplied by reverse-transcriptases encoded by endogenous retroviruses or retrotransposons commonly present in eukaryotic genomes and cellular machinery for homologous recombination would then be relied upon to insert the copy into the genome. It is difficult to assess the feasibility of such a mechanism since experiments employing PCR to detect rare reverse-splicing have so far not detected any such activity (C. Guthrie, personal communication) perhaps because not all introns can be reverse-spliced.

This passive mechanism of intron acquisition (Roger and Doolittle, 1993) is attractive because the gain of introns depends only on a side reaction of the spliceosome and does not propose the existence of an unknown transposing entity. But it suffers from the general problem that intron loss, by the replacement of intron-containing alleles with intron-less cDNAs, ought to be thousands of times more likely than incorporation of rarely reverse-spliced versions. Because introns are present in high density in many genomes of eukaryotic multicells, there must have been periods in the evolution of certain lineages during which intron gain was far more likely than loss; that is, periods when introns were actively transposing.

To account for this, we suggest that there is a subset of introns which are transposable and may encode a reverse-transcriptase/maturase-like protein which recognizes sequences in its intron parent as well as binding with spliceosomal components in such a way as to promote reverse-splicing. Such an element could effectively reproduce itself by spliceosomal reverse-splicing of parent introns into proto-splice sites in RNAs followed by catalyzing the reverse transcription of these messages into cDNA. In genomes where homologous recombination occurs frequently, the fecundity of this kind of retrotransposon would be high, causing the rapid turnover of introns in genes. However, if it is rare, as in vertebrate genomes, retroposing introns would die out (mutations in them would accumulate faster than new copies could be generated by transposition events) leaving an intron distribution that is phylogenetically stable. Consistent with this is the observation that there is a relative stasis of intron positions within vertebrate genes like tubulins and actin (Dibb and Newman, 1989).

Towards a Resolution to the Introns Debate?

The "introns-late" model we have presented rests largely on the mobility of spliceosomal introns, a phenomenon for which there is presently no direct evidence. It is often difficult to catch a transposon in the act, especially when it is assumed to be a rare event, nevertheless there are several strategies which may be employed to demonstrate transposition convincingly.

Because the actual constraints on intron sequence appear to be so lax, it is somewhat naive to hope to discover two closely related introns within a genome by chance. It is necessary to hedge our bets by first looking for an organism containing an intron at some position which is intronless in numerous closely related species; this is indicitive of a recent transposition. If the time of divergence between the organism and its intronless relatives is less than ~65 million years, we could expect to find a parental copy of the new intron to have maintained a detectable level of sequence identity (~60%). Similar introns in different sequence contexts provides good evidence of transposition.

Such a finding would happily provide concrete evidence in a debate where it is so desperately needed. As molecular biology is catching up with theory, we may hope for a settlement to this ancient question, or at least for more data to fuel another sixteen years of vigourous argument.

References

Augustin, S., M. W. Mueller, and R. J. Schweyen. 1990. Reverse self-splicing of group II intron RNA *in vitro. Nature.* 343:383–386.

Blake, C. C. F. 1983. Exons—present from the beginning? *Nature.* 306:535–537.

Bonen, L. 1993. *Trans*-splicing of pre-mRNA in plants, animals and protists. *FASEB Journal.* 7:40–46.

Cavalier-Smith, T. 1985. Selfish DNA and the origin of introns. *Nature.* 315:283.

Cavalier-Smith, T. 1993. Kingdom Protozoa and its 18 phyla. *Microbiological reviews.* 57:953–994.

Dibb, N. J., and A. J. Newman. 1989. Evidence that introns arose at proto-splice sites. *EMBO Journal.* 8:2015–2021.

Doolittle, W. F. 1991. The origins of introns. *Current Biology.* 1:145–146.

Dorit, R. L., L. Schoenbach, and W. Gilbert. 1990. How big is the universe of exons? *Science.* 250:1377–1382.

Duester, G., H. Jornvall, and G. W. Hatfield. 1986. Intron-dependent evolution of the nucleotide-binding domains within alcohol dehydrogenase and related enzymes. *Nucleic Acids Research.* 14:1931–1941.

Darnell, J. E., and W. F. Doolittle. 1986. Speculations on the early course of evolution. *Proceedings of the National Academy of Sciences, USA.* 83:1271–1275.

Ferat, J.-L., and F. Michel. 1993. Group II self-splicing introns in bacteria. *Nature.* 364:358–361.

Gilbert, W. 1978. Why genes in pieces? *Nature.* 271:501.

Gilbert, W., M. Marchionni, and G. McKnight. 1986. On the antiquity of introns. *Cell.* 46:151–154.

Gray, M. W. 1989. Origin and evolution of mitochondrial DNA. *Annual Review of Cell Biology.* 5:25–50.

Hickey, D. A., B. F. Benkel, and S. M. Abukashawa. 1989. A general model for the evolution of nuclear pre-mRNA introns. *Journal of Theoretical Biology.* 137:41–53.

Iwabe, N., K. Kuma, H. Kishino, M. Hasegawa, and T. Miyata. 1990. Compartmentalized isozyme genes and the origin of introns. *Journal of Molecular Evolution.* 31:205–210.

Jacquier, A. 1990. Self-splicing group II and nuclear pre-mRNA introns: how similar are they? *Trends in Biochemical Sciences.* 15:351–354.

Juretic, N., U. Mattes, M. Ziak, P. Christen, and R. Jaussi. 1990. Structure of the genes of two homologous intracellularly heterotopic isoenzyme. *European Journal of Biochemistry.* 192:119–126.

Kennel, J. C., J. V. Moran, P. S. Perlman, R. A. Butow, and A. M. Lambowitz. 1993. Reverse-transcriptase activity associated with maturase-encoding group II introns in yeast mitochondria. *Cell.* 73:133–146.

Kersenach, R., H. Brinkmann, M.-F. Liaud, D.-X. Zhang, W. Martin, and R. Cerff. 1994. Five identical intron positions in ancient duplicated genes of eubacterial origin. *Nature.* In press.

Knoop, V., and A. Brennicke. 1991. A mitochondrial intron sequence in the 5′-flanking region of a plant nuclear lectin gene. *Current Genetics.* 20:423–425.

Kwiatowski, J., D. Skarecky, and F. J. Ayala. 1992. Structure and sequence of the *Cu,Zn Sod* gene in the mediterranean fruit fly, *Ceratitis capitata:* intron insertion/deletion and evolution of the gene. *Molecular Phylogenetics and Evolution.* 1:72–82.

Lambowitz, A. M., and M. Belfort. 1993. Introns as mobile genetic elements. *Annual Review of Biochemistry.* 62:587–622.

Lamond, A. I. 1993. A glimpse into the spliceosome. *Current Biology.* 3:62–64.

Leipe, D. D., J. H. Gunderson, T. A. Nerad, and M. L. Sogin. 1993. Small subunit ribosomal RNA of *Hexamita inflata* and the quest for the first branch in the eukaryotic tree. *Molecular and Biochemical Parasitology.* 59:41–48.

Madhani, H., and C. Guthrie. 1992. A novel base-pairing interaction between U2 and U6 snRNAs suggests a mechanism for the catalytic activation of the spliceosome. *Cell.* 71:803–817.

Michel, F., and B. F. Lang. 1985. Mitochondrial class II introns encode proteins related to the reverse transcriptases of retroviruses. *Nature.* 316:641–643.

Mueller, M. W., M. Allmaier, R. Eskes, and R. J. Schweyen. 1993. Transposition of group II intron *aI1* in yeast and invasion of mitochondrial genes at new locations. *Nature.* 366:174–178.

Mowatt, M. R., E. C. Weinbach, T. C. Howard, and T. E. Nash. 1993. Complementation of an *Escherichia coli* glycolysis mutant by *Giardia lamblia* triosephosphate isomerase. *Experimental Parasitology.* In press.

Newman, A. J., and C. Norman. 1992. U5 snRNA interacts with exon sequences at 5′ and 3′ splice sites. *Cell.* 68:743–754.

Palmer, J. D., and J. M. Logsdon. 1991. The recent origins of introns. *Current Opinion in Genetics and Development.* 1:470–477.

Patthy, L. 1987. Intron-dependent evolution: preferred types of exons and introns. *FEBS Letters.* 214:1–7.

Patthy, L. 1991. Modular exchange principles in proteins. *Current Opinion in Structural Biology.* 1:351–361.

Roger, A. J., and W. F. Doolittle. 1993. Why introns-in-pieces? *Nature.* 364:289–290.

Rogers, J. H. 1989. How were introns introduced into nuclear genes? *Trends in Genetics.* 5:213–216.

Rogers, J. H. 1990. The role of introns in evolution. *FEBS Letters.* 268:339–343.

Sellem, C. H., G. Lecellier, and L. Belcour. 1993. Transposition of a group II intron. *Nature.* 366:176–178.

Sharp, P. A. 1985. On the origin of RNA splicing and introns. *Cell.* 42:397–400.

Shih, M. C., P. Heinrich, and H. M. Goodman. 1988. Intron existence predated the divergence of eukaryotes and prokaryotes. *Science.* 2423:1164–1166.

Stoltzfus, A., and W. F. Doolittle. 1993. Slippery introns and globin gene evolution. *Current Biology.* 3:215–217.

Suchy, M., and C. Schmelzer. 1991. Restoration of the self-splicing activity of a defective group II intron by a small *trans*-acting RNA. *Journal of Molecular Biology.* 222:179–187.

Ushida, C., and A. Muto. 1993. A small RNA of Mycoplasma capricolum that resembles eukaryotic U6 small nuclear RNA. *Nucleic Acids Research.* 21:2649–2653.

Weiner, A. M. 1993. mRNA splicing and autocatalytic introns: distant cousins or the products of chemical determinism? *Cell.* 72:161–164.

Woese, C. R. 1987. Bacterial evolution. *Microbiological Reviews.* 51:221–271.

Monophyletic Origin of Animals: A Shared Ancestry with the Fungi

Patricia O. Wainright,* David J. Patterson,‡ and Mitchell L. Sogin§

**Institute of Marine and Coastal Sciences, Rutgers University, Cook Campus, New Brunswick, New Jersey 08903-0231; ‡School of Biological Sciences, The University of Sydney, Sidney, New South Wales 2006, Australia; and §Marine Biological Laboratory, Woods Hole, Massachusetts 02543*

Phylogenetic frameworks inferred from comparisons of macromolecular sequences can provide insights into the origin of the Animalia and its major phyla. Structural and molecular data now offer corroborating evidence that animals are monophyletic with the choanoflagellates as a sister group. Within the metazoa, the divergence of sponges is followed by the Ctenophora, the Cnidaria plus the placozoan *Trichoplax adhaerans,* and finally, by an unresolved polychotomy of triploblastic animal phyla. Maximum likelihood analyses of small subunit ribosomal RNAs reveal animals and fungi share a unique evolutionary history. Their last common ancestor was a uni-flagellated protist similar to extant choanoflagellates or chytrids. This relationship between animal and fungal ribosomal RNAs agrees with molecular systematic studies of other gene families including α-tubulins, β-tubulins, elongation factors and actins.

Historical Theories

Most theories describing the origins of metazoa (Hanson, 1958, 1977) are based upon comparative morphology, developmental biology, and physiology. More recent studies of molecular evolution only underscore the complexity of deciphering the sudden radiation of animal phyla previously identified through paleontology. Preferred branching patterns for simple animals are not convincing in molecular and nonmolecular phylogenetic trees, yet there is general agreement that animals must have evolved from ancestral protists.

More than 100 years ago, Haeckel (1874), Lankester (1887), and Metschnikoff (1886) suggested that early animals evolved from a colonial flagellate resembling an embryological stage of extant metazoa. Because of similar cell structures identified through light microscopy, James-Clark (1867, 1868) proposed a direct relationship between the unicellular choanoflagellates and the choanocytes of sponges (James-Clark, 1867, 1868). Discovery of a colonial choanoflagellate, *Proterospongia haeckeli* Kent (1880–1882), reinforced the hypothesis linking choanoflagellates and simple animals (Kent, 1880–1882). This discovery also strengthened Metschnikoff's theory that a solid colony (parenchymula) was the first metazoan form. Lameere extended this idea to a colonial integration model in which colonies of choanoflagellates gave rise to sponges and in turn to the major animal phyla (Lameere, 1901, 1908).

A frequently cited competing idea placed the origins of metazoa within the ciliates (Hadzi, 1953). As described in Hadzi's Syncytial Theory, a multinucleated ciliate with seemingly bilateral symmetry might have been the evolutionary precursor to flat bilateral acoel flatworms. Cnidarians and ctenophores would have been derived secondarily from these earliest multicellular animals (Barnes, 1986). Although many other theories have been proposed, there is no overwhelming body of comparative morphological nor embryological data to support any single hypothesis (Brien, 1971; de Saedeleer, 1930; Grassé, 1971; Hanson, 1958, 1977; Laval, 1971; Morris, 1993; Nielsen, 1985, 1987; Salvini-Plawen, 1978; Tuzet, 1973; reviewed in Hyman, 1940; and Willmer, 1990).

Precambrian and Cambrian Fossil History

The inconsistent fossilization of organisms lacking mineralized hard parts or degradation-resistant organic materials constrains interpretation of Precambrian biological history (Knoll, 1992). Even so, Knoll (1992) described evidence of eukaryotes in the late paleoproterozoic and documented eukaryotic diversification in the mesoproterozoic from preserved organic compressions and impressions in sandstones and shales. The earliest known multicellular protist, a red alga, is preserved in silicified carbonate rocks of the 1250 to 750 million-year-old hunting formation (Butterfield, Knoll, and Swett, 1990). The presence of metazoa at that time is debatable as a molecular clock calibration places the initial radiation of metazoa at ~1,000 million years ago (mya) (Runnegar, 1982) while the first appearance of unskeletonized multicellular animals (Ediacaran faunas) was closer to the pre-Cambrian-Cambrian boundary (late Neoproterozoic) ~620 to 700 ma (Conway Morris, 1985, 1987, 1989*a;* Knoll and Walter, 1992). This was followed by the abrupt appearance of a diverse array of predominately skeletonized invertebrate phyla during the early Cambrian (Bergström, 1990) and a non skeletonized faunal component exemplified by those preserved in the Burgess Shale formation (Conway Morris, 1989*b*). This fauna may in fact be the survivors of the pre-Cambrian Ediacaran assemblages (Conway Morris, 1989*b*). Recent geochemical evidence suggests that the proterozoic-Cambrian boundary occurred 544 ma rather than 570 ma and the transition occurred over a 5–10 million year period (Bowring, Grotzinger, Isachsen, Knoll, Pelechaty, Kolosov, 1993). While geologic history provides information on the estimated time of emergence for many animals, it lends few clues about pre-Cambrian protists ancestral to contemporary metazoa. In part this represents lack of consensus about phenotypic characteristics that can be assigned to a protist ancestor for animals.

Molecular (Biochemical) Analyses to Date

Comparisons of ribosomal RNA sequences (rRNA) have established the phylogenetic relationships within the prokaryotic and protist worlds (Sogin, 1991; Woese, 1987), but similar procedures have given contradictory topologies for early diverging metazoan lineages (Christen, Ratto, Baroin, Perasso, Grell, and Adoutte, 1991; Erwin, 1991; Field, Olsen, Lane, Giovannoni, Ghiselin, Raff, Pace, and Raff, 1988; Lake, 1990). To date, distance and parsimony analyses of rRNA data portray protists as a series of independent branches that precede the nearly simultaneous separation of plants, animals, fungi, plus two novel and complex evolutionary assemblages described as stramenopiles (some chromophyte algae, diatoms, oomycetes, labyrinthulids, and other heterokont protists) and alveolates (ciliates, dinoflagellates, and apicomplexans) (Patterson and Sogin, 1992). Collectively these assemblages plus several other protist lineages define the crown of the eukaryotic tree (Knoll, 1992).

Relative branching order for the crown groups has been difficult to establish as the nodes separating each group are represented by fewer than five nucleotide changes per thousand positions. The absence of sequence data from early branching lineages within each of the crown groups and inadequate representation of eukaryotic microbial diversity has weakened the utility of molecular data in resolving the origins of animals (Christen et al., 1991; Field et al., 1988; Lake, 1990). In some studies, difficulties in reconstructing organismal phylogeny was further exacerbated by incomplete sequence information. Most analyses of animal origins have been based upon partial sequences and hence lack a statistically meaningful number of independently variable sites. Furthermore, these sequences may be subject to systematic errors which are not detected when only single stranded rRNA templates are available for sequence analyses. One percent sequence errors (Clark and Whittam, 1992) and inadequate methods for inferring molecular frameworks that include both slowly and rapidly evolving lineages (Christen et al., 1991; Field et al., 1988) can lead to major rearrangements of deep interior nodes or terminal taxa separated by short branches.

Complete small subunit ribosomal RNA (16S-like rRNA) sequences are now available for several lineages that diverged early in the metazoan radiation (Wainright, Hinkle, Sogin, and Stickel, 1993). Together with an expanded ribosomal sequence data base (Olsen, Overbeek, Larsen, Marsh, McCaughey, Maciukenas, Kuan, Macke, Xing, and Woese, 1992) this study employed maximum likelihood (Felsenstein, 1981) and distance techniques (Olsen, 1988) to explore relationships between animals and other eukaryotes. In these analyses, only unambiguously aligned nucleotide positions of complete 16S-like rRNA sequences were considered. The alignments were achieved using a computer assisted method that took into account conservation of both primary and secondary structural features in rRNA molecules (Elwood, Olsen, and Sogin, 1985; Olsen, 1988). An accelerated method for maximum likelihood analyses of DNA sequences (Olsen, Matsuda, Hagstrom, and Overbeek, 1994) using the generalized two parameter model of evolution (Kishino and Hasegawa, 1989) produced the phylogenetic tree in Fig. 1. With the use of jumbled orders for taxa addition and allowing global swapping to cross three branches, the search for an optimal tree was repeated until the best log likelihood score was reached in at least three independent searches.

Bootstrap techniques (Felsenstein, 1985) provided relative confidence levels for topological elements in the phylogenetic trees. Bootstrap values that exceed fifty percent in fastDNAml analyses and the corresponding values from resamplings for neighbor-joining methods (Saitou and Nei, 1987) are indicated in Fig. 1. Although the use of bootstrap values has gained widespread acceptance for measuring reliability in phylogenetic tree reconstructions, we do not consider them to be reliable estimates of statistical confidence. Bootstrap values are affected by the number of resamplings, the topology of the tree, and underlying models of molecular evolution. Lengthy computer runs for larger data sets can lead to lower confidence levels associated with fewer bootstrap resamplings (Hedges, 1992; Li and Gouy, 1990). Observed bootstrap values in molecular phylogeny reconstructions can be influenced by the number of deep interior nodes separated by short evolutionary distances as well as the fraction of sites that change on a given segment separating two nodes (Hillis and Bull, 1993). Even taxon representation in phylogenetic inferences can have profound effect on observed bootstrap values. For example, in the data set described (Fig. 1), bootstrap values uniting fungi and animals approach 98 percent when the protist *Acanthamoeba castellani* is excluded from the analysis. Rather than selectively excluding taxa (e.g., *Acanthamoeba*) that result in high bootstrap values, we included representatives from all the crown groups.

Consensus among topologies when different taxa and methods are selected for analyses may be a preferred method for measuring confidence in phylogenetic

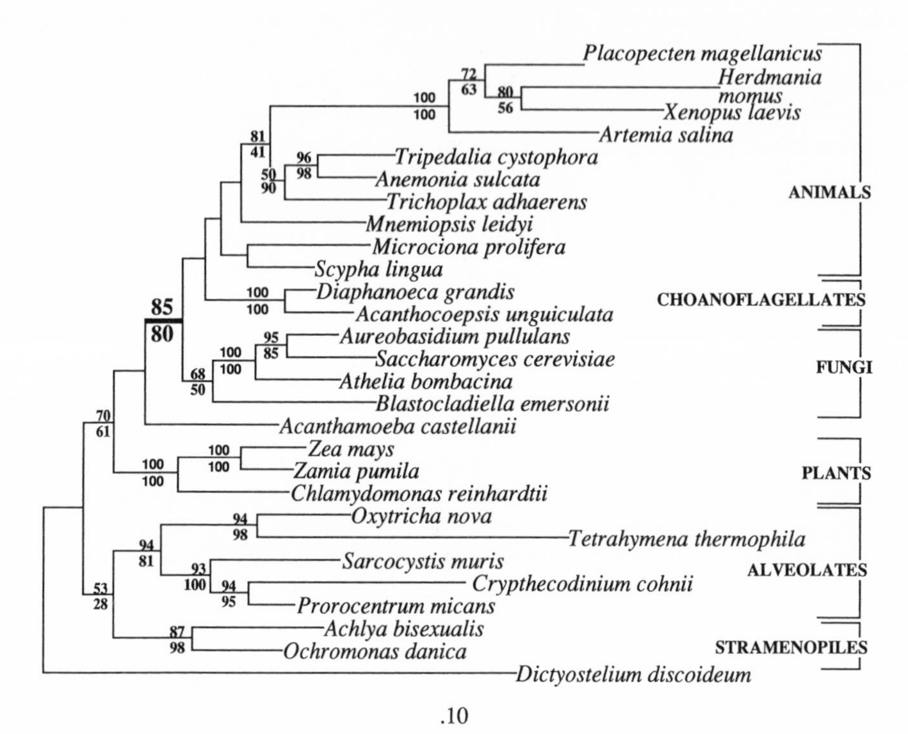

Figure 1. Relationships between animals, fungi, plants, and other eukaryotic groups inferred from complete 16S-like rRNAs. The percentage of 110 bootstrap resamplings that support topological elements in maximum likelihood inferences are shown above the branches: each resampling required 48 h to compute on a SUN Microsystems SPARC 2 Workstation. Values below the branches represent the fraction of 200 bootstrap neighbor joining replicates (Saitou and Nei, 1987) that support the evolutionary hypothesis. Confidence levels below fifty percent are not indicated in the figure. The root of the tree is placed within the *Dictyostelium discoideum* lineage.

reconstructions (Leipe, Gunderson, Nerad, and Sogin, 1993). A series of maximum likelihood trees were constructed for different combinations of fungal, ciliate, chlorophyte, and chrysophyte taxa with the choanoflagellates and diploblastic and triploblastic animals represented in Fig. 1. Maximum likelihood trees were also constructed with single representatives of choanoflagellates, sponges, and cnidaria. In all cases, the best tree topologies as judged by optimal log likelihood scores displayed branching patterns that were consistent with the phylogeny presented. The relationships between major groups in the maximum likelihood and neighbor joining trees were also found in maximum parsimony using the computer package PAUP 3.0. (Swofford, 1990).

Monophyletic Origins of Metazoa

The metazoa appear as a monophyletic lineage that shares a recent common ancestry with choanoflagellates. Surprisingly, the animal and fungal lineages share a more recent common ancestor than either does with the plant, alveolate, or stramenopile lineages. Within the metazoan subtree the divergence of the choanoflagellates, sponges, and the ctenophore were followed by the placozoan and cnidarians which directly preceded the triploblastic animal phyla. Although the bootstrap values do not substantiate any single branching order for choanoflagellates, sponges, and ctenophores, the topology in Fig. 1 is consistent when alternative protist, fungal, and lower metazoan taxa are included in maximum likelihood analyses.

Similarities of Choanoflagellates and Sponges

The early metazoan branching pattern inferred from the rRNA phylogeny confirms hypotheses about the relationship of choanoflagellates and sponges to diploblastic and triploblastic animals (James-Clark, 1867, 1868; Lameere, 1901, 1908). The cellular similarities shared between choanoflagellates and the choanocytes of sponges are well known (Hollande, 1952; Laval, 1971; Tuzet, 1973). The most obvious structures shared are the microvilli arranged in a collar-like structure surrounding a single flagellum (de Saedeleer, 1930; James-Clark, 1867, 1868; Laval, 1971) with the collar functioning as a filtering organelle (Fjerdingstad, 1961*a,b*). The microfilaments found within the microvilli are arranged in a similar manner in both cell types (Leadbeater, 1983*a*).

Additional ultrastructural features between choanocytes and choanoflagellates lend further support for their evolutionary relatedness. Choanoflagellates and metazoa have flat nondiscoid cristae in their mitochondria (Hibberd, 1975; Leadbeater and Manton, 1974). Similarity in the mitotic division process has recently been reported in choanoflagellates and animals (Karpov and Mylnikov, 1993). The flagellum of choanoflagellates and choanocytes both have a bilateral vane on either side of the axis (Brill, 1973; Hibberd, 1975). However, the flagellar vane found in sponges appear more robust and may not be comparable to those in choanoflagellates (Mehl and Relswig, 1991). There is a detailed report of the flagellar rootlet system in choanoflagellates (Hibberd, 1975) but without a comparable account of this architecture in choanocytes, the degree of identity between these cells remains unclear.

The Evolution of Multicellularity

Multicellularity is here regarded as that state where coexisting cells with the same genome (with differential expression) carry out different but complementary functions. Cellular differentiation of this nature has been acquired on many occasions (for example: red algae, brown algae, and slime moulds) and is usually accompanied by an increase in organismal size. Together, differentiation and large size can only be achieved through the emergence of cellular support and intercellular communication. The molecular data presented in Fig. 1 is largely consistent with our traditional and emerging understanding of the structural basis for the acquisition of multicellularity in animals. It is consistent with information on nonepitheliate sponges and placozoa, for the diploblastic ctenophores and cnidaria, and for the triploblasts. The status of the mesozoa still awaits clarification.

The Extracellular Matrix

Structural support for animal cells has been achieved through the elaboration of an intercellular matrix (ECM, Morris, 1993). The ECM is composed of several substances but principally includes fibrillar collagens, nonfibrillar collagens, and noncollagenous materials such as laminin and fibronectin (Pedersen, 1991). These may be present as a matrix with little organization (as in mesohyl and mesogloea) or may exist in a complex assemblage, the basement membrane. The assembly of a basement membrane is one of the primary features responsible for organization of cells into the sheetlike arrays (epithelia) from which higher levels of complexity have developed.

The origins of this matrix may lie within the choanoflagellates. But some of these flagellates (acanthocoecids) exploit siliceous elements as structural devices (Buck, 1990). One genus, the colonial *Proterospongia,* has cells embedded within a central jellied matrix of unknown composition (Kent, 1880–1882; Leadbeater, 1983*b*). Within the sponges, the extracellular matrix takes the form of the mesohyl. A major component of the mesohyl is fibrillar collagen or spongin (Bergquist, 1978; De Vos, 1977; Garrone, 1984; Garrone, 1985; Gross, 1985; Harrison and De Vos, 1991; Pedersen, 1991). The presence of fibrillar collagen leads to the assertion that mesohyl is homologous to the ECM of other animals (Bergquist, 1978). Additional components of the mesohyl include a fibronectinlike protein (Labat-Robert, Robert, Auger, Lethias, and Garrone, 1981) and a nonfibrillar collagen similar to Type IV (Exposito, Ouazana, and Garrone, 1990). However, unlike more complex animals, sponge cells are not arranged in an epithelia. The ECM components together with siliceous and calcareous spicules, provided sponges with a system for support for an integrated array of sophisticated and specialized cells (Wood, 1991). Sponge cells within this matrix do not form tight connections (such as desmosome structures found between true epithelial cells) with the mesohyl or to each other. Some cells, such as the pinacocytes, form less organized layers: the pinacoderm.

The integration of collagen with other noncollagenous materials for the formation of an ECM creates a structural argument that the sponges and animals are a monophyletic group (Morris, 1993; Pedersen, 1991). However, collagen alone should not be regarded as the synapomorphy of the animals as this material has also been reported in foraminifera (Hedley and Wakefield, 1967). This report needs to be confirmed and further efforts made to establish if collagen is produced by other protists.

Within the remaining animals (placozoa and mesozoa excepted), the ECM takes the form of a complex basement membrane which intercedes between epithelia and connective tissues. The basement membrane has three distinct layers: lamina lucida, basal lamina, and lamina reticularis (Pedersen, 1991). This pattern of organization has not been reported in sponges, but structures resembling the lamina reticularis (Harrison, Kaye, and Kaye, 1990) and basal lamina (Pedersen, 1991) have been reported. Other features which some or all sponges share with the remaining animals are the occurrence of spermatogenesis (Harrison and De Vos, 1991), acrosomate sperm (Ax, 1989), polar bodies (Ax, 1989), yolky eggs (Kaye, 1990), aspects of early embryonic development (Misevic, Schlup, and Burger, 1990), and septate junctions (see below) (Green and Bergquist, 1979; Ledger, 1975). The concordance of molecular and structural data is consistent with placing the sponges as a group diverging from the base of the animal lineage (Brien, 1971; Tuzet, 1970, 1973), and provides no support for the argument that the sponges are an independent or dead-end lineage (Delage, 1892; Sollas, 1884).

Cell junctions. Junctional complexes (=cell junctions) are differentiations of the surface of cell and underlying cytoplasm. Desmosomes and septate junctions permit adhesion of cells to each other or to the extracellular matrix. Tight and septate junctions prevent the passage of fluids between cells and gap junctions enhance communication between cells (Staehelin, 1974). The evolution of these complexes was integral to the acquisition of animal multicellularity. Among the animals, sponges show lesser variety of junctional complexes. Intercellular structures include: simple parallel membrance structures (Felge, 1969; Green and Bergquist, 1979, 1982) as well as structures reported as septate junctions (Green and Bergquist, 1979; Ledger, 1975; Mackie and Singla, 1983). Organized particle arrays in membranes have been described from freeze-fracture studies (Garrone and Lethias, 1990).

In addition, structures which are reported as being 'desmosome-like' have been reported (Lethias, Garrone, and Mazzorana, 1983; Pavans de Ceccaty, 1986). Desmosomes may be defined critically as structures occurring between two parallel membranes of adjacent cells consisting of pentalaminate structures. Two layers include thickened plasma membranes with the outer layer of each with an electron-lucent layer. A central electron-dense layer is sandwiched between the plasma membranes. The cytoplasmic side of the desmosome gives rise to tonofilaments or cytokeratins which penetrate into the cell. This type of structure has been observed in placozoans, ctenophores, cnidaria, and other higher animals as well as in Myxozoan sporozoa. The evolution of these structures has apparently accompanied the transformation of the ECM into a more organized layer. Precisely this type of structure has yet to be reported in sponges, and the use of the term 'desmosome-like' is potentially misleading.

Ctenophores have desmosomes (arrayed as spots or belts) and gap junctions (Hernandez-Nicaise, 1991), but tight and septate junctions have yet to be reported. In cnidaria, a more comprehensive array of junctional complexes have been reported (Filshie and Flower, 1977; Hand and Globel, 1972; King and Spencer, 1979; Wood, 1959). Three types of septate junctions have been observed in cnidarians; one, found in hydrozoans, compares more closely to those found in the triploblasts (pleated septate junctions with narrow septa). Two different types with broad septa occur in the anthozoans (Green and Bergquist, 1982). Although there are several reports of gap junctions in the hydrozoans (Hand and Gobel, 1972; Flower, 1977; Mackie, Nielsen, and Singla, 1989), none have been found in anthozoans or scyphozoans (Mackie, Anderson, and Singla, 1984). Although, from the perspective of junctional complexes, ctenophores appear less complex than cnidaria, both groups are diploblastic and exploit relatively disorganized ECM (mesogloea) as a supporting material. Basement membranelike structures are found in cnidarians (Pedersen, 1991) and true basal laminae in ctenophores (Hernandez-Nicaise, 1991). Ctenophora have relatively well developed musculature, nervous and sensory systems, and excretory systems (Harbison, 1985; Hyman, 1940). The exact relationship between the ctenophores and the coelenterates remains in dispute. Because both groups are diploblastic, and because of developmental similarities (Hernandez-Nicaise, 1991) we regard them as closely related.

One group of organisms normally assigned to the protozoa, the myxozoa also have desmosomes that satisfy the definition above. They are multicellular organisms in which the spore filaments show developmental, structural, and functional similarities with the cnidarian nematocysts. For this reason, it has been argued that this group is assignable to the cnidaria (Patterson, 1994; Patterson and Sogin, 1993).

The position of the placozoa is not clear. Placozoa lack basal lamina and have been reported as having belt desmosomes (Grell and Ruthmann, 1991; Ruthmann, Behrendt, and Wahl, 1986) a piece of structural evidence which corroborates the molecular evidence that they are a bona fide member of the animal lineage. If the branching pattern in Fig. 1 is valid, then this group must have lost a number of junctional complexes during their evolutionary history.

Together molecular and structural studies provide a robust concept which endorses the monophyly of the group (with Mesozoa remaining of uncertain status). A progressive sophistication of the ECM and of junctional complexes in sponges, ctenophores, and cnidaria (embracing Myxozoa) provides evidence of the monophyly of the diploblastic and triploblastic animals. This process has achieved the tissue grade of organization as distinguished by cells organized into epithelia found in the more complex animals.

Relationship of the Animals and Fungi

The maximum likelihood analysis in Fig. 1 shows a close phylogenetic relationship between animals and fungi. Independent confirmation of a common evolutionary history for animals and fungi is provided by maximum likelihood analyses of amino acid sequences for elongation factors (Hasegawa, 1993) and from molecular systematic studies of other gene families including α-tubulins, β-tubulins, elongation factors and actins (Baldauf and Palmer, 1993). At different times, fungi have been considered to be plants, members of other protist groups, or in more contemporary schemes, worthy of kingdom level status (Whittaker, 1959). Most conventional phylogenies place the origin of animals among protists. Towe (1981) and Cavalier-Smith (1987) argue that similarities in complex biosynthetic pathways including syntheses of hydroxyproline, chitin, and even ferritin suggest a specific evolutionary relationship between fungi and animals. And, even though cellulose which is most prevalent in plants has been detected in some tunicates, chitin is the most abundant structural polysaccharide in fungi and animals (Brown, 1990). Molecular phylogenies based upon 16S-like rRNA sequence comparisons (Bowman, Taylor, Brownlee, Lee, Lu, and White, 1992; Bruns, Vilgalys, Barns, Gonzalez, Hibbett, Lane, Simon, Stickel, Szaro, Weisberg, and Sogin, 1992) and ultrastructural features observed during mitosis (Fuller, 1976) describe the higher fungi as a monophyletic group. The chytrids with a flagellated zoospore represent the earliest diverging lineage of the fungi. Because chytrids and choanoflagellates both have flattened mitochondrial cristae and a single posterior flagellum, the hypothetical protist representing the most recent common ancestor to the animals and fungi in the rRNA phylogeny very likely was a unicellular protist with these ultrastructural characteristics. This new rRNA phylogeny that recognizes animal origins and their relationship with the fungi provides us with a framework and a rational basis through which we may explore the origins and diversifications of metazoan phenotypes.

Acknowledgements

We thank S. C. Wainright for comments on this manuscript.

This manuscript is Rutgers University contribution number 93-39. This work was supported by NIH GM32964 to M. L. Sogin and the G. Unger Vetlesen Foundation to the Marine Biological Laboratory at Woods Hole.

References

Ax, P. 1989. Basic phylogenetic systematization of the metazoa. *In* The Hierarchy of Life. H. Fernholm, K. Bremer, and J. Jornwall, editors. Elsevier Science Publishers, Cambridge, England. 229–245.

Baldauf, S. L., and J. D. Palmer. 1993. Animals and fungi are each other's closest relatives: evidence from multiple proteins. *Proceedings of the National Academy of Sciences, USA.* 90:11558–11562.

Barnes, R. D. 1986. Introduction to the metazoa. *In* Invertebrate Zoology, Saunders College Publishing, Philadelphia, PA. 63–65; 71–91.

Bergquist, P. R. 1978. Sponges. University of California Press, Berkeley, CA. 84–90; 140–143.

Bergström, J. 1990. Precambrian trace fossils and the rise of bilaterian animals. *Ichnos.* 1:3–14.

Bowman, B. H., J. W. Taylor, A. G. Brownlee, J. Lee, S-D. Lu, and T. J. White. 1992. Molecular evolution of the fungi: relationship of the Basidiomycetes, Ascomycetes, and Chytridiomycetes. *Molecular Biology and Evolution.* 9:285–296.

Bowring, S. A., J. P. Grotzinger, C. E. Isachsen, A. H. Knoll, S. M. Pelechaty, and P. Kolosov. 1993. Calibrating rates of early Cambrian evolution. *Science.* 261:1293–1298.

Brien, P. 1971. Phylogenese des metazoaires diploblastiques. Passage phylogenetique de la diploblastie a la triploblastie. *In* Actas del I Simposio Internacional de Zoofilogenia. R. Alvarado, and E. Gadea y A. de Haro, editors. *Acta Salmanticensia.* 193–202.

Brill, B. 1973. Untersuchungen zur ultrastruktur der choanocyte von *Ephydatia fluviatilis* L. *Zeitschrift für Zellforschung und Mikroskopische Anatomie.* 144:231–245.

Brown, R. M. Jr. 1990. Algae as tools in studying the biosynthesis of cellulose, nature's most abundant macromolecule. *In* Experimental Phycology: Cell Walls and Surfaces, Reproduction, Photosynthesis. W. Wiessner, D. G. Robinson, and R. C. Starr, editors. Springer-Verlag, Berlin. 20–39.

Bruns, T. D., R. Vilgalys, S. M. Barns, D. Gonzalez, D. S. Hibbett, D. J. Lane, L. Simon, S. Stickel, T. M. Szaro, W. G. Weisburg, and M. L. Sogin. 1992. Evolutionary relationships within the fungi: analyses of nuclear small subunit rRNA sequences. *Molecular Phylogenetics and Evolution.* 1:231–241.

Buck, K. R. 1990. Phylum zoomastigina, class choanomastigotes (choanoflagellates). *In* Handbook of Protoctista. L. Margulis, J. O. Corliss, M. Melkonian, and D. J. Chapman, editors. Jones and Bartlett Publishers, Boston, MA. 194–199.

Butterfield, N. J., A. Knoll, and K. Swett. 1990. A bangiophyte red alga from the Proterozoic of Arctic Canada. *Science.* 250:104–107.

Cavalier-Smith, T. 1987. The origin of fungi and pseudofungi. *In* Evolutionary Biology of the Fungi. British Mycological Society Symposium 12. A. D. M. Rayner, C. M. Brasier, and D. Moore, editors. Cambridge University Press, Cambridge, UK. 339–353.

Christen, R., A. Ratto, A. Baroin, R. Perasso, K. G. Grell, and A. Adoutte. 1991. An analysis of the origin of metazoans, using comparisons of partial sequences of the 28S RNA reveals an early emergence of triploblasts. *EMBO Journal.* 10:499–503.

Clark, A. G., and T. S. Whittam. 1992. Sequencing errors and molecular evolutionary analysis. *Molecular Biology and Evolution.* 9:744–752.

Conway-Morris, S. 1985. Non-skeletalized lower invertebrates: a review. *In* The Origins and Relationships of Lower Invertebrates. S. Conway Morris, J. D. George, R. Gibson, and H. M. Platt, editors. Clarendon Press, Oxford, UK. 343–359.

Conway Morris, S. 1987. The search for the Precambrian-Cambrian boundary. *American Scientist.* 75:156–167.

Conway Morris, S. 1989*a*. Early metazoans. *Science Progress.* 73:81–99.

Conway Morris, S. 1989*b*. Burgess shale faunas and the cambrian explosion. *Science.* 246:339–346.

de Saedeleer, H. 1930. Recherches sur les choanocytes; l'origine des Spongiarires. *Annales de la Société Royale Zoologique de Belgique.* 60:16–21.

Delage, Y. 1892. Embryongénie des éponges: développement post-larvaire des éponges siliceuses et fibreuses marines et d'eau douce. *Archives de Zoologie Expérimentale et Générale.* 10:345–498.

De Vos, L. 1977. Morphogenesis of the collagenous shell of the gemmules of a fresh-water sponge *Ephydatia fluviatilis. Archives de Biologie.* 88:479–494.

Elwood, H. J., G. J. Olsen, and M. L. Sogin. 1985. The small-subunit ribosomal RNA gene sequences from the hypotrichous ciliates *Oxytrica nova* and *Stylonychia pustulata. Molecular Biology and Evolution.* 2:399–410.

Erwin, D. H. 1991. Metazoan phylogeny and the Cambrian radiation. *Trends in Ecology and Evolution.* 6:131–134.

Exposito, J-Y., R. Ouazana, and R. Garrone. 1990. Cloning and sequencing of a Porifera partial cDNA coding for a short-chain collagen. *European Journal of Biochemistry.* 190:401–406.

Felge, V. W. 1969. Die feinstruktrur der epithellen von *Ephydatia fluviatills. Zoologische Jahrbücher Abteilung für Anatomie und Ontogenie der Tiere.* 86:177–237.

Felsenstein, J. 1981. Evolutionary trees from DNA sequences: a maximum likelihood approach. *Journal of Molecular Evolution.* 17:368–376.

Felsenstein, J. 1985. Confidence limits on phylogenies: an approach using the bootstrap. *Evolution.* 39:783–791.

Field, K. G., G. J. Olsen, D. J. Lane, S. J. Giovannoni, M. T. Ghiselin, E. C. Raff, N. R. Pace, and R. A. Raff. 1988. Molecular phylogeny of the animal kingdom. *Science.* 239:748–753.

Filshie, B. K., and N. E. Flower. 1977. Junctional structures in *Hydra. Journal of Cell Science.* 23:151–172.

Flower, N. E. 1977. Invertebrate gap junctions. *Journal of Cell Science.* 25:163–171.

Fjerdingstad, E. J. 1961*a*. The ultrastructure of choanocyte collars in *Spongilla lacustris* (L.). *Zeitschrift für Zellforschung.* 53:645–657.

Fjerdingstad, E. J. 1961*b*. Ultrastructure of the collar of the choanoflagellate *Codonosiga* botrytis (Ehrenb.). *Zeitschrift für Zellforschung.* 54:499–510.

Fuller, M. S. 1976. Mitosis in fungi. *International Review of Cytology.* 45:113–153.

Garrone, R. 1984. Formation and involvement of extracellular matrix in the development of sponges. A primitive multicellular system. *In* The Role of Extracellular Matrix in Development. R. L. Treland, editor. Alan R. Liss, Inc., NY. 461–477.

Garrone, R. 1985. The collagen of the porifera. *In* Biology of the Invertebrates. A. Bairati and R. Garrone, editors. Plenum Publishing Corp., NY. 157–175.

Garrone, R., and C. Lethias. 1990. Freeze-fracture study of sponge cells. *In* New Perspectives in Sponge Biology. Third International Conference on Biology of Sponges. Woods Hole, MA, 1985. K. Kutzler, editor. Smithsonian Institution Press, Washington, DC. 121–128.

Grassé, P-P. 1971. Les protozoaires sont-ils les ancetres des metazoaires? *In* Actas del I Simposio Internacional de Zoofilogenia. R. Alvarado and E. Gadea y A. de Haro, editors. *Acta Salmanticensia.* 65–92.

Green, C. R., and P. R. Bergquist. 1979. Cell membrane specialisations in the Porifera. *In* Colloques Internationaux du Centre National de la Recherche Scientific, No. 291. Biologie des Spongiaires, Paris, 18–22 Décembre, 1978. Quai Anatole, Paris, France. 153–158.

Green, C. R., and P. R. Bergquist. 1982. Phylogenetic relationships within the invertebrata in relation to the structure of septate junctions and the development of occluding junctional types. *Journal of Cell Science.* 53:279–305.

Grell, K. G., and A. Ruthman. 1991. Placozoa. *In* Microscopic Anatomy of Invertebrates. Vol. 2. F. W. Harrison and J. A. Westfall, editors. Wiley-Liss, NY. 14–17.

Gross, G. 1985. Invertebrate collagens in the scheme of things. *In* Biology of the Invertebrates. A. Bairati and R. Garrone, editors. Plenum Publishing Corp., NY. 1–28.

Hadzi, J. 1953. An attempt to reconstruct the system of animal classification. *Systematic Zoology.* 2:145–154.

Haeckel, E. 1874. The Gastraea-theory, the phylogenetic classification of the animal kingdom and the homology of the germ-lamellae. *Quarterly Journal of Microscopical Science.* 14:142–165; 14:223–247.

Hand, A. R., and S. Globel. 1972. The structural organization of septate and gap junctions of *Hydra. The Journal of Cell Biology.* 52:397–408.

Hanson, E. D. 1958. On the origin of the Eumetazoa. *Systematic Zoology.* 7:16–47.

Hanson, E. D. 1977. The Origin and Early Evolution of Animals. Wesleyan University Press, Middletown, CT. 670 pp.

Harbison, G. R. 1985. On the classification of the Ctenophora. *In* The Origins and Relationships of Lower Invertebrates. S. Conway Morris, J. D. George, R. Gibson, and H. M. Platt, editors. Clarendon Press, Oxford, England. 78–100.

Harrison, F. W., and L. De Vos. 1991. Porifera. *In* Microscopic Anatomy of Invertebrates. Vol. 2, F. W. Harrison and J. A. Westfall, editors. Wiley-Liss, NY. 29–87.

Harrison, F. W., N. W. Kaye, and G. W. Kaye. 1990. The dermal membrane of *Eunapius fragilis. In* New Perspectives in Sponge Biology. K. Kutzler, editor. Third International Conference on Biology of Sponges. Woods Hole, MA, 1985. Smithsonian Institution Press, Washington, DC. 223–227.

Hasegawa, M. 1993. Origin and evolution of eukaryotes as inferred from protein sequence data. *In* The Origin and Evolution of Prokaryotic and Eukaryotic Cells. H. Hartmann and K. Matsuno, editors. World Scientific, Singapore. 107–130.

Hedley, R. H., and J. St. J. Wakefield. 1967. A collagen-like sheath in the arenaceous foraminifer *Haliphysema* (Protozoa). *Journal of the Royal Microscopical Society.* 87:475–481.

Hedges, S. B. 1992. The number of replications needed for accurate estimation of the bootstrap *P* value in phylogenetic studies. *Molecular Biology and Evolution.* 9:366–369.

Hernandez-Nicaise, M.-L. 1991. Ctenophora. *In* Microscopic Anatomy of Invertebrates. Vol. 2, F. W. Harrison and J. A. Westfall, editors. Wiley-Liss, NY. 364–367.

Hibberd, D. J. 1975. Observations on the ultrastructure of the choanoflagellate *Codosiga botrytis* (EHR.) Saville-Kent with special reference to the flagellar apparatus. *Journal of Cell Science.* 17:191–219.

Hillis, D. M., and J. J. Bull. 1993. An empirical test of bootstrapping as a method for assessing confidence in phylogenetic analysis. *Systematic Biology.* 42:182–192.

Hollande, A. 1952. Ordre des Choanoflagellés ou Craspédomonadines. *In* Traité de Zoologie, Introduction (Protozaires) Generalités, Flagellés. Grassé, P-P, editor. Masson et Cie, Paris, France. 1:579–598.

Hyman, L. H. 1940. The Invertebrates: Protozoa through Ctenophora. McGraw-Hill, NY. 248–283.

James-Clark, H. 1867. Conclusive proofs of the animality of the ciliate sponges, and of their affinities with the infusora flagellata. *The Annals Magazine of Natural History.* 19:13–18.

James-Clark, H. 1868. On the spongiae ciliatae as infusoria flagellata; or observations on the structure, animality, and relationship of *Leucosolenia botryoides,* Bowerbank. *The Annals Magazine of Natural History.* 1:133–142; 1:188–215; 1:250–264.

Karpov, S. A., and A. P. Mylnikov. 1993. Preliminary observations on the ultrastructure of mitosis in choanoflagellates. *European Journal of Protistology.* 29:19–23.

Kaye, H. R. 1990. Reproduction in West Indian commercial sponges: oogenesis, larval development, and behavior. *In* New Perspectives in Sponge Biology. Third International Conference on Biology of Sponges, Woods Hole, MA, 1985. K. Kutzler, editor. Smithsonian Institution Press, Washington, DC. 161–169.

Kent, W. Saville-. 1880–1882. In Manual of the Infusoria. D. Bogue, London. 1:363–365; Plate X, Figure 3.

King, M. G. and A. N. Spencer. 1979. Gap and septate junctions in the excitable endoderm of *Polyorchis penicillatus* (Hydrozoa, Anthomedusae). *Journal of Cell Science.* 36:391–400.

Kishino, H., and M. Hasegawa. 1989. Evaluation of the maximum likelihood estimate of the evolutionary tree topologies from DNA sequence data, and the branching order in Hominodea. *Journal of Molecular Evolution.* 29:170–179.

Knoll, A. 1992. The early evolution of eukaryotes: a geological perspective. *Science.* 256:622–627.

Knoll, A., and M. R. Walter. 1992. Latest proterozoic stratigraphy and earth history. *Science.* 356:673–678.

Labat-Robert, J., L. Robert, C. Auger, C. Lethias, R. Garrone. 1981. Fibronectin-like protein in Porifera: its role in cell aggregation. *Proceedings of the National Academy of Sciences, USA.* 78:6261–6265.

Lake, J. A. 1990. Origin of the metazoa. *Proceedings of the National Academy of Sciences, USA.* 87:763–766.

Lameere, M. 1901. De l'origine des éponges. *Annales de la Société Royale Malacologique de Belgique.* 36:7–8.

Lameere, M. 1908. Eponge et polype. *Annales de la Société Royale Zoologique et Malacologique de Belgique.* 43:107–124.

Lankester, E. R. 1877. Notes on embryology and classification of the animal kingdom: comprising a revision of speculation relative to the origin and significance of germ layers. *Quarterly Journal of Microscopical Science.* 17:399–354.

Laval, M. 1971. Ultrastructure et mode de nutrition du choanoflagellé *Salpingoeca pelagica,* sp. nov. comparaison avec les choanocytes des spongiaires. *Protistologica.* 7:325–336.

Leadbeater, B. S. C. 1983*a*. Distribution and chemistry of microfilaments in choanoflagellates, with special reference to the collar and other tentacle systems. *Protistologica.* 19:157–166.

Leadbeater, B. S. C. 1983*b*. Life-history and ultrastructure of a new marine species of *Proterospongia* (Choanoflagellida). *Journal of the Marine Biological Association.* 63:135–160.

Leadbeater, B. S. C., and I. Manton. 1974. Preliminary observations on the chemistry and biology of the lorica in a collared flagellate (*Stephanoeca diplocostata* Ellis). *Journal of the Marine Biological Association.* 54:269–276.

Ledger, P. W. 1975. Septate junctions in the calcareous sponge *Sycon ciliatum. Tissue and Cell.* 7:13–18.

Leipe, D., J. H. Gunderson, T. A. Nerad, and M. L. Sogin. 1993. Small subunit ribosomal RNA^+ of *Hexamita inflata* and the quest for the first branch in the eukaryotic tree. *Molecular and Biochemical Parasitology.* 59:41–48.

Lethias, C., R. Garrone, and M. Mazzorana. 1983. Fine structure of sponge cell membranes: comparative study with freeze-fracture and conventional thin section methods. *Tissue and Cell.* 15:523–535.

Li, W.-H., and M. Gouy. 1990. Statistical tests of molecular phylogenies. *Methods in Enzymology.* 183:645–659.

Mackie, G. O., P. A. V. Anderson, and C. L. Singla. 1984. Apparent absence of gap junctions in two classes of Cnidaria. *Biological Bulletin.* 167:120–123.

Mackie, G. O., C. Nielsen, and C. L. Singla. 1989. The tentacle cilia of *Aglantha digitale* (Hydrozoa: Trachylina) and their control. *Acta Zoologica.* 70:133–141.

Mackie, G. O., and C. L. Singla. 1983. Studies on Hexactinellid sponges. I. Histology of *Rhabocalyptus dawsonii* (Lambei, 1873). *Philosophical Transactions of the Royal Society of London.* 301:365–400.

Mehl, D., and H. M. Relswig. 1991. The presence of flagellar vanes in choanomeres of Porifera and their possible phylogenetic implications. *Zeitschrift für Zoologische Systematik und Evolutionsforschung.* 29:312–319.

Metschnikoff, E. 1886. Embryologische Studien an Medusen, mit Atlas. Ein Beitrag zür Genealogie der Primitiv-Organe. A Hölder, Wein, Austria. 1–159.

Misevic, M. N., V. Schlup, and M. M. Burger. 1990. Larval metamorphosis of *Microciona prolifera:* evidence against the reversal of layers. *In* New Perspectives in Sponge Biology. Third International Conference on Biology of Sponges. Woods Hole, MA, 1985. K. Kutzler, editor. Smithsonian Institution Press, Washington, DC. 182–187.

Morris, P. J. 1993. The developmental role of the extracellular matrix suggests a monopnylectic origin for the kingdom Animalia. *Evolution.* 47:152–165.

Nielsen, C. 1985. Animal phylogeny in the light of the trochaea theory. *Biological Journal of the Linnean Society.* 25:243–299.

Nielsen, C. 1987. Structure and function of metazoan ciliary bands and their phylogenetic significance. *Acta Zoologica.* 68:205–262.

Olsen, G. J. 1988. Phylogenetic analysis using ribosomal RNA. *Methods in Enzymology.* 164:793–812.

Olsen, G. M. Matsuda, R. Hagstrom, and R. Overbeek. 1994. FastDNAml: a tool for construction of phylogenetic trees of DNA sequences using maximum likelihood. *Cabios.* In press.

Olsen, G. J., R. Overbeek, N. Larsen, T. L. Marsh, M. J. McCaughey, M. A. Maciukenas, M.-M. Kuan, T. J. Macke, Y. Xing, and C. R. Woese. 1992. The ribosomal database project. *Nucleic Acids Research.* 20:2199–2200.

Patterson, D. J. 1994. Protozoa: evolution and systematics. *European Journal of Protistology.* In press.

Patterson, D. J., and M. L. Sogin. 1992. Eukaryote origins and protistan diversity. *In* The Origin and Evolution of the Cell. H. Hartmann and K. Matsuno, editors. World Scientific Publishers, Singapore. 13–46.

Pavans de Ceccaty, M. 1986. Cytoskeletal organization and tissue patterns of epithelia in the sponge *Ephydatia mulleri. Journal of Morphology.* 189:45–65.

Pedersen, K. J. 1991. Invited review: structure and composition of basement membranes and other basal matrix systems in selected invertebrates. *Acta Zoologica.* 72:181–201.

Ruthmann, A., G. Behrendt, and R. Wahl. 1986. The ventral epithelium of *Trichoplax adhaerens* (Placozoa): cytoskeletal structures, cell contacts, and endocytosis. *Zoomorphology.* 106:115–122.

Runnegar, B. 1982. A molecular-clock date for the origin of the animal phyla. *Lethaia.* 15:199–205.

Saitou, N., and M. Nei. 1987. The neighbor-joining method: a new method for reconstructing phylogenetic trees. *Molecular Biology and Evolution.* 4:406–425.

Salvini-Plawen, L. v. 1978. On the origin and evolution of the lower metazoa. *Zeitschrift für Zoologische Systematik und Evolutionsforschung.* 16:40–88.

Sogin, M. L. 1991. Early evolution and the origin of eukaryotes. *Current Opinion in Genetics and Development.* 1:457–463.

Sollas, W. J. 1884. On the development of *Halisarca lobularis. Quarterly Journal of Microscopical Science.* 24:603–621.

Staehelin, L. A. 1974. Structure and function of intercellular junctions. *International Review of Cytology.* 39:191–283.

Swofford, D. L. 1990. PAUP. Phylogenetic analysis using parsimony, Version 3.0. Computer program distributed by the Illinois Natural History Survey, Champaign, IL.

Towe, K. M. 1981. Biochemical keys to the emergence of complex life. *In* Life in the Universe. J. Billingham, editor. MIT Press, Cambridge, MA. 297–306.

Tuzet, O. 1970. La signification des porifères pour l'évolution des Métazoaires. *Zeitschrift für Zoologische Systematik und Evolutionsforschung.* 8:119–128.

Tuzet, O. 1973. Introduction et place des spongiaires dans la classification. In Traité de Zoologie, Anatomie, Systématique, Biologie. P-P. Grassé, editor. Masson et Cie, Paris, France. 3:1–26.

Wainright, P. O., G. Hinkle, M. L. Sogin, and S. K. Stickel. 1993. Monophylectic origins of the metazoa: an evolutionary link with fungi. *Science.* 260:340–342.

Whittaker, R. H. 1959. On the broad classification of organisms. The Quarterly Review of Biology. 34:210–226.

Willmer, P. 1990. The origin of the metazoa. *In* Invertebrate Relationships: Patterns in Animal Evolution. Cambridge University Press, Cambridge, UK. 162–197.

Woese, C. R. 1987. Bacterial evolution. *Microbiological Reviews.* 51:221–271.

Wood, R. L. 1959. Intercellular attachment in the epithelium of *Hydra* as revealed by electron microscopy. *Journal of Biophysical and Biochemical Cytology.* 6:343–352.

Wood, R. L. 1991. Problematic reef-building sponges. *In* The Early Evolution of Metazoa and the Significance of Problematic Taxa. Proceedings of the International Symposium, University of Camerino, 1989. A. M. Simonetta and S. Conway Morris, editors. Cambridge University Press, Cambridge, UK. 99–105.

To Tree the Truth: Biological and Numerical Simulations of Phylogeny

David M. Hillis and John P. Huelsenbeck

Department of Zoology, The University of Texas at Austin, Austin, Texas 78712

The reconstruction of phylogenetic history has become an integral part of all comparative biological studies over the past few decades (e.g., see Brooks and McLennan, 1991; Harvey and Pagel, 1991; Hillis and Moritz, 1990; Maddison and Maddison, 1992). The range of applications of phylogenetic inference is immense: phylogenies are used for everything from tracking infections of viruses within human populations (e.g., Ou et al., 1992) to studying the evolution of sex determining mechanisms across hundreds of millions of years (e.g., Hillis and Green, 1990) to tracing the earliest lineages of life billions of years ago (e.g., Olsen, 1987). However, it is obviously not possible to go back in time and directly observe any of these phylogenies, so how can we know if phylogenetic methods are finding the correct phylogenies? As with most scientific theories and methods, there are two choices to evaluate the validity of phylogenetic techniques: empirical and theoretical experimentation (or to put it in other terms, biological and numerical simulation). The purpose of this review is to examine the results of both types of studies with regard to performance of phylogenetic methods, and then to make general recommendations about selecting a method for use.

Numerical Versus Biological Simulations

To date, most evaluations of phylogenetic methods have involved numerical simulations: an investigator defines a simple model of evolution, creates some sequences (perhaps at random), and specifies a tree or some rules for generating a tree (e.g., a Markov process of speciation). The investigator then applies the model of evolution to the given sequences and tree, and a computer program carries out the tasks of assigning mutations and recording successive generations of new sequences. After the sequences have "evolved" in computer memory, the various methods of phylogenetic inference can be tested to see which ones perform the most effectively. The advantages of simulating phylogenies in this manner include the ability to generate a sample of thousands or millions of phylogenies with great ease and the ability to generate any conceivable phylogeny. Thus, we can choose some aspect of trees to investigate, define the parameter space of interest, and then examine samples of trees from throughout this parameter space. The limitations of the approach lie only in our ability to identify relevant problems and in computational limitations in analyzing the simulated data sets.

Given the flexibility of the numerical simulation approach, why would we ever turn to experiments with real organisms? There are two principal reasons: we are painfully ignorant about the details of molecular evolution, and computer simulations, by necessity, incorporate gross simplifications of evolutionary processes. The most complex of computer simulations still make sweeping generalizations and simplifying assumptions about how organisms evolve. This will likely always be the case, because a computer simulation that did not simplify evolutionary processes would have to be as complex as a real functioning organism. As an example, the vast majority of simulations of molecular evolution to date have defined one or two parameters associated with mutation rates; typically, there is either a single mutation rate or two different rates for transitions and transversions, respectively. The most complex simulations may specify as many as twelve different mutation rates for the twelve possible changes that can occur among nucleotides. However, in a real sequence, these rates are likely to differ in ways we are yet to understand across every position in a given gene. Many simulations also assume a constant rate of change across all positions, a situation we know to be very different from what actually occurs in real sequences. There are also complexities that we can imagine but are difficult to model: there may be complex interactions among different nucleotide positions (e.g., having to do with RNA or protein secondary structure, or related to the binding of control sequences), or there may be fluctuating kinds and levels of selection at various developmental stages. Of course, given that we could thoroughly understand such complexities, we could incorporate them into simulations. Unfortunately, our knowledge of molecular evolution is far too rudimentary to develop any but the simplest models at present, and even if we had complete knowledge, it would be computationally intractable to develop such detailed models. Therefore, we need some check on the simulations to see to what extent our simplifications have led us astray, as well as to suggest ways in which the models need to be modified to make them more realistic. This is the role of experimental phylogenies, also known as biological simulations.

When we create an experimental phylogeny, we would like to control some aspects of evolution while we let the experimental organisms control the rest. For instance, if we are interested in the effects of differing branch lengths on the performance of phylogenetic methods, we might design a series of experimental trees in which we systematically vary branch lengths across trees for some experimental lineages, while we hold such factors as population size and environmental conditions constant. The biological constraints of the organisms are established by the organisms themselves, rather than modeled by an investigator. If we model the same trees and show that the results are consistent with the experimental lineages, then we can begin to conclude that the simplifications of the numerical simulations are not adversely affecting our conclusions. On the other hand, as will often be the case, we may see a difference between the simulated trees and the experimental trees. In this situation, we can evaluate the two data sets and determine why they are different. After doing so, not only will we now know more about the processes of molecular evolution, but we can also incorporate this information into new and better simulations. The new simulations may suggest new conditions to test experimentally, and the process can be repeated indefinitely, with the investigators learning more about the behavior of phylogenetic methods and the processes of molecular evolution with every cycle.

Of course, the scenario above assumes three things. First, we need to be able to define what we mean by "good performance" of a phylogenetic method. Second, we need to be able to identify and define relevant "parameter space" to explore in the numerical and biological simulations. Third, we need to identify organisms that evolve quickly enough that we can create experimental phylogenies in reasonable periods of time (hopefully measured in weeks or months rather than years).

How Do We Know a Good Method When We See One?

An ideal phylogenetic method would be fast, powerful, consistent, robust, discriminating, and versatile (see Penny, Hendy, and Steel, 1992). Unfortunately, there are trade-offs involved in optimizing these criteria, so that it is usually necessary to rank their importance in selecting a method for a given problem. Below we consider each criterion in turn.

Computational speed. If everything else were equal, computational speed would be very important. In general, however, the methods that rank the best for speed rank among the worst for some of the other criteria, and "quick-and-dirty" approaches are not often favored in science except as a way to get a first approximation. However, there is a wide spectrum of computational speeds among the methods from very fast single-tree clustering algorithms, to the character and distance-based methods that identify an optimality criterion, to methods such as maximum likelihood that require enormous computational efforts. Even though we may not want to select a method based on speed considerations alone, we still must select a method that is fast enough to give an answer without having to wait across geological time, and for some applications, a fast approximation may be appropriate.

Power. In the real world, we have a finite number of data that we can analyze for a given problem. If two methods are otherwise equal, but one correctly estimates a phylogeny from sequences 100 bp long whereas the other requires sequences 1,000 bp long to achieve the same success rate, then we would obviously prefer the one that requires fewer data to get the correct answer. Methods may differ in power because they consider different kinds of variation among the sequences to be informative, or because they give different weights to different kinds of variable characters. In the latter case, power may be a function of the model of evolution and the degree to which the assumptions of the method are matched.

Consistency. A method is consistent if it converges on the correct answer as more data are examined. All methods of phylogenetic analysis proposed to date are consistent under some conditions but are inconsistent under others. Some methods explicitly state a set of assumptions, which if violated may lead to inconsistency; for other methods, the assumptions are implicit and the conditions that lead to inconsistency have to be determined empirically. Many methods are based on a stated model of evolution, and as long as the organisms are evolving under the model conditions the method is consistent. Ideally, we would like to have a method that is consistent for the most general possible models of evolution.

Robustness. Even if we know the model conditions under which a method is consistent, it does not necessarily follow that deviations from the model will automatically lead to inconsistency. A robust method is insensitive to deviations from the ideal (model) conditions. This is obviously an important attribute, because it is unlikely that any real organisms ever evolve precisely in accord with any but the most general of models.

Discriminating ability. Methods should be able to return no answer if certain basic assumptions are violated (e.g., if there is no underlying tree), and they should be capable of comparing and ranking alternative hypotheses. Some methods (e.g., clustering algorithms like unweighted pair-group method of averages [UPGMA] and neighbor-joining) will always return one tree with no means of comparing alternatives, although tests can be applied to ask if any of the branch-lengths are significantly different from zero. Most other methods show limited ability to reject a tree-like structure but do specify an optimality criterion that can be used to compare and test alternative trees. Separate methods have been developed to identify data sets that contain no more structure than would be expected at random (e.g., Hillis and Huelsenbeck, 1992), and in principle such methods can be applied before deciding that it is appropriate to proceed to phylogenetic analysis. Once it has been determined that phylogenetic analysis is appropriate, a discriminating method should identify a range of potential solutions and provide a means of evaluating their optimality.

Versatility. The objectives of a phylogenetic analysis are usually more than simply finding the branching structure of a tree, although that is a universal first step. Some methods do little more than specify a branching structure, however. A versatile method would also provide such properties as estimates of branch lengths and estimates of the character states of the ancestral nodes in the tree. A method is also versatile if it is applicable to a wide range of character types (e.g., both molecular and morphological data); some methods are not versatile because they are applicable only to DNA sequences or incorporate only information about substitutional changes (e.g., information on insertion-deletion events is ignored).

Given that we accept the above criteria as important, how can we rate a given method for each criterion? Some of the rankings are straightforward, such as computational speed, but others (such as power and robustness) are harder to evaluate. Two main approaches can be used to compare the methods, namely numerical simulations and experimental phylogenies.

Numerical Simulations: Examining Conceivable Limits

A common objection to numerical simulations is that the conclusions of a typical simulation study invariably seem to support the investigator's a priori views on the relevant methods. For instance, one investigator who likes the neighbor-joining method (for whatever reason) may simulate phylogenies that indicate its superiority, whereas another investigator who prefers the UPGMA method may conduct simulations to show support for that approach. The reason for such discrepancies is that each of the methods has conditions under which it performs optimally, so a method looks best if trees are simulated under only those conditions. As an example, UPGMA performs best if rates of evolution are exactly the same in all lineages in the tree. Under such conditions, UPGMA can be shown to estimate the correct trees as well as or better than many other methods. However, the conclusions from a study that only includes such conditions are not very general and do not present a fair comparison of different methods. If simulations are to be used effectively, then, we need to define a specific problem for investigation and then examine the potential parameter space for that problem as exhaustively as possible.

As an example of defining a problem exhaustively, consider the simple and often-simulated "four-taxon tree with two rates" problem (Figs. 1 and 2). Felsenstein (1978) discussed this problem to demonstrate that some methods of phylogenetic analysis are inconsistent for some trees of this type; if the lineages represented by two opposing peripheral branches are evolving at a very high rate compared to the other three branches, then the parallel changes in the two long branches can overwhelm

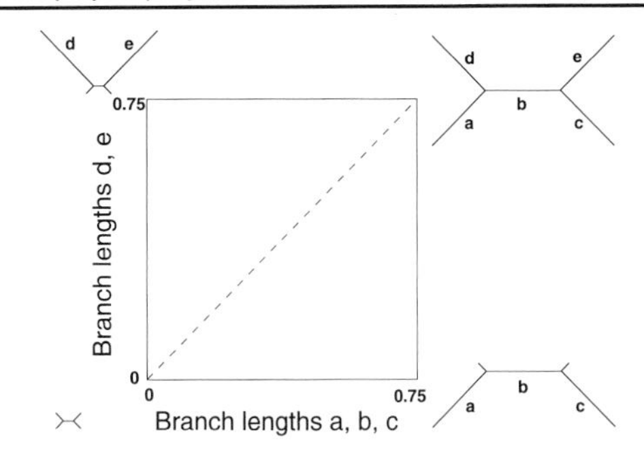

Figure 1. The four-taxon, two-rates problem. Consider a four-taxon tree in which two of the lineages (represented by the opposing branches *d* and *e*) have a different rate of change than the other three lineages (represented by branches *a, b,* and *c*). The instantaneous rate of change varies from 0 to infinity along the two axes of the graph, so that in the upper right corner of the graph all the characters of the taxa are as divergent as if chosen at random (e.g., the probability that there is a difference in any given nucleotide position in a DNA sequence is 0.75). Different areas of the resulting parameter space represent trees with different shapes. Along the diagonal (*dashed line*) the lengths of all five branches are equal. Near the top left corner of the graph, branches *a–c* are short and branches *d–e* are long, which produces conditions that are inconsistent for several methods of phylogenetic inference (see Fig. 2).

any signal in the small internal branch, leading one to be positively misled in estimating the tree (Fig. 2). Several methods are inconsistent under such conditions: the more data that are applied to the problem, the more likely the incorrect tree will be estimated. Because of this well-known behavior, and because of the simplicity of simulating and evaluating such trees, there have been numerous simulation studies of this problem (see summaries by Nei, 1991; and Huelsenbeck and Hillis, 1993). However, it is also simple to identify specific types of four-taxon trees that are particularly amenable to most of the common phylogenetic methods. Therefore, given that we have identified a particular problem (namely a four-taxon tree with two different rates), there is no reason not to examine the problem exhaustively for any given model of evolution. This is relatively easy to do in this case: we can graph out the two rates along two axes, and vary the instantaneous rate of evolution from zero to infinity along both axes (Fig. 1). We can then partition the graph space as finely as our computational limitations will permit, and simulate trees from throughout the entire possible parameter space. Using this approach, we can compare any set of methods for all potential conditions simultaneously, rather than only examining a biased set of trees that tends to support an a priori preference. If we are interested in consistency, we can calculate the expectations for infinitely large data sets; if we are interested in power, we can examine a regular series of finite data sets. A power analysis using this approach is illustrated in Fig. 3, with colors used to show the probability that a given method will find the correct tree in different areas of the parameter space.

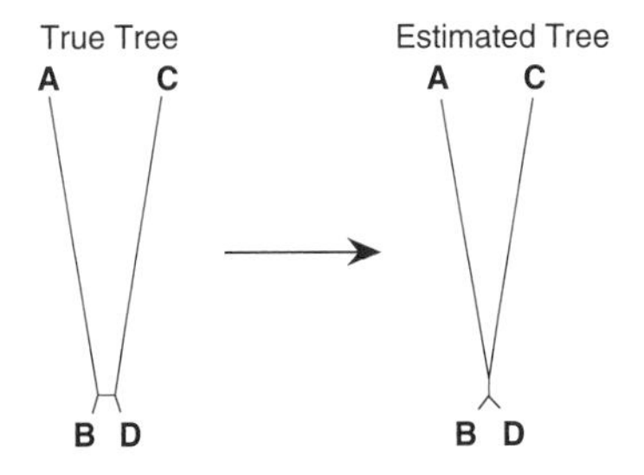

Figure 2. A tree from the Felsenstein zone: two of the opposing branches are long, and the other three branches (including the internal branch) are short. Parallel changes in the two long branches can be confused as phylogenetic signal, which leads some methods to estimate the incorrect tree on the right.

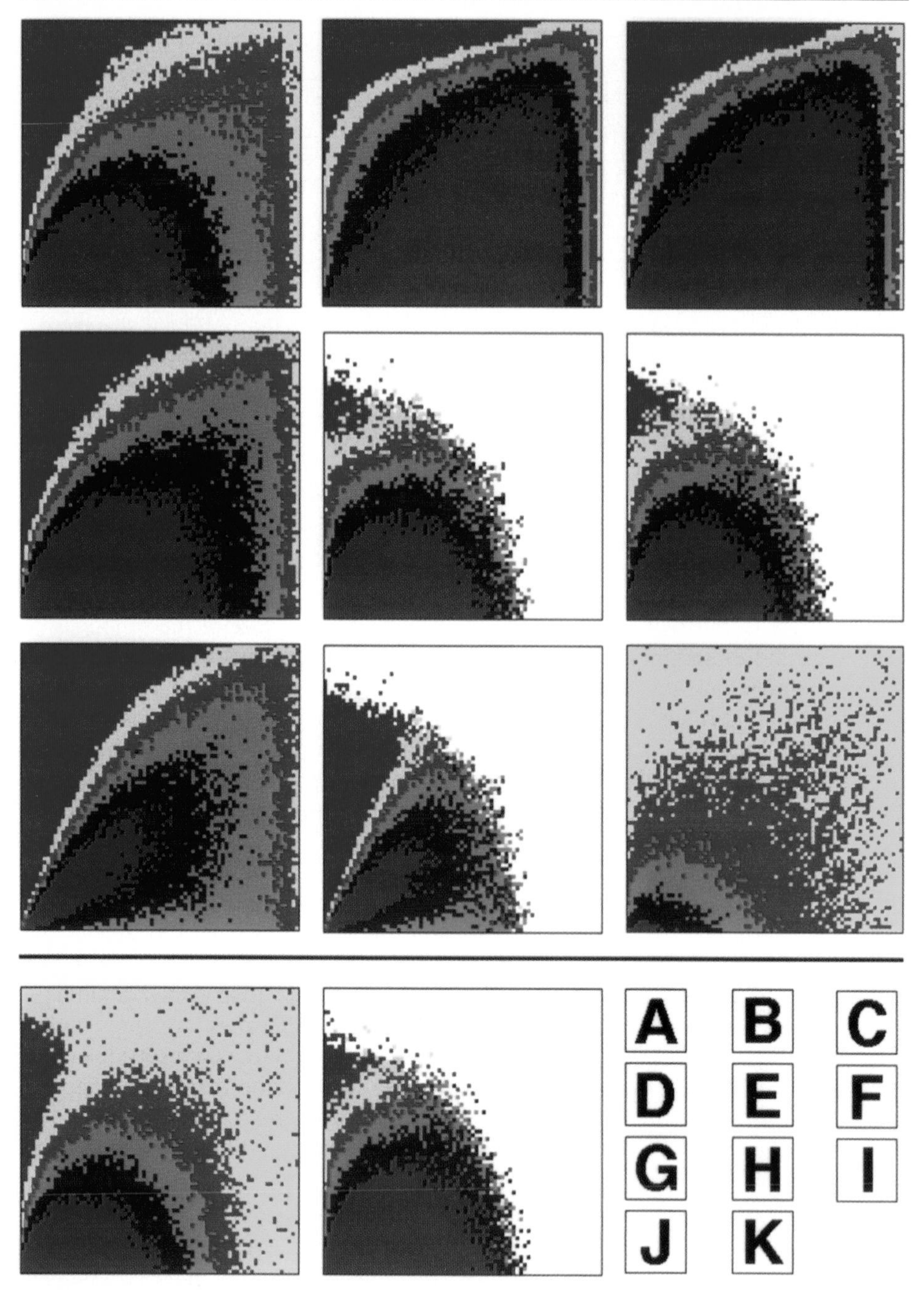

Figure 3. Comparison of the power of several methods of phylogenetic inference. The colors represent the probability of correctly estimating the phylogeny: green (>95%), dark blue (80–95%), light blue (60–80%), magenta (40–60%), yellow (20–40%), and red (<20%). White areas represent conditions in which over 90% of the data sets include undefined pairwise distances (so no tree can be constructed). In graphs *A–I,* DNA sequence evolution followed the Kimura model, with a 5:1 transition:transversion ratio. In graphs *J* and *K,* there was a 5:1 ratio of G–C and A–T changes compared to G–A, G–T, A–C, or C–T changes. 100

The model of evolution used to simulate the data sets analyzed in Fig. 3, *a–i,* is the simple Kimura model of nucleotide substitutions: there is one mutation rate for transitions and another for transversions, and in this case transitions are five times as common as transversions. This simple model is widely used because it approximates the pattern of evolution observed for many genes, in particular mammalian mitochondrial genes (e.g., see Brown, Prager, Wang, and Wilson, 1982). Parsimony (Fig. 3 *a*), UPGMA (Fig. 3 *g*), and neighbor-joining using uncorrected distances (Fig. 3 *d*) are inconsistent under these conditions for trees in the upper left-hand corner of the graph, a region sometimes called the Felsenstein zone (for details of the regions of inconsistency for most major methods under these conditions, see Huelsenbeck and Hillis, 1993). The performance of most of the methods shown falls off at high rates of change and near the Felsenstein zone. The performance of parsimony is increased dramatically at higher rates of evolution by weighting the transversions more heavily than the transitions, or by simply ignoring the transitions altogether (Fig. 3, *b* and *c*). Distance methods and parsimony can be made consistent by correcting the data in accord with the model of evolution. For this model, pairwise distance methods like neighbor-joining can be made consistent by correcting the distances as suggested by Kimura (1980); parsimony can be made consistent through a Hadamard transformation (see Penny et al., 1992). If the model of evolution matches the correction exactly, as shown in Fig. 3 *e* for neighbor-joining with Kimura distances, then many methods are consistent throughout the parameter space (Penny et al., 1992). However, note that these corrections may have a minimal effect on the power of the technique, except within the region of former inconsistency (e.g., compare Fig. 3, *d* with *e*). At high rates of change, the area in which neighbor-joining with Kimura-corrected distances finds the correct tree is still small compared to the weighted parsimony method (compare Fig. 3, *c* with *e*). Moreover, essentially the same level of performance can be achieved using the Kimura distances with the Fitch-Margoliash method as with neighbor-joining, with the added advantage of discriminating ability with the Fitch-Margoliash approach (compare Fig. 3, *e* with *f*). Using the Kimura corrections actually decreases the performance of the UPGMA method, which has little power in any case (Fig. 3, *g* and *h*). Lake's method of invariants (Fig. 3 *i*) shows an extreme trade-off between consistency and power: the method is consistent over the entire parameter space under these conditions, but has very low power. Interestingly, if we modify the model slightly by changing the classes of common versus rare substitutions, thereby violating the assumptions of Lake's method of invariants, the power of the method actually increases, even though it also become inconsistent in upper left corner of the graph (Fig. 3 *j*). In contrast, a similar change of models has a

---

variable nucleotide positions were included in all data sets. (*A*) Parsimony, all changes weighted equally; (*B*) transversion parsimony (transitions weighted zero); (*C*) weighted parsimony (transversions weighted five times more heavily than transitions); (*D*) neighbor-joining with uncorrected distances; (*E*) neighbor-joining with Kimura distances; (*F*) Fitch-Margoliash method with Kimura distances; (*G*) UPGMA with uncorrected distances; (*H*) UPGMA with Kimura distances; (*I*) Lake's method of invariants, all assumptions met; (*J*) Lake's method of invariants, mutation assumptions violated (see above); (*K*) neighbor-joining with Kimura distances, mutation assumptions violated (see above). For a description of all these methods, see Swofford and Olsen (1990).

comparatively little effect on the power of neighbor-joining (Fig. 3 *k;* although it too becomes inconsistent in part of the parameter space under these conditions).

The challenge in the field of numerical simulations is to identify ways that more complex problems can be explored in a similarly unbiased manner. This requires that we specify the purposes of a set of simulations explicitly: once the problem is defined, the relevant parameter space will often be obvious. Unfortunately, for many of the problems we face, the necessary computations will be much more difficult than in the simple four-taxon problem discussed above. As the number of taxa increases, the number of possible phylogenies increases rapidly, even if we ignore the lengths of branches. For rooted bifurcating trees, the number of distinct, labeled topologies for *n* species is equal to

$$\prod_{k=2}^{n}(2k - 3)$$

(Cavalli-Sforza and Edwards, 1967). Therefore, for just fifty taxa, there are 2.8×10^{76} possible rooted bifurcating trees. Obviously, we can not examine every possible solution in such cases; even with a computer that could examine 10^{50} trees a second, it would take much more time to examine all the trees than has existed in the history of the universe! We can get around this problem to a large extent by developing efficient heuristic searching methods (see Swofford and Olsen, 1990; Swofford, 1993), but the harder problems will still require major increases in computational power, such as those afforded by massively parallel computing (see W. D. Hillis and Boghosian, 1993).

Experimental Phylogenies: Observing Evolution in Action

The principal limitation in experimental phylogenies is producing the relevant levels of evolutionary change within a human life span (or more realistically, within the life span of a funded research project). This requires an organism with a short generation time and high mutation rate. Ideally, we would like to be able to control the spectrum of potential mutations, have the ability to examine a large fraction of the organism's genome (to avoid or explore problems of sampling bias), control population size over a wide spectrum, control and manipulate the organism's environment with ease (to examine the effects of selection), and culture the organism without great difficulty. All of these conditions are met most closely by viruses, and particularly bacteriophages, the viruses of bacteria.

Fig. 4 shows an actual and two estimated trees for an experimental phylogeny based on cultures of the bacteriophage T7 (see Hillis, Bull, White, Badgett, and Molineux, 1992). The T7 cultures were grown in the presence of a mutagen (in this case, nitrosoguanidine) to increase the rate of mutations. The mutational spectrum produced in such an environment is biased in favor of certain substitutions, but deletions occur as well. With nitrosoguanidine, many kinds of mutations occur, but the majority are transitions, especially G $\rightarrow$ A and C $\rightarrow$ T. This spectrum of mutations is similar to that observed in some natural systems that have been examined (e.g., mammalian genes and pseudogenes: Gojobori, Li, and Graur, 1982; Li, Wu, and Luo, 1984; human immunodeficiency viruses: Moriyama, Ina, Ikeo, Shimizu, and Gojobori, 1991). To create the phylogeny, an initial culture of bacteria plus bacteriophage is divided into three lineages (one of which will become the outgroup for purposes of rooting the tree), and then the lineages are redivided after a predetermined number of lytic cycles. Stocks of ancestral cultures are saved at each cycle for later analysis. After the phylogeny has been created, DNA is isolated from the mutated lineages of T7 and analyzed by restriction digestion, sequencing, or DNA-DNA hybridization. Different methods of phylogenetic analysis can then be used to estimate the phylogeny, the branch lengths, and the genotypes of the ancestors, and in each case evaluated against the true tree and actual ancestors.

What are the results of such experiments? An analysis of a set of just over 200 restriction sites that are variable among the lineages produces the correct phylogeny with almost every method tested to date. However, if one uses the presence or absence of restriction fragments as the primary data, instead of mapping out the sites for analysis, then every method returns the incorrect phylogeny. Restriction fragment analyses are troublesome because restriction fragments are not independent characters. There are two reasons for this. First, a single site gain causes one fragment to be lost and two other fragments to be gained, and second, a single deletion can cause changes to multiple fragments. However, it is often argued that mapping and aligning restriction sites may not be worth the extra effort (e.g., Bremer,

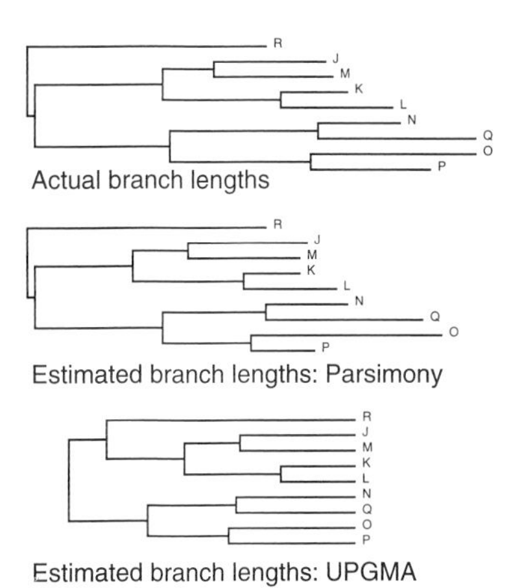

Figure 4. An actual and two estimated trees for an experimental phylogeny derived from bacteriophage T7 (Hillis et al., 1992). The branch lengths are drawn proportional to the number of actual or estimated restriction-site changes. Both estimated trees produce the correct unrooted topology (the root is ambiguous, because the ancestral node is a trichotomy), but parsimony produces much better estimates of branch lengths than does UPGMA.

1991). Yet, in the seemingly ideal situation of the experimental T7 phylogeny, restriction sites always give the correct tree and restriction fragments always give the wrong tree, indicating that the problem with nonindependence is a severe one. This result has not been obvious from simulations, probably because simulations rarely include insertions or deletions in the models of nucleotide change. This result indicates that we need to improve our simulation models to include realistic frequencies of insertion-deletion events.

As discussed earlier, the branching structure is not the only aspect of the tree we would like to estimate. Branch lengths are also of interest, as are the estimation of ancestral genotypes. For this phylogeny, as well as for simulations based on the kinds and distributions of mutations observed across the tree (Bull, Cunningham, Molineaux, Badgett, and Hillis, 1993), the character-based parsimony method produces significantly better estimates of branch lengths (estimated number of restriction-site changes) than do pairwise distance methods such as neighbor-joining or the Fitch-Margoliash method, and these latter methods produce significantly better estimates than does UPGMA. Fig. 4 compares the actual branch lengths (in number of restriction-site changes) with the estimated branch lengths from parsimony and UPGMA.

Of the common methods of phylogenetic analysis, only parsimony provides estimates of the ancestral character states at internal nodes in the tree (see Maddison and Maddison [1993] for an excellent discussion of the reconstruction of ancestral character states). How well does this method fair in this example? For any given restriction site, the estimated state in an ancestor may be (*a*) correctly inferred, (*b*) incorrectly inferred, or (*c*) ambiguous (i.e., the presence or absence of the site is equally parsimonious). When we compared the inferred restriction maps to the actual restriction maps for the seven ancestral nodes in the model phylogeny, we found that the inferred restriction maps were $>98\%$ correct. This finding indicates that the method for reconstructing ancestral character states is working well under the conditions tested.

To what extent can we generalize these findings from a study of bacteriophage to other organisms? Phylogenetic methods are purported to be general for living organisms, so any organism should therefore provide a test of the fallibility of the methods. Some findings, such as the efficiency of reconstructing ancestral genotypes, will be dependent on the level of homoplasy (reversals, parallelisms, convergences) present across the phylogeny. In the case presented above, the level of homoplasy present in the tree was almost exactly the average of that seen in real examples in the literature for the same number of taxa and characters (Hillis, Bull, White, Badgett, and Molineux, 1993; data on other organisms in Sanderson and Donoghue, 1989). Obviously, there is a need to examine other tree topologies, the effects of highly unequal branch lengths on the performance of the various methods, the effects of parallel selection pressures on different parts of the tree, the effects of other mutational spectra, the effects of population size, and many other aspects of evolution that potentially affect our ability to reconstruct phylogenies. This is a strength of experimental phylogenies: all of these aspects of phylogenies can be studied, without detailed a priori knowledge of the underlying mechanisms. In fact, the studies also can provide first-hand information on the processes of molecular evolution. Experimental phylogenies can (and do) provide actual examples of the failure of certain methods, as in the use of restriction fragment analysis in the example above. Such findings can be used to improve the methods or to identify conditions under which they should not be applied.

How Do the Major Methods Rate?

It should be clear that no single method is optimal for all of the criteria we have identified. As with most things in life, there are trade-offs. Thus, the methods that are fastest produce just a single estimate of the tree, without any obvious way to compare or rank the alternatives (i.e., they have very low discriminating ability); fast methods also tend to be less powerful. Methods that are fully consistent under a specific model of evolution are usually less versatile (e.g., they only apply to certain kinds of data). We have attempted to rank the most commonly used methods for the assessment criteria identified earlier (see Table I). These rankings are somewhat subjective, and we recommend that interested readers refer to the primary literature on simulations and experimental phylogenies to make their own rankings. Also note that these methods are constantly being modified and refined, so the rankings may change over time. This is particularly true for the maximum likelihood methods, which have increased dramatically in speed and versatility in recent years (our rankings for maximum likelihood reflect the current common implementations rather than the full potential of the method).

Which method is finally selected will depend to a large extent on the goals of the study: neighbor-joining is perfectly adequate for a fast, initial estimate of a tree, even though it is useless for comparing and ranking alternative solutions. For most studies, one of three classes of methods will usually be most appropriate: Fitch-Margoliash and related methods (including the minimum evolution method), parsimony methods, or maximum likelihood. Fitch-Margoliash and related methods specify an optimality criterion for fitting observed pairwise distances to path-length distances on trees. The optimality criterion can be assessed for any tree, and the methods are relatively insensitive to variations in branch lengths across the tree. They are currently somewhat limited in their power and versatility by the simplistic models that are used to compute evolutionary distances (e.g., it is difficult to incorporate different weights for different characters or types of character change into the

TABLE I
Comparison of the Most Commonly Used Methods of Phylogenetic Analysis

Method	Criterion					
	Speed	Power	Consistency	Robustness	Discrimination	Versatility
UPGMA	++	−	−	−	−	−
Neighbor-joining	++	+	++	+	−	+
Fitch-Margoliash (and related methods)	+	+	++	+	++	+
Parsimony methods	+	++	+	+	++	++
Methods of invariants	+	−	++	+	++	−
Maximum likelihood (as currently implemented)	−	++	++	±	++	±

Key: ++, excellent; +, good; and −, poor.

analyses, and it is difficult to combine analyses of different kinds of data into a single analysis). Parsimony methods correct these deficiencies, but at a cost of lower overall consistency for simple models of evolution. Parsimony methods are the most versatile approaches: they can be applied to all kinds of data, can easily incorporate differential weights for different character-state changes or for entire characters, are amenable to combination of results among studies, and provide accurate reconstructions of branch lengths and ancestral character states. As shown in Fig. 3, this versatility (e.g., character-state weighting) can lead to increased power. Although parsimony methods are consistent under more limited conditions than some other approaches, this limitation has been lifted to some extent by recent developments (for more information, see discussion of the Hadamard transformation in Penny et al., 1992). Finally, maximum likelihood methods are likely to undergo the most development in coming years. Their use has been limited because they currently are too slow to be applied to many real world data sets, and because they are not very versatile (for instance, they currently are limited primarily to substitutional changes in DNA sequences). However, their versatility is being constantly improved, and they are useful methods when they can be applied because of their high power, consistency, and discriminating ability.

References

Bremer, B. 1991. Restriction data from chloroplast DNA for phylogenetic reconstruction: Is there only one accurate way of scoring? *Plant Systematics and Evolution*. 175:39–54.

Brooks, D. R., and D. A. McLennan. 1991. Phylogeny, Ecology, and Behavior: A Research Program in Comparative Biology. University of Chicago Press, Chicago. 434 pp.

Brown, W. M., E. M. Prager, A. Wang, and A. C. Wilson. 1982. Mitochondrial DNA sequences in primates: tempo and mode of evolution. *Journal of Molecular Evolution*. 18:225–239.

Bull, J. J., C. W. Cunningham, I. J. Molineux, M. R. Badgett, and D. M. Hillis. 1993. Experimental molecular evolution of bacteriophage T7. *Evolution*. 47:993–1007.

Cavalli-Sforza, L. L., and A. W. F. Edwards. 1967. Phylogenetic analysis: models and estimation procedures. *Evolution*. 21:550–570.

Felsenstein, J. 1978. Cases in which parsimony and compatibility methods will be positively misleading. *Systematic Zoology*. 27:401–410.

Gojobori, T., W.-H. Li, and D. Graur. 1982. Patterns of nucleotide substitution in pseudogenes and functional genes. *Journal of Molecular Evolution*. 18:360–369.

Harvey, P. H., and M. D. Pagel. 1991. The Comparative Method in Evolutionary Biology. Oxford University Press, Oxford. 239 pp.

Hillis, D. M., and D. M. Green. 1990. Evolutionary changes of heterogametic sex in the phylogenetic history of amphibians. *Journal of Evolutionary Biology*. 3:49–64.

Hillis, D. M., and C. Moritz, editor. 1990. Molecular Systematics. Sinauer Associates, Inc., Sunderland, MA. 588 pp.

Hillis, D. M., J. J. Bull, M. E. White, M. R. Badgett, and I. J. Molineux. 1992. Experimental phylogenetics: generation of a known phylogeny. *Science*. 255:589–592.

Hillis, D. M., and J. P. Huelsenbeck. 1992. Signal, noise, and reliability in molecular phylogenetic analyses. *Journal of Heredity*. 83:189–195.

Hillis, D. M., J. J. Bull, M. E. White, M. R. Badgett, and I. J. Molineux. 1993. Experimental approaches to phylogenetic analysis. *Systematic Biology*. 42:90–92.

Hillis, W. D., and B. M. Boghosian. 1993. Parallel scientific computation. *Science*. 261:856–863.

Huelsenbeck, J. P., and D. M. Hillis. 1993. Success of phylogenetic methods in the four-taxon case. *Systematic Biology*. 42:247–264.

Kimura, M. 1980. A simple method for estimating evolutionary rate of base substitutions through comparative studies of nucleotide, sequences. *Proceedings of the National Academy of Sciences, USA*. 78:454–458.

Li, W.-H., C.-I. Wu, and C.-C. Luo. 1984. Nonrandomness of point mutation as reflected in nucleotide substitutions in pseudogenes and its evolutionary implications. *Journal of Molecular Evolution*. 21:58–71.

Maddison, W. P., and D. R. Maddison. 1992. MacClade: Analysis of Phylogeny and Character Evolution. Sinauer Associates, Inc., Sunderland, MA. 398 pp.

Moriyama, E. N., Y. Ina, K. Ikeo, N. Shimizu, and T. Gojobori. 1991. Mutation pattern of human immunodeficiency virus genes. *Journal of Molecular Evolution.* 32:360–363.

Nei, M. 1991. Relative efficiencies of different tree-making methods for molecular data. *In* Phylogenetic Analysis of DNA Sequences. M. M. Miyamoto and J. Cracraft, editors. Oxford University Press, Oxford.

Olsen, G. J. 1987. Earliest phylogenetic branchings: comparing rRNA-based evolutionary trees inferred from various techniques. *Cold Spring Harbor Symposia in Quantitative Biology.* 52:825–837.

Ou, C.-Y., C. A. Ciesielski, G. Myers, C. I. Bandea, C.-C. Luo, B. T. M. Korber, J. I. Mullins, G. Schochetman, R. L. Berkelman, A. N. Economou, J. J. Witte, L. J. Furman, G. A. Satten, K. A. MacInnes, J. W. Curran, and H. W. Jaffe. 1992. Molecular epidemiology of HIV transmission in a dental practice. *Science.* 256:1165–1171.

Penny, D., M. D. Hendy, and M. A. Steel. 1992. Progress with methods for constructing evolutionary trees. *Trends in Ecology and Evolution.* 7:73–79.

Sanderson, M. J., and M. J. Donoghue. 1989. Patterns of variation in levels of homoplasy. *Evolution.* 43:1781–1795.

Swofford, D. L. 1993. PAUP 3.1: Phylogenetic Analysis Using Parsimony. Illinois Natural History Survey, Champaign, IL.

Swofford, D. L., and G. J. Olsen. 1990. Phylogeny reconstruction. *In* Molecular Systematics. D . M. Hillis and C. Moritz, editors. Sinauer Associates, Inc., Sunderland, MA.

Chapter 2

Genome Organization and the Generation of Diversity

Transposable Elements and the Evolution of Heterochromatin

Allan C. Spradling

Howard Hughes Medical Institute Research Laboratories, Department of Embryology, Carnegie Institution of Washington, Baltimore, Maryland 21210

Drosophila P elements were shown to insert frequently into telomeric and centromeric heterochromatin, and to prefer a region associated with efficient copy number regulation. Upon excision, P elements frequently altered the number of repeats in a tandem array of heterochromatic sequences, by inducing unequal gene conversion. These studies suggest that a flux of transposable element insertions and excisions has the capacity to rapidly and nonrandomly modify heterochromatic sequences dispersed at multiple chromosomal sites. We propose that transposable elements maintain genomic heterochromatin in a state of dynamic equilibrium and drive its rapid evolution.

Introduction

It is fashionable to view evolution as occuring so slowly that it can only be understood in historical rather than mechanistic terms. However, genomes sometimes change relatively rapidly in response to well understood forces. The evolution of domesticated plants (Doebley, 1993), of drug-resistant bacteria (Falkow, 1975) and the adaptation of mosquitos to new microenvironments associated with human habitation (Coluzzi, Sabatini, Petrarca, and DiDeco, 1979) all provide instructive examples. Surprisingly, rapid but small evolutionary changes are frequently associated with disproportionately large modifications in chromosome structure. Even closely related species often differ substantially in repetitive DNAs that make up visible heterochromatic blocks on metaphase chromosomes. Such changes in repetitive DNAs have largely been ignored, and are usually assumed to be irrelevant to underlying selective forces. In this paper we consider how transposable elements, the known agents of some selectively important changes, may concomitantly control seemingly insignificant modifications in genome and chromosome structure.

Heterochromatin constitutes a poorly defined and enigmatic component of virtually all eucaryotic genomes (reviewed in Verma, 1988). Heterochromatic regions are rich in repetitive sequences, contain genes only infrequently, and often constitute large blocks near chromosome centromeres and telomeres. While heterochromatic sequences have been viewed as junk (Orgel and Crick, 1980), they include nucleolus organizers, telomeres, centromeres, and elements that interact with specific euchromatic genes (reviewed in Pardue and Hennig, 1990; Gatti and Pimpinelli, 1992). Many heterochromatic regions evolve rapidly (White, 1973), and are frequently polymorphic within populations. Genetic drift (see Kimura, 1983) and transposition (see Steinemann and Steinemann, 1992) are among the process that have been proposed to contribute to changes within heterochromatin.

Our studies have been motivated by the idea that transposable elements play a crucial role in creating, maintaining and modifying the heterochromatic regions of eucaryotic plant and animal genomes. In principle, if one could remove all the heterochromatin from a eucaryotic genome and provide only a single active copy of each transposable element family, we would expect that within a few hundreds of generations, complete and defective elements would have proliferated, and their rate of transposition would have declined greatly due to element-encoded and cellular regulatory processes. Concomitantly, heterochromatin-like regions would have begun to appear on the chromosomes as a consequence of transposon-catalyzed insertions and rearrangements. Subsequent perturbation of the regulatory balance, for example by chromosome rearrangement, would transiently reactivate affected transposons, however their activity would decline as soon as a new regulatory equilibrium was approached. Dynamic responses of eucaryotic genomes to chromosome rearrangements, including the activation of dormant transposable elements, have long been documented (reviewed in McClintock, 1984).

Approximately 25% of the *Drosophila melanogaster* genome falls into the category of heterochromatin, and nearly 10% is composed of transposable elements (see Rubin and Spradling, 1981; Finnegan and Fawcett, 1986). Two transposons have recently been shown to protect the ends of *Drosophila* chromosomes from progressive shortening (Biessmann, Valgeirsdottir, Lofsky, Chin, Ginther, Levis, and Pardue, 1992; Levis, Ganesam, Houtchens, Tolar, and Sheen, 1993). Both the HeT and TART elements are located exclusively at chromosome termini and on the Y chromosome. The structure of *Drosophila* telmomeres can be explained by frequent retrotranspositions of HeT and TART elements onto free chromosome ends in a fixed orientation. HeT addition events of this kind have been observed onto the ends of terminally deleted chromosomes, strongly supporting their proposed role in telomere maintenance. Therefore, *Drosophila* telomeres provide an instructive model of the sort of actions we propose transposons exert on heterochromatin generally.

The P element has provided particular insight into how transposable elements behave and how they are regulated. The introduction of single active P elements into naive *Drosophila* strains results in the proliferation and eventual stabilization of element copy number (Daniels, Clark, Kidwell, and Chovnik, 1987; Preston and Engels, 1989). Before transposition, P element excision generates a double stranded break that recombines with homologous sequences located on the sister strand, the homolog or at an ectopic site (Engels, Johnson-Schlitz, Eggleston, and Sved, 1990; Gloor, Nassif, Johnson-Schlitz, Preston, and Engels, 1991). The properties of P elements have made them powerful tools for characterizing and manipulating the *Drosophila* genome (Cooley, Kelley, and Spradling, 1988). We have utilized genetically marked P element to learn more about the structure of *Drosophila* heterochromatin and to provide specific information on how this element interacts with heterochromatic sequences. Our results support the idea that transposable elements play a major part in the evolution of this ubiquitous genomic component. Of course, a full understanding of the roles played by transposons will require that the structure, function and regulation of each transposable element family within eucaryotic genomes, like each stable gene, be determined.

Experimental Procedures

Genetic Strains

P element insertions into the *Dp1187* minichromosome were obtained as described in Karpen and Spradling (1992). Terminal deficiencies were subsequently recovered by remobilizing insertion *0801* (Tower, Karpen, Craig, and Spradling, 1993) or *8002* (Zhang and Spradling, 1993). All strains were propagated at room temperature on standard *Drosophila* medium (Ashburner, 1990).

In Situ Hybridization to Mitotic Chromosomes

P element insertions were localized by in situ hybridization using biotinylatyed probes complementary to the PZ transposon. Mitotic chromosomes were prepared from the brains of third instar larvae and hybridized as described by Zhang and Spradling (1994).

DNA Blots

Southern transfers and hybridization with DNA probes was as described in Tower et al. (1993).

Results

Transposable Elements and Heterochromatin Structure

The *Drosophila melanogaster* genome contains at least 50 families of transposable elements that together comprise an estimated 5–10% of genomic DNA (reviewed in Berg and Howe, 1989). The heterochromatic chromosome regions are particularly rich in transposable element sequences and transposon-specific probes frequently label the polytene chromocenter in situ. Based on the structure and apparent immobility of some resident elements, heterochromatin has been thought of as a "sink" or "graveyard" for transposons that became trapped and subject to degeneration (see for example, Steinemann and Steinemann, 1992). Attempts to investigate alternative possibilities are limited by the poorly characterized structure of heterochromatic chromosome regions and the lack of methods to mutate (all copies of) dispersed, repetitive sequences. Consequently, we have attempted to simplify the problem by studying the behavior of a single family of transposons, *Drosophila* P elements, under conditions where the activity of other elements was minimized.

P Elements Insert Readily into Diverse Heterochromatic Regions

Transposable elements that influenced the structure and evolution of heterochromatin would be expected to have the ability to insert within heterochromatic regions. In the past, heterochromatic P element insertions have proved difficult to document (see Engels, 1989). However, using a P element marked with the *rosy*$^+$ eye color marker (called PZ), we have observed recently that insertions into both centromeric and telomeric heterochromatin occur frequently (Figs. 1 and 2; Karpen and Spradling, 1992; Zhang and Spradling, 1994).

Several obstacles made it difficult to document P element insertion events in centromeric heterochromatin. These genomic regions are underrepresented in polytene salivary gland cells (reviewed in Spradling and Orr-Weaver, 1987), so mitotic chromosomes must be used when localizing insertions by situ hybridization.

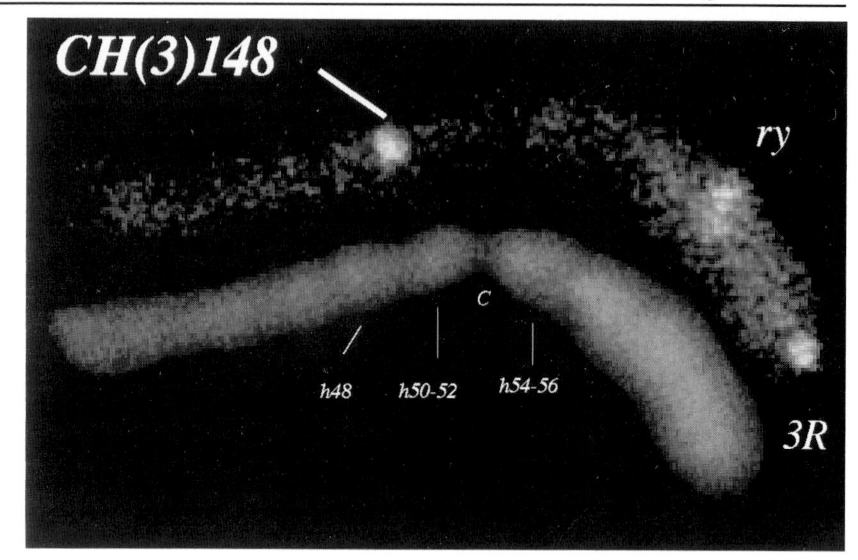

B

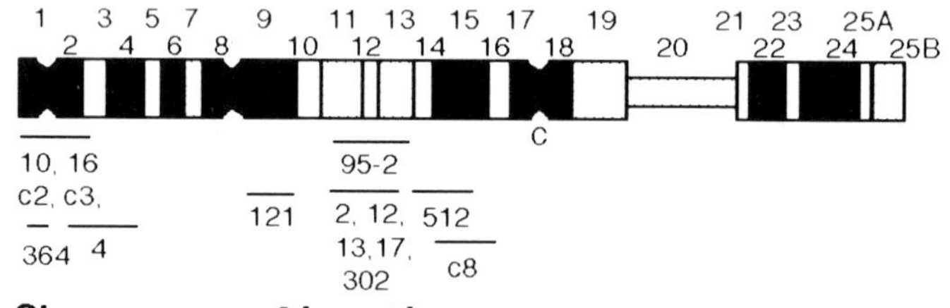

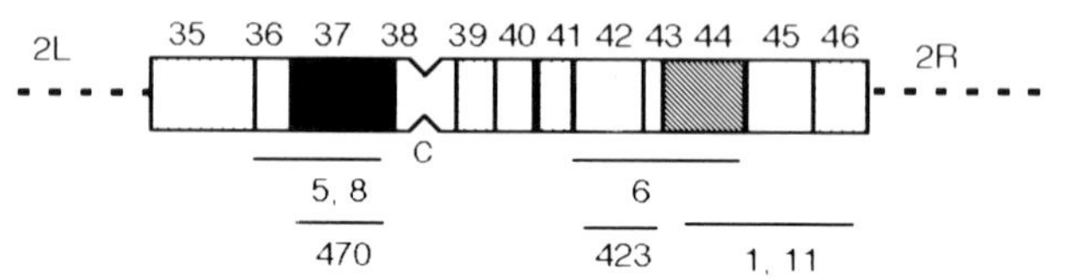

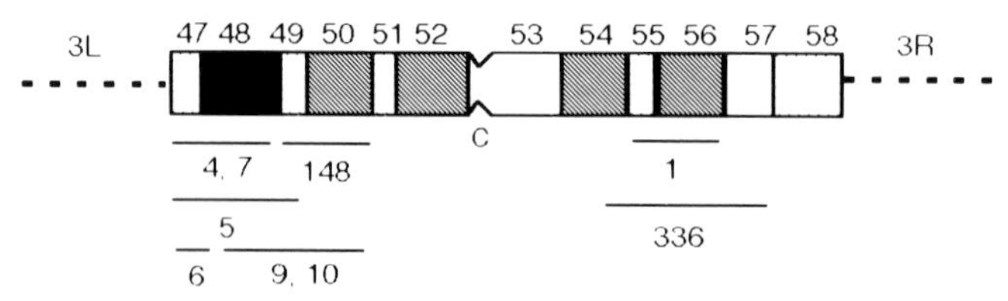

Figure 1. P elements insert in *Drosophila* heterochromatin. (*A*) Localization of the CH(3)148 insertion in chromosome three centromeric heterochromatin by in situ hybridization to mitotic chromosomes. The hybridization of the PZ probe in region h49–h50 is shown on the upper chromosome, along with cross-hybridization to the *rosy* locus (*ry*), and hybridization from a control probe that labels the tip of the right arm (*3R*). Below, the staining pattern of the same chromosome with the dye DAPI was used to align the hybridization signal with the major heterochromatic bands (i.e., h48–h56). (*B*) Summary of sites of heterochromatic P element insertion. The diagrams show the heterochromatic regions of three major *Drosophila*

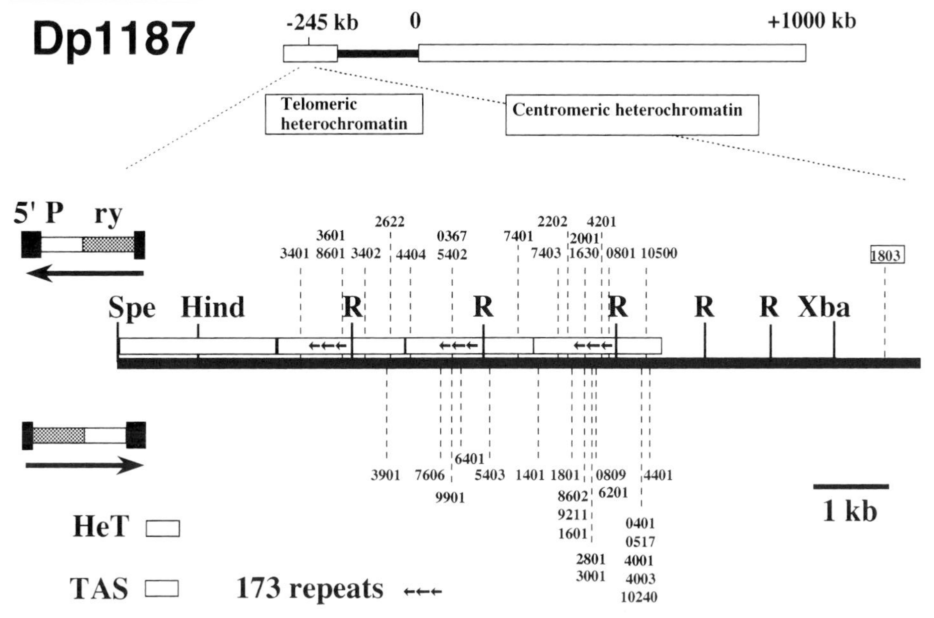

Figure 2. (*A*) A P element insertion hotspot in subtelomeric heterochromatin. The structure of the minichromsome Dp1187 is shown above (heterochromatin, *open boxes;* euchromatin, *thick line*), and a 10-kb region located 40–50 kb from the distal (*left*) telomere is expanded below. The positions of 37 independent insertions of the PZ element are indicated by dashed lines. The orientation of the PZ insertions (which differ for those shown above and below the line) is diagrammed at the left. The insertions lie within 3.1 tandem repeats of a subtelomeric repetitive element called *TAS.* The location of three 173-bp subrepeats within each *TAS* element is indicated by arrows (*173 repeats*), and of HeT DNA by a shaded box. Modified from Karpen and Spradling (1992).

Furthermore, genes juxtaposed within or near to centromeric heterochromatin undergo position-effect variegation (reviewed in Karpen, 1994), so the expression of marker genes carried into heterochromatin by integrating P elements may be repressed. Table I summarizes experiments that monitored the movement of *rosy*$^+$ P elements, and revealed that position effects are the primary reason insertions into centromeric and Y chromosome heterochromatin have rarely been recovered. When jumping was carried out without suppressing position-effect variegation, no lines were recovered in autosomal centromeric heterochromatin or on the Y chromosome (except near its telomeres). In contrast, when position-effect variegation was suppressed by the presence of an extra Y chromosome, 2–7% of all transpositions were recovered at a wide variety of different heterochromatic sites (Zhang and Spradling, 1994; Fig. 1).

---

chromosomes: 2, 3, and Y. Dashed lines indicate junctions with euchromatin, while the open, shaded and black regions represent the mitotic chromosome banding pattern within heterochromatin (see Ashburner, 1990). The locations, determined by in situ hybridization as in *A,* of the single transposed P elements in lines containing heterochromatic insertions are indicated below (Zhang and Spradling, 1994).

These experiments also contradicted the common view that heterochromatic insertions represent a dead-end pathway that traps elements and leads to their ultimate degeneration. Genes inserted in heterochromatin are thought to become compacted and unavailable for binding by transcription factors and other proteins necessary for expression. A highly compacted state would be expected to preclude access by transposase, and immobilize elements inserted in heterochromatin. Insertion 95-2 (Fig. 1) exhibited an extremely strong position effect on *rosy*$^+$ expression; even the addition of an extra Y chromosome was insufficient to produce a *rosy*$^+$ phenotype. However, the 95-2 insertion and those at seven other heterochromatic sites transposed at essentially normal rates (Zhang and Spradling, 1994; Tower et al., 1993). Therefore, the mechanisms that suppress gene transcription from heterochromatic regions do not block P element transposition. Perhaps compaction does not occur in the germline cells where transposition takes place, or is insufficient to impede access by transposase. Alternatively, some mechanism other than generalized compaction may cause heterochromatic position effects.

TABLE I
Transposition of PZ Elements Located at Different Starting Sites*

Starting site‡	Suppression§	Total‖	Telomeric Y het¶	Centromeric or internal Y heterochromatic**
X (euchromatic)	–	7,825	24 (0.3%)	0 (0%)
X (euchromatic)	+	496	2 (0.4%)	8 (2%)
Y (heterochromatic)	+	279	3 (1%)	19 (7%)

*Data from Karpen and Spradling (1992) and Zhang and Spradling (1994).
‡X (euchromatic) a euchromatic site on the X; Y (heterochromatic) the heterochromatic 95-2 site (see Fig. 1).
§Transpositions scored in the presence(+) or absence (–) of an extra Y chromosome to suppress position effect variegation.
‖Number of independent transpositions analyzed.
¶Number of insertions that mapped near Y chromosome telomeres by in situ hybridization.
**Number of insertions that mapped to autosomal centromeric heterochromatin, or internally on the Y chromosome by in situ hybridization.

Preferential Insertion into a Heterochromatic Region Mediating P Element Repression

Additional studies of P element transposition utilized the Dp1187 minichromosome, a 1300 kb X-chromosome derivative rich in heterochromatin (Karpen and Spradling, 1990, 1992). PZ elements inserted at high frequency into a 4.7-kb subtelomeric region of Dp1187 located ~40 kb from the chromosome terminus (Fig. 2). Insertions within this same "hotspot" have been seen to occur frequently in wild Drosophila populations (Ajioka and Eanes, 1989; Ajoika, 1987). The presence of only two complete P elements in this site can strongly suppress P element transposition in trans (Ronsseray, Lehman, and Anxolabéhère, 1991). Comparable repression has never been observed when P elements were inserted at euchromatic sites, even when the elements were engineered to express high levels of P element proteins (Rio,

1991). Thus, P elements insert efficiently into telomeric as well as centromeric heterochromatin, and prefer at least one site that is unusually active in repressing their own transposition.

The association of a heterochromatic P element insertion hotspot with repression was consistent with the dynamic equilibrium model discussed previously. If the 1A site plays an important role in controlling P element transposition, rearrangements that disrupt this site might lead to increased P element activity. Equilibrium would be restored when a new transposition reversed the damage to the 1A site, or caused an element to insert into another repressive location. It will be particularly interesting to test the ability of complete P elements to mediate repression when located at defined sites within centromeric heterochromatin. Evidence that heterochromatic elements are involved in the repression of several other transposable elements, including the I factor (Bucheton, 1990) has been reported.

Transposable Element Activity May Drive the Concerted Evolution of Tandemly Repetitive DNA Sequences

The presence of tandemly repeated sequences is a hallmark of heterochromatin. However, relatively little is known at present what factors are responsible for generating and maintaining this type of sequence organization. The observation that repetitive DNA evolves in a concerted fashion (Brown, Wensink, and Jordan, 1972) argues that individual repeat copies within tandem arrays are corrected against one another by unequal sister chromatid exchanges (Smith, 1973; Petes, 1981) or gene conversions (Dover, 1982). Although extensive theoretical studies have been carried out, actual correction events within the repetitive DNA sequences of multicellular eukaryotes had not been detected. Consequently, the times in germline development when corrections take place (meiosis?), and the possible existence of specific biases that might drive array lengths in particular directions remain unknown (reviewed by Dover, 1993).

Studies of P element behavior suggested that tandem arrays could be generated and maintained by transposable elements. P elements have a nonrandom tendency to insert very close to their starting site, a property called "local jumping" (Tower et al., 1993; Zhang and Spradling, 1993). Very frequently, such local transposition events create tandem or inverted repeats of the starting transposon. Slippage during gap repair after excision has also been observed to give rise to partially duplicated segments (Gloor et al., 1991). Once initiated, tandem repeats might expand by a variety of recombination mechanisms. A large tandem array of P element sequences has arisen in *Drosophila guanche* (Miller, Hagemann, Reiter, and Pinsker, 1992).

We investigated whether P element excision from a short tandem array could alter its repeat copy number (Thompson-Stewart, Karpen, and Spradling, 1994). PZ elements inserted within 1.8 kb "TAS" elements that make up the Dp1187 hotspot (Fig. 2) were mobilized. The structures of the TAS arrays in 400 progeny that retained a PZ element were analyzed by Southern blotting, and by cloning and sequencing the 5′ P element junctions. In 12 cases, only two 1.8-kb repeats remained whereas in another line, five copies had been generated. Changes in the copy number of a 173-bp subrepeat located at the insertion site were also observed (see Fig. 2, *173 repeats*). Unexpectedly, a large fraction of the derivatives had undergone a local transposition as well as a change in TAS repeat copy number.

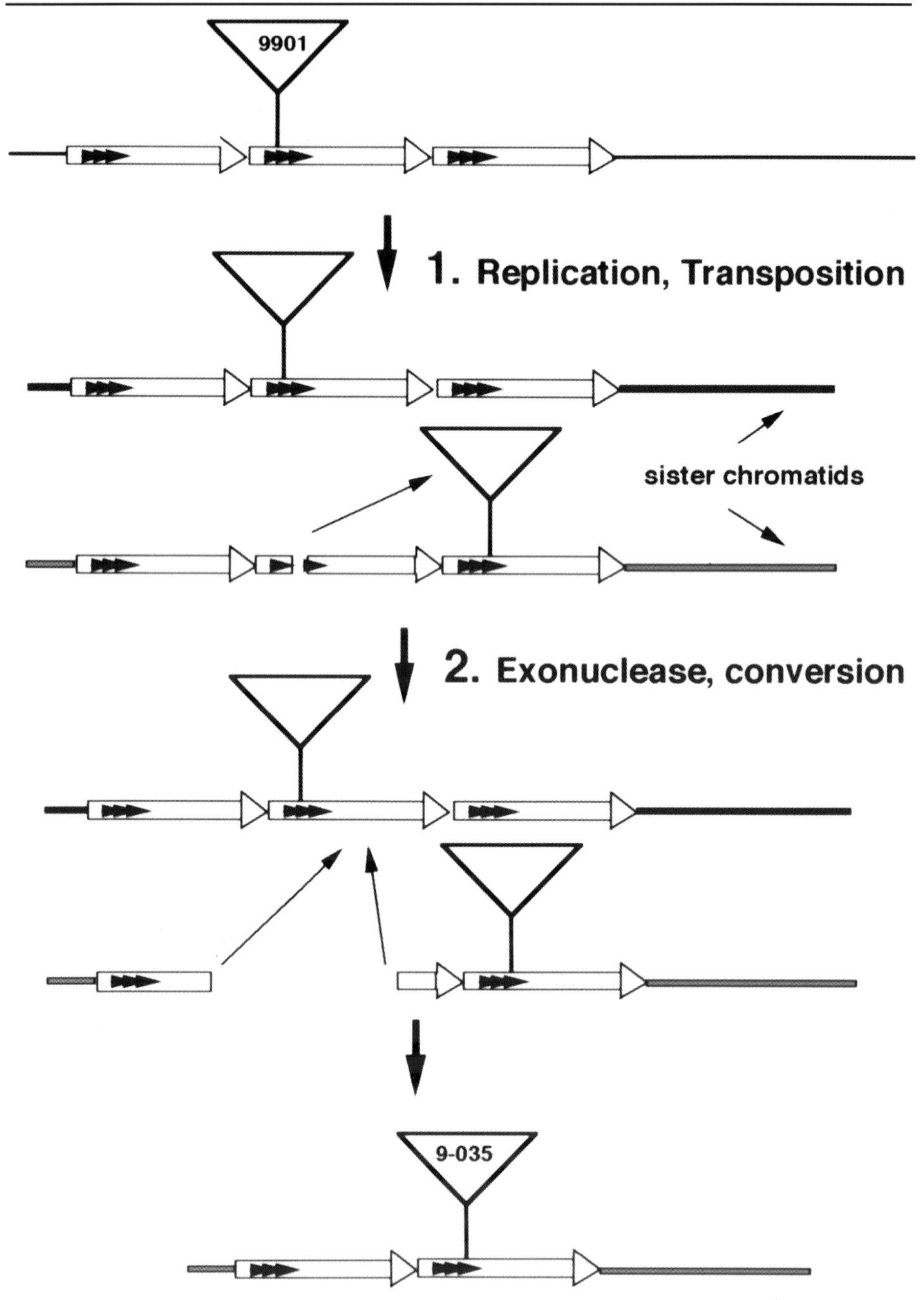

Figure 3. P element activity alters the copy number of tandemely repeated TAS sequences. A PZ element located within the subtelomeric hotspot in strain 9901 (see *upper line,* Tower et al., 1993) was activated and derivatives were examined to determine if the copy number of the TAS repeats had been altered. The figure shows the structure (*bottom line*) and the proposed derivation of one such line, 9-035. The model involves the following steps: (*1*) the starting 9901 chromatid undergoes replication to generate two identical sister chromatids. The P element excises from one of the sisters leaving a double strand break (shown below as a gap) and

These derivatives could be explained by a gap repair model in which the broken ends created by P element excision invade homologous sequences on the sister strand (Fig. 3). However, the absence of more arrays with copy number increases, and the high frequency of local transpositions among the recovered products suggested that the two broken ends usually invaded the same repeat copy. Exonuclease trimming before invasion was probably responsible for the reductions in TAS copy number. These results support the idea that transposon activity plays a role in generating and maintaining the many tandemly repeated sequences associated with heterochromatin (Thompson-Stewart et al., 1994).

Discussion

Transposable Elements May Play a Major Role in Heterochromatin Evolution

The properties of P elements discussed above are consistent with the proposal that transposable elements generate and maintain genomic heterochromatin. Such a mechanism would explain why heterochromatin is a nearly ubiquitous component of eucaryotic genomes, why it is rich in transposon sequences, and why heterochromatic chromosome regions are frequently polymorphic in populations and evolutionarily labile.

The variability of heterochromatic regions probably arises from several different sources. Even in the repressed state characteristic of wild P strains, a low level of P element transposition still takes place. Thus, sequences molded by transposon activity would continue to diverge at a significant basal rate. Such events would increase following depression of one or more transposon families by chromosome rearrangement, gene mutation or other factors. Under conditions of strong selection, this variability might prove highly advantageous. Transposons are known to be of selective value in bacteria (Hartl, Dykhuizen, Miller, Green, and DeFramond, 1983; Chao and McBroom, 1992).

The introduction of new transposon families represents an unpredictable and essentially irreversible source of additional variation. Transposons appear to be transferred between insect species at high frequency. For example, a variety of evidence shows that *Drosophila melanogaster* has acquired at least three new families of transposable elements (P, hobo, and I) within the last 50 yr (see Kidwell, 1994). Phylogenetic studies of the mariner family reveal many examples of horizontal transmission (Robertson and MacLeod, 1993). Because the incoming elements are likely to carry with them intrinsic site specificites, tissue specificities and regulatory properties, they would be expected to drive the evolution of the resident genome in nonrandom directions.

Subtelomeric Heterochromatin

The subterminal regions of different *Drosophila* chromosomes vary in length, and also frequently differ between individuals (Roberts, 1979). The X chromosome

transposes locally into the third 173-bp repeat of the third TAS repeat (*thin arrow*). (*2*) The double strand break is expanded by exonuclease, and the two ends subsequently invade homologous sequences (*arrows*) within the second TAS repeat. After the completion of repair, the observed structure of the region in line 9-035 would result.

contains the largest block of subtelomeric heterochromatin of any characterized tip region, as much as 100 kb on Dp1187 (Karpen and Spradling, 1992; our unpublished observations). In contrast, the first gene lies only ~ 15 kb from the tip of chromosome 3R (Levis et al., 1993). Terminal retrotransposon additions generate a significant fraction of telomeric heterochromatin. They make up the first 10 kb of 3R DNA and may account for the first 40 kb of Dp1187. It is not known how elements like HeT and TART recognize chromosome termini, or whether the rate of addition is controlled in any manner. Besides adding directly at the chromosome end, transpositions of other elements can introduce DNA sequences at subterminal sites (Fig. 2), and transposon-catalyzed changes in the copy number of subterminal tandem arrays may also help mold the structure of subterminal heterochromatin.

Large differences in telomeric heterochromatin within populations (*Atractomorpha similis:* John and King, 1983) or between closely related species (*Parascaris equorum* vs *Parascaris univalens;* Goday and Pimpinelli, 1986; *Cyclops divulsus* vs *Cyclops strenuus;* Beermann, 1977) might reflect changes in any of these mechanisms. *Anopheles melas* contains an X chromosome heterochromatic arm at least three times longer than that in any of the other members of the *Anopheles gambiae* complex (Coluzzi et al., 1979). The introduction of a new telomere-preferring transposable element into a species by horizontal transfer might be responsible for particularly dramatic growth in telomeric blocks. In *Drosophila,* at least one phenotypic effect, the suppression of meiotic recombination, is associated with telomeres, although the mechanism of inhibition is unknown. Interestingly, the larger block of subtelomeric heterochromatin present on the X chromosome is correlated with suppression of recombination over a longer distance compared to autosomal termini.

Concerted Evolution of Repetitive DNA

Our experiments indicated that transposable element activity might be an important factor driving the concerted evolution of repetitive DNA. Tandemly repeated DNAs are frequently found within recognizable heterochromatic blocks. For example, nucleolus organizers are almost invariably located within heterochromatin, while *Xenopus* 5S gene clusters are associated with the telomeres (Callan, Gall, and Berg, 1987). The copy number and chromosomal location of the repeated genes for ribosomal and 5S RNA vary widely between species. Transposons may generate variations in copy number that can be acted upon by selection, and even drive copy number changes in a directed manner.

The ribosomal DNA of insects provides a particularly interesting example of tandemly repeated DNA that may be subject to these mechanisms. Most insect species contain several hundred rDNA genes that are invariably located in heterochromatic tandem arrays. Two families of non-LTR retrotransposons, R1 and R2, are present at specific sites within the 28S gene of many rDNA repeats; genes containing insertions are inactive. The number of rDNA genes containing and lacking insertions varies widely between species and within populations (Jakubczak, Burke, and Eickbush, 1991; Jakubczak, Zenni, Woodruff, and Eickbush, 1992). The induction of double-strand breaks by the R2 integrase protein (Xiong and Eickbush, 1988) may be associated with new transpositions as well as gene conversions changing both rDNA repeat copy number, and the fraction of genes that can become active. Variation in the number of active genes appears to be of adaptive value in *D. mercatorum* populations (Templeton, Hollocher, Lawler, and Johnston, 1989). Flies with a low number of active rDNA genes develop more slowly and would be expected to lay eggs at a reduced rate.

The presence of elements with extremely high site specificity such as R1 and R2 is particularly interesting. Changes in the regulatory circuits affecting these elements might affect the number of active rDNA genes specifically. Transposons have been observed in several other tandemly repetitive DNAs, including *Drosophila* histone genes (Matsuo and Yamazaki, 1989). It is not known if these elements contribute to variations in gene number and activity like the R1 and R2 elements.

Transposon activity is likely to influence the evolution of euchromatic gene families as well. DNA sequencing has revealed evidence of frequent gene conversion events among multigene family members, such as those encoding immunoglobulin or MHC proteins (Kawamura, Saitou, and Ueda, 1992; Kuhner and Peterson, 1992). Our results suggest that transposable elements capable of inserting and excising from the chromosome regions in question may play a role in their concerted evolution. Those segments subject to frequent transposon insertion would be expected to show the highest amounts of conversion.

Don't We Know Already That Transposable Elements Are Unimportant for Evolution?

It has been argued that transposable elements play a relatively minor role in evolution (reviewed in Charlesworth and Langely, 1989). However, these studies have primarily addressed only one possible role for one class of transposable elements (the copialike retroelements), i.e., that they integrate near euchromatic genes and modify their regulation. Failure to identify evolutionarily successful integrations as sites of transposon in situ hybridization at fixed chromosomal positions can only eliminate the existence of very recent events of this type. Many other possible functions of transposons can never be addressed simply by mapping the sites of transposons within euchromatin or by analyzing their DNA sequences. HeT and TART transposons now appear to maintain *Drosophila* telomeres (see Levis et al., 1993), and heterochromatic elements have been proposed to modify chromosome and nuclear structure (see Spradling, 1993). Because the dispersed, multicopy nature of most transposons renders it is virtually impossible to eliminate their RNAs and proteins by mutation, we have no effective way to determine whether transposon-encoded products function during the organismal life cycle. Consequently, it remains plausible that transposons play a significant role in organismal development and evolution.

Are Changes in Heterochromatin Directly Selected or Side Effects?

This raises two possible reasons for the success of organisms bearing new transposon-mediated changes in heterochromatin. Selection might favor an organism bearing a new or reactivated transposon family that modified the copy number or expression of useful genes. Changes in heterochromatin structure might have resulted as well, simply as a byproduct of increased transposition. For example, the introduction of new transposon families at various times during the evolution of the *Anopheles gambiae* species complex may have generated the multiple chromosome inversions that characterize the six morphologically indistinguishable species (Coluzzi et al., 1979). These may have led to physiological changes that permitted adaptation to specific microenvironments. As a byproduct of these events, a transposon specific to

A. melas may have induced the large increase in its X chromosome heterochromatin. Alternatively, different configurations of heterochromatic sequences may modify chromatin structure in such a manner as to have been directly selected.

Further Studies of Transposable Elements and Heterochromatin Are Needed

We have shown that P elements have a variety of properties that are consistent with a role in generating heterochromatin and maintaining it in a state of dynamic equilibrium. Obviously, the functions of transposable elements need to be tested further and more directly. For example, it would be worthwhile to create strains containing one or more marked defective "responder" P elements that are maintained in a repressed state by a small number of differentially marked "repressor" elements located at special heterochromatic sites such as the X chromosome tip. The effects of chromosome rearrangements and of mutating cellular genes could then be determined by looking for activation of transposition and altered marker gene expression. The genetic interactions between multiple families of transposable elements and the host genome are undoubtedly complex. However, they are within the reach of a systematic analysis that builds on the knowledge that is already being gathered about the behavior of individual elements using systems of gradually increasing complexity.

Acknowledgements

The author thanks Drs. Gary Karpen, Ping Zhang, John Tower, and Nancy Craig for stimulating discussions.

This work was supported by grant GM27875 from the United States Public Health Service and by the Howard Hughes Medical Institute.

References

Ajioka, J. W. 1987. The molecular natural history of the P element family in *Drosophila melanogaster.* PhD thesis. State University of New York, Stony Brook, NY. 251 pp.

Ajioka, J. W., and W. F. Eanes. 1989. The accumulation of P elements on the tip of the X chromosome in populations of *Drosophila melanogaster. Genetic Research.* 53:1–6.

Ashburner, M. 1990. Drosophila: A Laboratory Handbook. Cold Spring Harbor Laboratory, Cold Spring Harbor, NY. 1331 pp.

Beermann, S. 1977. The diminution of heterochromatic chromosomal segments in Cyclops (Crustacea, Copepoda). *Chromosoma.* 60:297–344.

Berg, D. E., and M. M. Howe, editors. 1989. Mobile DNA. American Society of Microbiology Publishers, Washington, DC. 972 pp.

Biessmann, H., K. Valgeirsdottir, A. Lofsky, C. Chin, B. Ginther, R. W. Levis, and M.-L. Pardue. 1992. HeT-A, a transposable element specifically involved in "healing" broken chromosome ends in *Drosophila melanogaster. Molecular and Cellular Biology.* 12:3910–3918.

Brown, D. D., P. C. Wensink, and E. Jordan. 1972. Comparison of the ribosomal DNA's of *Xenopus laevis* and *Xenopus mulleri:* the evolution of tandem genes. *Journal of Molecular Biology.* 63:57–73.

Bucheton, A. 1990. I transposable elements and I-R hybrid dysgenesis in Drosophila. *Trends in Genetics.* 6:16–21.

Callan, H. G., J. G. Gall, and C. A. Berg. 1987. The lampbrush chromosomes of *Xenopus laevis:* preparation, identification and distribution of 5S DNA sequences. *Chromosoma.* 95:236–250.

Chao, L., and S. McBroom. 1992. Evolution of transposable elements: an IS10 insertion increases fitness in *E. coli. Journal of Molecular Evolution.* 2:359–369.

Charlesworth, B., and C. Langley. 1989. The population genetics of Drosophila transposable elements. *Annual Review of Genetics.* 23:251–287.

Coluzzi, M., A. Sabatini, V. Petrarca, and M. A. DiDeco. 1979. Chromosomal differentiation and adaptation to human environments in the *Anopheles gambiae* complex. *Transactions of the Royal Society of Tropical Medicine and Hygiene.* 73:483–497.

Cooley, L., R. Kelley, and A. Spradling. 1988. Insertional mutagenesis of the Drosophila genome with single P elements. *Science.* 239:1121–1128.

Daniels, S. B., S. H. Clark, M. G. Kidwell, and A. Chovnick. 1987. Genetic transformation of *Drosophila melanogaster* with an autonomous P element: phenotypic and molecular analyses of long-established transformed lines. *Genetics.* 115:711–723.

Doebley, J. 1993. Genetics, development and plant evolution. *Current Opinion in Genetics and Development.* 3:865–872.

Dover, G. A. 1982. Molecular drive: a cohesive model of species evolution. *Nature.* 299:111–117.

Dover, G. A. 1993. Evolution of genetic redundancy for advanced players. *Current Opinion in Genetics and Development.* 3:902–910.

Engels, W. R. 1989. P elements in Drosophila. *In* Mobile DNA. D. Berg and M. Howe, editors. ASM Publications, Washington, DC. 437 pp.

Engels, W. R., D. M. Johnson-Schlitz, W. B. Eggleston, and J. Sved. 1990. High-frequency P element loss in Drosophila is homolog dependent. *Cell.* 62:515–525.

Fakow, S. 1975. Infectious multiple antibiotic resistance. Pion, London, UK. 400 pp.

Finnegan, D. J., and D. H. Fawcett. 1986. Transposable elements in *Drosophila melanogaster. In* Oxford Surveys on Eukaryotic Genes. Vol. 3. N. Maclean, editor. Oxford University Press, Oxford, UK. 1–62.

Gatti, M., and S. Pimpinelli. 1992. Functional elements within *Drosophila melanogaster* heterochromatin. *Annual Review of Genetics.* 26:239–275.

Gloor, G. B., N. A. Nassif, D. M. Johnson-Schlitz, C. R. Preston, and W. R. Engels. 1991. Targeted gene replacement in Drosophila via P element-induced gap repair. *Science.* 253:1110–1117.

Goday, C., and S. Pimpinelli. 1986. Cytological analysis of chromosomes in the two species *Parascaris univalens* and *P. equorum. Chromosoma.* 94:1–10.

Hartl, D. L., D. R. Dykhuizen, R. D. Miller, L. Green, and J. DeFramond. 1983. Transposable element improves growth rate of *E. coli* without transposition. *Cell.* 35:503–510.

John, B., and M. King. 1983. Population cytogenetics of *Atractomorpha similis.* I. C. band variation. *Chromosoma.* 88:57–68.

Jakubczak, J. L., W. D. Burke, and T. H. Eickbush. 1991. Retrotransposable elements R1 and R2 interrupt the rRNA genes of most insects. *Proceedings of the National Academy of Sciences, USA.* 88:3295–3299.

Jakubczak, J. L., M. K. Zenni, R. C. Woodruff, and T. H. Eickbush. 1992. Turnover of R1 (typeI) and R2 (typeII) retrotransposable elements in the ribosomal DNA of *Drosophila melanogaster. Genetics.* 131:129–142.

Karpen, G. H. 1994. Position-effect variegation and the new genetics of heterochromatin. *Current Opinion in Genetics and Development.* In press.

Karpen, G. H., and A. C. Spradling. 1990. Reduced DNA polytenization of a minichromosome region undergoing position-effect variegation in Drosophila. *Cell.* 63:97–107.

Karpen, G. H., and A. C. Spradling. 1992. Analysis of subtelomeric heterochromatin in the Drosophila minichromosome *Dp1187* by single P element insertional mutagenesis. *Genetics.* 132:737–753.

Kawamura, S., N. Saitou, and S. Ueda. 1992. Concerted evolution of the primate immunoglobulin a-gene through gene conversion *Journal of Biological Chemistry.* 267:7359–7367.

Kidwell, M. G. 1994. Horizontal transfer of P elements and other short inverted repeat transposons. *In* Transposable Elements and Evolution. J. F. Macdonald, editor. Kluwer Academic Publishers, Dordrecht, The Netherlands. In press.

Kimura, M. 1983. The neutral theory of molecular evolution. Cambridge University Press, London, UK. 363 pp.

Kuhner, M. K., and M. J. Peterson, 1992. Genetic exchange in the evolution of the human MHC class II loci. *Tissue Antigens.* 39:209–215.

Levis, R. W., R. Ganesan, K. Houtchens, L. A. Tolar, and F. Sheen. 1993. Transposons in place of telomeric repeats at a Drosophila telomere. *Cell.* 75:1083–1093.

Matsuo, Y., and T. Yamazaki. 1989. tRNA derived insertion element in histone gene repeating unit of *Drosophila melanogaster. Nucleic Acids Research.* 17:225–238.

McClintock, B. 1984. The significance of responses of the genome to challenge. *Science.* 226:792–796.

Miller, W. J., S. Hagemann, E. Reiter, and W. Pinsker. 1992. P-element homologous sequences are tandemly repeated in the genome of *Drosophila guanche. Proceedings of the National Academy of Sciences, USA.* 89:4018–4022.

Orgel, L., and F. Crick. 1980. Selfish DNA, the ultimate parasite. *Nature.* 284:604–607.

Pardue, M. L., and W. Hennig. 1990. Heterochromatin: junk or collector's item? *Chromosoma.* 100:3–7.

Petes, T. 1981. Unequal meiotic recombination within tandem arrays of yeast ribosomal DNA genes. *Cell.* 19:765–774.

Preston, C. R., and W. R. Engels. 1989. Spread of P transposable elements in inbreed lines of *Drosophila melanogaster. Progress in Nucleic Acids Research and Molecular Biology.* 36:71–85.

Rio, D. C. 1991. Regulation of Drosophila P element transposition. *Trends in Genetics.* 7:282–287.

Roberts, P. A. 1979. Rapid change of chromomeric and pairing patterns of polytene chromosome tips in D. melanogaster: migration of polytene-nonpolytene transition zone? *Genetics.* 92:861–878.

Robertson, H. M., and E. G. MacLeod. 1993. Five major subfamilies of mariner transposable elements in insects, including the Mediterranean fruit fly, and related arthropods. *Insect Molecular Biology.* 2:125–139.

Ronserray, S., M. Lehmann, and D. Anxolabéhère. 1991. The maternally inherited regulation of P elements in *Drosophila melanogaster* can be elicited by two P copies at cytological site 1A on the X chromosome. *Genetics.* 129:501–512.

Rubin, G. M., and A. C. Spradling. 1981. Drosophila genome organization: conserved and dynamic aspects. *Annual Review of Genetics.* 15:219–264.

Smith, G. P. 1973. Unequal crossover and the evolution of multigene families. *Cold Spring Harbor Symposium on Quantitative Biology.* 38:507–513.

Spradling, A. C., and T. Orr-Weaver. 1987. Regulation of DNA replication during Drosophila development. *Annual Review of Genetics.* 21:373–403.

Spradling, A. C. 1993. Position-effect variegation and genomic instability. Cold Spring Harbor Symposium on Quantitative Biology. Cold Spring Harbor Laboratories, Cold Spring Harbor, NY. In press.

Steinemann, M., and S. Steinemann. 1992. Degenerating *Y* chromosome of *Drosophila miranda:* a trap for retrotransposons. *Proceedings of the National Academy of Sciences, USA.* 89:7591–7595.

Templeton, A. R., H. Hollocher, S. Lawler, and J. S. Johnston. 1989. Natural selection and ribosomal DNA in Drosophila. *Genome.* 31:296–303.

Thompson-Stewart, D., G. Karpen, and A. Spradling. 1994. A transposable element can drive the concerted evolution of tandemly repetitious DNA. *Proceedings of the National Academy of Sciences, USA.* In press.

Tower, J., G. Karpen, N. Craig, and A. Spradling. 1993. Preferential transposition of Drosophila P elements to nearby chromosome sites. *Genetics.* 133:347–359.

Verma, R. S., editor. 1988. Heterochromatin, molecular and structural aspects. Cambridge University Press: Cambridge, UK. 301 pp.

White, M. J. D. 1973. Animal cytology and evolution. Third edition. Cambridge University Press. London, UK. 961 pp.

Xiong, Y. E., and T. H. Eickbush. 1988. Functional expression of a sequence-specific endonuclease encoded by the retrotransposon R2Bm. *Cell.* 55:235–246.

Zhang, P., and A. Spradling. 1993. Efficient and dispersed local P element transposition from Drosophila females. *Genetics.* 133:361–373.

Zhang, P., and A. Spradling. 1994. Insertional mutagenesis of Drosophila heterochromatin with single P elements. *Proceedings of the National Academy of Sciences, USA.* 91:3539–3543.

Gene Conversion and the Evolution of Gene Families in Mammals

John C. Schimenti

The Jackson Laboratory, Bar Harbor, Maine 04609

Introduction

Gene duplication is a critical process in the evolution of higher organisms. It recruits preexisting genetic material as a substrate for the formation of novel functional units, thereby catalyzing rapid genetic changes. Extra gene copies created through duplication may ultimately diverge to perform related but specialized developmental and biochemical function. The mammalian globin family, which has evolved a highly coordinated process of tissue and stage specific expression of developmentally specialized genes from a single ancestral gene, exemplifies this process.

Genes can duplicate by a variety of mechanisms. Most dramatic is whole genome duplication, which happens very rarely during evolution. Duplication via unequal recombination is probably the most common mechanism. This can occur after pairing of nonallelic, homologous sequences. Initial duplication events of single copy genes can be catalyzed by repetitive elements, such as Alu or L1 (Cross and Renkawitz, 1990), and subsequent events can occur within the duplicated genes themselves, creating new hybrid genes. Depending on the number of genes in the family and the location of the cross-over events, duplication by this mechanism can dramatically increase the gene copy number of a family.

Immediately after duplication via unequal recombination, a new gene or DNA fragment would be identical to the preexisting genetic information. These duplicated sequences are susceptible to gene conversion, a fundamental recombination mechanism.

Gene conversion is the nonreciprocal transfer of genetic information between two related genes or DNA sequences. It was first discovered in fungi, manifested as the non-Mendelian segregation of alleles in a diploid heterozygous at a particular locus. While most meioses create tetrads (in yeast) with two spores of each genotype, there are frequent exceptions which demonstrate 3:1 segregation of alleles. This type of inheritance is called gene conversion. Explaining the mechanism of gene conversion lies at the cornerstone of general recombination models (reviewed in Orr-Weaver and Szostak, 1985).

Studies of gene conversion in fungi have shown that conversion occurs not only between alleles, but intrachromosomally between duplicated sequences on either the same chromatid or the sister chromatid, and ectopically between sequences on nonhomologous chromosomes. Essentially, any homologous sequences in the genome are potential participants in gene conversion.

Fungi are particularly well-suited for studying gene conversion because all the products from a meiosis can be isolated (Orr-Weaver and Szostak, 1985; Petes and

Hill, 1988). Gene conversion is extremely active in these organisms. Intrachromosomal meiotic conversions between duplicated genes generally occurs in 1–10% of yeast tetrads (Klein, 1984; Klein and Petes, 1981; Nicolas, Treco, Schultes, and Szostak, 1989), and conversion between unlinked sequences falls within that range (Haber, Leung, Borts, and Lichten, 1991). In mammals, much of the evidence for germline gene conversion is based upon comparative sequence analysis of duplicated genes. A duplication unit containing a patch of near sequence identity within a larger stretch of considerable divergence is the kind of data best explained by gene conversion.

Results and Methods

A comprehensive investigation into mammalian gene conversion frequency faces two major technical and logistical difficulties. The first is that large numbers of progeny must be scored. Secondly, the events must be readily detectible. We have devised a

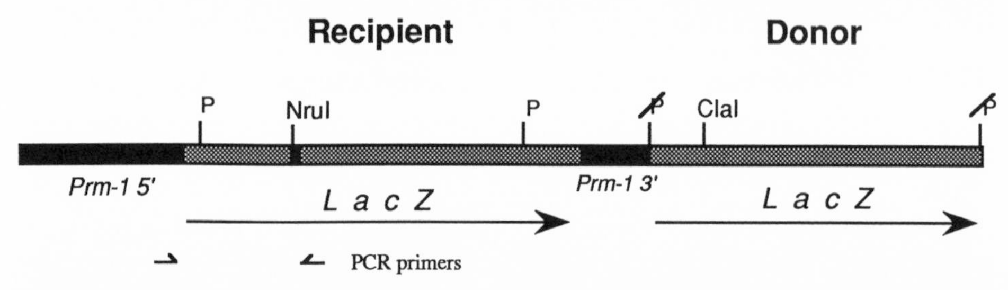

Figure 1. As indicated, the black boxes represent mouse protamine 1 (*Prm-1*) sequences, and hatched boxes are *lac*Z sequences. Transcriptional orientations of the *lac*Z genes are indicated by the arrows. The black vertical stripe in the Recipient *lac*Z gene is a 2 bp insertion mutation, created by filling in a *Cla*I site to make a *Nru*I site. P = *Pvu*II. The crossed-out "P"s correspond to the *Pvu*II sites in the Recipient gene, but were destroyed during cloning. The Donor *lac*Z gene is deleted for the first 36 and last 136 amino acids of the enzyme. The construct is 7.5 kb in length. The location of PCR primers for amplification of the Recipient gene from sperm DNA is indicated.

novel strategy to detect germline gene conversion which eliminates these constraints (Murti, Bumbulis, and Schimenti, 1992). The crux of the system is to generate transgenic mice in which a planned conversion event can be visualized in sperm, rather than offspring, by histochemical staining. This strategy allows the rapid quantification of thousands of meiotic events for gene conversion at the transgene locus.

The basic constructs for detecting intrachromosomal gene conversion consist of two *lac*Z genes with different mutations (Fig. 1; Murti et al., 1992). One of these genes, called the "recipient," is under the transcriptional control of the mouse protamine-1 (*Prm*-1) promoter, and contains an internal 2-bp insertion that abolishes β-galactosidase activity by frameshift mutation. Linked to it is a promoterless "donor" *lac*Z gene that is truncated at both the amino and carboxy termini. Male mice carrying this construct transcribe the recipient gene specifically in haploid germ cells, but do not produce functional *lac*Z unless a spontaneous mutation or recombination event corrects the frameshift mutation.

To quantitate the percentage of spermatids that have undergone gene conversion, testicular cells from adult transgenic hemizygotes are fractionated in a BSA gradient at unit gravity. They are then fixed and stained for *lac*Z activity by 5-bromo-4-chloro-3-indolyl-β-D-galactopyranoside (X-gal) staining, followed by photography and manual quantitation. In a second method, a crude testicular cell mixture is analyzed by flow cytometry, whereby haploid cells are distinguished in the basis of DNA content, and a fluorescent *lac*Z substrate (Nolan, Fiering, Nicolas, and Hertzenberg, 1988) is used to detect converted spermatids.

As we have previously reported (Murti et al., 1992), no *lac*Z staining is observed in wild-type mice or transgenics harboring a construct that contains the recipient but no donor gene. Therefore, interaction between donor and recipient sequences must be responsible for the gene correction. Regardless of copy number or genomic insertion site, 1–2% of spermatids from hemizygous animals are *lac*Z positive. PCR amplification of mature sperm using the primers shown in Fig. 1, followed by digestion of the products with *Cla*I (which is diagnostic for a conversion event in the recipient gene) confirms the frequencies observed by *lac*Z staining (Murti, Schimenti, and Schimenti, 1994). Histological analysis indicated that both meiotic and mitotic events contributed to this percentage (Murti et al., 1992). Conversion events in mitotically proliferating spermatogonia would be inherited by half their progeny. Therefore, the percentage of converted cells does not equal the frequency of conversion events.

A variation of this system has been used to detect ectopic gene conversion-recombination between homologous sequences on nonhomologous chromosomes. In this experimental paradigm, the Donor and Recipient *lac*Z genes are placed on different chromosomes, followed by detection of converted spermatids by the methods outlined above. Remarkably, gene conversion is also quite frequent in this situation, although at frequencies (percentages of converted spermatids) 2–10-fold lower than intrachromosomally (J. R. Murti, M. Bumbulis, and J. Schimenti, unpublished observations).

Discussion

Questions Unique to Recombination in Mammals

Whereas details and mechanics of recombination are best studied in fungal systems, recombination occurs in situations that are unique to mammals, and must ultimately be addressed in that context. For example, in addition to the implications of mitotic recombination in the germ line, somatic recombination is an important process which has no parallel in fungi. Immunoglobulin gene rearrangement and mitotic conversion/crossing over causing loss of heterozygosity at tumor suppressor genes are examples of recombinational mechanisms that are distinct between mammals and fungi. Because recombination activities appear to differ from tissue to tissue (germ cells probably being the highest), and that specialized rearrangement activities exist (immune cells), it is likely that either multiple recombination enzymes exist with different cellular specificities, or that a common group of recombination enzymes is differentionally regulated in different tissues.

Aside from specialized somatic recombination systems, the existence of general recombination in somatic cells is probably (in part) related to DNA repair (Padmore, Cao, and Kleckner, 1991). Recombination induction appears to be a broad indicator of an agent's mutagenicity, reflecting repair of induced DNA damage (Zimmermann, 1971). In meiosis, recombination is required for proper disjunction of chromosomes. Furthermore, meiotic recombination has the important role of creating genetic diversity (although it is unknown whether the evolution of efficient recombination was selected for on that basis). Mitotically dividing somatic cells do not have any apparent requirement for recombination in fidelity of replication, and recombination events are not transmitted to progeny. The role of recombination in DNA repair will require investigations into the recombinational capacity of various tissue or cell types, and whether recombination is induced by DNA damage. Notably, cells from Ataxia-Telangiectasia patients (Meyn, 1993), which are difficient in DNA repair, exhibit greatly elevated levels of intrachromosomal recombination.

Gene Conversion and the Evolution of Gene Families

We have found that up to 2% of gametes reflect an intrachromosomal conversion event in our transgenic mice. This approaches the meiotic gene conversion frequency in yeast (Orr-Weaver and Szostak, 1985). It is unclear whether the gene conversion frequency observed in this system is generally representative of random duplicated sequences elsewhere in the genome. In yeast, meiotic gene conversion frequency shows great locus variation; values have been observed to range from 0.5 to 30% of meioses with different genes (Fogel, Mortimer, Lusnak, and Travares, 1979; Nag, White, and Petes, 1989). Because the percentage of corrected spermatids between the transgenic lines used in our experiments was similar, it is unlikely that the numbers are unusually high due to chromosomal context. Rather, the intrachromosomal conversion activity of the transgenes must be an inherent property of the construct.

Gene conversion can play paradoxical roles in the evolution of a gene family. On one hand, related gene family members can undergo sequence homogenization by gene conversion, thereby inhibiting divergence and potential adaptation. Several theoretical studies have addressed the confounding effects of gene conversion-mediated homogenization upon the evolutionary analysis of repeated genes, presenting mathematical models on the role of gene conversion in evolution (Dover, 1982; Gutz and Leslie, 1976; Lamb and Helmi, 1982; Nagylaki and Petes, 1982; Walsh, 1987). On the other hand, micro-gene conversions can rapidly generate diversity by introducing multiple, simultaneous sequence changes into a member of a large gene family (Baltimore, 1981). A gene conversion-like mechanism appears to be responsible for the generation of novel MHC alleles in human populations (Belich, Madrigal, Hildebrand, Zemmour, Williams, Luz, Petzl, and Parham, 1992; Kuhner, Lawler, Ennis, and Parham, 1991; Parham and Lawlor, 1991). Our results showing that ectopic gene conversion is also relatively frequent in mice indicates that the divergence of duplicated genes is not guaranteed by genomic dispersal. The significance of gene conversion upon diversification or homogenization during evolution is directly related to its frequency in the germ line.

The frequency of gene conversion we have observed greatly exceeds values used in theoretical studies on the influence of this process upon the evolution of duplicated genes (Nei, 1987; Walsh, 1987). According to the formula derived by Walsh (1987), the frequency of gene conversion we have observed in our transgenic system would be incompatible with the divergence of genes. Because genes do in fact diverge despite being susceptible to such frequent homogenization indicates that mechanisms must exist to diminish unabated gene conversion. Selection against conversion should not be a factor immediately after a duplication event, because the newly created gene would not have an unique, critical function. Although the frequency of recombination decreases as the length and degree of homology decreases (Ayares, Chekuri, Song, and Kucherlapati, 1986; Liskay, Letsou, and Stachelek, 1987; Rubnitz and Subramani, 1984; Singer et al., 1982; Waldman and Liskay, 1988), how do duplicated genes escape conversion in the first place? It has been suggested that structural events which interrupt homology and heteroduplex formation, such as insertion of repetitive elements, could substantially inhibit gene conversion between duplicated genes (Cross and Renkawitz, 1990; Fisher, Pecht, and Hood, 1989; Hess, Fox, Schmid, and Shen, 1983; Michelson and Orkin, 1983; Schimenti and Duncan, 1984), allowing them to diverge freely. Another potential mechanism to promote divergence of genes is "RIPping" (Kricker, Drake, and Radman, 1992). This is a phenomenon whereby duplicated sequences undergo cytosine methylation at a high frequency, which catalyzes $C > T$ mutations. As random events such as these occur to interrupt stretches of perfect or high homology between duplicated gene pairs, the gene conversion frequency would decrease locally. This would increase the average time between conversion events, and hence permit the accumulation of base changes. If duplicated genes evolve differential function, then selection may become a strong force against maintainance of conversion events.

Future Directions

Our results indicate that gene conversion is a significant force in the genome. We are conducting additional lines of experimentation to further characterize the nature and properties of gene conversion in mice. In particular, variations of the system described here will be used to compare the frequency of premeiotic versus meiotic conversion, measure conversion tract lengths, and study the effect of introducing heterologies.

References

Ayares, D., L. Chekuri, K.-Y. Song, and R. Kucherlapati. 1986. Sequence homology requirements for intermolecular recombination in mammalian cells. *Proceedings of the National Academy of Sciences, USA.* 83:5199–5203.

Baltimore, D. 1981. Gene conversion: some implications for immunoglobulin genes. *Cell.* 24:592–594.

Belich, M. P., J. A. Madrigal, W. H. Hildebrand, J. Zemmour, R. C. Williams, R. Luz, E. M. Petzl, and P. Parham. 1992. Unusual HLA-B alleles in two tribes of Brazilian Indians. *Nature.* 357:326–329.

Cross, M., and R. Renkawitz. 1990. Repetitive sequence involvement in the duplication and divergence of mouse lysozyme genes. *European Molecular Biology Organization Journal.* 9:1283–1288.

Dover, G. 1982. Molecular drive: a cohesive mode of species evolution. *Nature.* 299:111–116.

Fisher, D., M. Pecht, and L. Hood. 1989. DNA sequence of a class I pseudogene from the *Tla* region of the murine MHC: recombination at a B2 Alu repetitive sequence. *Journal of Molecular Evolution.* 28:306–312.

Fogel, S., R. K. Mortimer, K. Lusnak, and F. Travares. 1979. Meiotic gene conversion: a signal of the basic recombination event in yeast. *Cold Spring Harbor Laboratory Symposium on Quantitative Biology.* 43:49–53.

Gutz, H., and J. F. Leslie. 1976. Gene conversion: a hitherto overlooked parameter in population genetics. *Genetics.* 83:861–866.

Haber, J., W.-Y. Leung, R. Borts, and M. Lichten. 1991. The frequency of meiotic recombination in yeast is independent of the number and position of homologous donor sequences: implications for chromosome pairing. *Proceedings of the National Academy of Sciences, USA.* 88:1120–1124.

Hess, J. F., M. Fox, C. Schmid, and C. K. Shen. 1983. Molecular evolution of the human adult alpha-globin-like gene region: insertion and deletion of Alu family repeats and non-Alu DNA sequences. *Proceedings of the National Academy of Sciences, USA.* 80:5970–5974.

Klein, H. 1984. Lack of association between intrachromosomal gene conversion and reciprocal exchange. *Nature.* 310:748–752.

Klein, H. L., and T. D. Petes. 1981. Intrachromosomal gene conversion in yeast. *Nature.* 289:144–148.

Kricker, M. C., J. W. Drake, and M. Radman. 1992. Duplication-targeted DNA methylation and mutagenesis in the evolution of eukaryotic chromosomes. *Proceedings of the National Academy of Sciences, USA.* 89:1075–1079.

Kuhner, M. K., D. A. Lawlor, P. D. Ennis, and P. Parham. 1991. Gene conversion in the evolution of the human and chimpanzee MHC class I loci. *Tissue Antigens.* 38:152–164.

Lamb, B. C., and S. Helmi. 1982. The extent to which gene conversion can change allele frequencies in populations. *Genetic Research.* 39:199–217.

Liskay, R. M., A. Letsou, and J. L. Stachelek. 1987. Homology requirement for efficient gene conversion between duplicated chromosomal sequences in mammalian cells. *Genetics.* 115: 161–7.

Meyn, M. S. 1993. High spontaneous intrachromosomal recombination rates in ataxia-telangiectasia. *Science.* 260:1327–30.

Michelson, A. M., and S. H. Orkin. 1983. Boundaries of gene conversion within in the duplicated human α-globin genes. *Journal of Biological Chemistry.* 258:15245–15254.

Murti, J. R., M. Bumbulis, and J. Schimenti. 1992. High frequency germline gene conversion in transgenic mice. *Molecular and Cellular Biology.* 12:2545–2552.

Murti, J. R., K. J. Schimenti, and J. C. Schimenti. 1994. A recombination-based transgenic mouse system for genotoxicity testing. *Mutation Research.* In press.

Nag, D. K., M. A. White, and T. D. Petes. 1989. Palindromic sequences in heteroduplex DNA inhibit mismatch repair in yeast. *Nature.* 340:318–320.

Nagylaki, T., and T. D. Petes. 1982. Intrachromosomal gene conversion and the maintainance of sequence homogeneity among repeated genes. *Genetics.* 100:315–337.

Nei, M. 1987. Molecular Evolutionary Genetics. Columbia University Press, New York.

Nicolas, A., D. Treco, N. Schultes, and J. Szostak. 1989. An initiation site for meiotic gene conversion in the yeast *Saccaromyces cerevesiae. Nature.* 338:35–39.

Nolan, G., S. Fiering, J.-F. Nicolas, and L. Hertzenberg. 1988. Fluorescence-activated cell analysis and sorting of viable mammalian cells based in β-D-galactosidase activity after transduction of *E. coli lacZ*. *Proceedings of the National Academy of Sciences, USA*. 85:2603–2607.

Orr-Weaver, T. L., and J. W. Szostak. 1985. Fungal recombination. *Microbiological Reviews*. 49:33–58.

Padmore, R., L. Cao, and N. Kleckner. 1991. Temporal comparison of recombination and synaptonemal complex formation during meiosis in *S. cerevisiae*. *Cell*. 66:1239–1256.

Parham, P., and D. A. Lawlor. 1991. Evolution of class I major histocompatibility complex genes and molecules in humans and apes. *Human Immunology*. 30:119–128.

Petes, T. D., and C. W. Hill. 1988. Recombination between repeated genes in microorganisms. *Annual Review of Genetics*. 22:147–168.

Rubnitz, J., and S. Subramani. 1984. The minimum amount of homology required for homologous recombination in mammalian cells. *Molecular and Cellular Biology*. 4:2253–2258.

Schimenti, J., and C. Duncan. 1984. Ruminant globin gene structures suggest an evolutionary role for Alu-type repeats. *Nucleic Acids Research*. 12:1641–1655.

Singer, B., L. Gold, P. Gauss, and D. H. Doherty. 1982. Determination of the amount of homology required for recombination in bacteriophage T4. *Cell*. 31:25–33.

Waldman, A. S., and R. M. Liskay. 1988. Dependence of intrachromosomal gene conversion in mammalian cells on uninterrupted homology. *Molecular and Cellular Biology*. 8:5350–5357.

Walsh, J. 1987. Sequence-dependent gene conversion: can duplicated genes diverge fast enough to escape conversion? *Genetics*. 117:543–557.

Zimmermann, F. 1971. Induction of mitotic gene conversion by mutagens. *Mutation Research*. 11:327–337.

Functional Polymorphism in Class I Major Histocompatibility Complex Genes

Peter Parham

Departments of Cell Biology and Microbiology and Immunology, Sherman Fairchild Building, Stanford University, Stanford, California 94305–5400

Introduction

Targets for the vertebrate immune system are diverse organisms capable of rapid change to evade the host response. In return, lymphocytes use four families of variable antigen binding molecules: immunoglobulins (Ig), T cell receptors (TcR) and the class I and II antigen presenting molecules of the Major Histocompatibility Complex (MHC). Whereas the repertoire of Ig and TcR is played out by rearranging genes within a single individual, diversity in MHC antigen presenting molecules involves genetic polymorphism seen at the level of populations.

Class I and II MHC molecules present short peptides of degraded antigenic proteins to $CD8^+$ and $CD4^+$ T cells. Their antigen binding sites bear similarities to the substrate binding domains of the heat shock 70 family of chaperones. To a first approximation, class I molecules present peptides from proteins made within the cell whereas class II molecules present peptides from proteins made without.

Phylogenetic comparison show the class I genes of the MHC evolve more rapidly than their class II counterparts. The number of class I genes and their organization varies greatly between species and only for closely related species, e.g., humans and apes, can orthologous genes be identified. The number of functional genes is always small, 1–3, and in crude terms functionality can be correlated with the extent of allelic polymorphism.

Alleles of class I genes vary by 1–77 nucleotide substitutions. Variation concentrates on residues of the peptide binding site, effecting changes in the specificity of peptide binding. Although many features of class I molecules in humans and apes are shared, no single allele is held in common by any two species and thus, no antigen-presenting specificity that existed in the common ancestor has survival unchanged along both the human and ape lineages.

For humans, we estimate there are 100–300 alleles of the class I HLA-A,B,C loci, many of which exhibit localized geographic distribution. Sequence comparisons indicate that interallelic recombination involving short segments of DNA (10–100 nucleotides) creates most new alleles. The mechanism of this process is unknown.

Disease-specific selection, heterozygote advantage, frequency-dependent selection and genetic drift may all play a role in the generation of class I HLA polymorphism. For urban populations, admixture resulting from recent migrations is also a major factor.

Immunological Functions in Antigen Presentation

As its name implies, the Major Histocompatibility Complex, or MHC, was discovered through observation of tissue transplantation. Experiments to propagate tumors from one mouse to another failed unless the mice were closely related. The same was true for normal tissues. Breeding studies revealed the influence of a single major locus—the MHC—and a number of minor ones. Luckily, immunization of one mouse with the cells from another produced antibodies against class I and II MHC proteins, providing simple tools for analysis of the mouse MHC (H-2 region). Similarly, antibodies for studying the human MHC (HLA region) were obtained first from patients who had been immunized through repeated blood transfusion and then from multiparous women who were repeatedly stimulated by the paternal HLA antigens expressed by their babies. Everything we know of the MHC stems from systematic application of the serological approach.

That mice reject each other's tissues is due to two fundamental properties of class I and II MHC antigens: extensive genetic polymorphism and a physiological function that is central to the stimulation of T lymphocytes.

Antigen receptors of T cells are cell surface molecules, the products of rearranging genes that are clonally expressed. The antigens with which T cell receptors interact comprise short peptides bound "presented" either by class I or class II MHC molecules. Presentation by class I molecules leads to cytotolytic T cell responses that can kill cells infected by viruses or parasites, whereas presentation by class II molecules leads to delayed type hypersensitivity (the classical cellular immunity) or to the release of cytokines which help B lymphocytes into antibody production. During development in the thymus and the peripheral circulation those T cells with receptor specificity for peptides derived from normal cellular proteins, "self," are either eliminated or inactivated. Thus, the T lymphocyte system is poised to respond to cells presenting foreign antigens.

The probability of two unrelated individuals having precisely the same set of class I and II HLA alleles is low. On transplantation of tissue the HLA differences are perceived in a similar fashion to the foreign antigens of parasites, viruses and bacteria. They trigger clones of T and B lymphocytes to respond and the resulting immune response eliminates the grafted cells in much the same way that virally infected cells are destroyed.

Class I and class II HLA (human leukocyte antigen) molecules are closely related in structure and function and they clearly represent variations upon an evolutionary theme. Both classes of molecules have a binding site for peptides and sites for interaction with the receptors and coreceptors of T cells. Differences between the class I and II molecules are found in the length of peptides they bind, the intracellular compartments in which binding takes place and the type of T cell coreceptor engaged. Class I MHC molecules bind peptides of 8–10 amino acid residues in the endoplasmic reticulum and engage T cells with the CD8 coreceptor whereas class II MHC molecules bind peptides of 15–25 residues in membrane compartments of the endocytic pathway and engage T cells with the CD4 coreceptor.

Flexibility and Efficiency in The Immune Response

Although the immune system has elaborated a series of elegant mechanisms for generating receptor diversity, notably in the immunoglobulin and the T cell receptors, it is nowhere close to being optimized for the tasks in hand. Many of its opponents, the virus, parasites and microorganisms, evolve faster than vertebrate organisms and can thereby evade the immune response. As a consequence of its flexibility, the immune system is slow to react and infections are often established before the immune system makes a contribution. Influenza and cold viruses are always getting ahead of the human immune with the result that most years we are all reminded individually of their effects. Inadequacy in immune recognition is unequivocal in the case of HIV, a virus which sabotages the whole enterprise. That the immune system benefits from a helping hand is illustrated by the efficacy of vaccination, which in effect allows the immune system to rehearse its act. It is perhaps not surprising that the immune system is this way, because those molecular mechanisms that provide flexibility in the immune response, in that it can have a go at anything, also prevent a swift targeted assault on any particular foe. In this and other respects, there are parallels with other organizations involved in defense.

Self preservation has its shortcomings too. The diseases bothering many in the over-developed countries are degenerative (diabetes, arthritis, multiple sclerosis) and automimmune in nature. In these diseases the slow erosion of the target tissue, pancreatic β cells in diabetes for example, is thought to be mediated by T cells responding to self peptides. A common idea suggests that the breaking of tolerance is a secondary consequence of some traumatic event: perhaps physical damage resulting in the exposure of self proteins which are normally sequestered from the immune system, or an acute infection which stimulates and expands T cell clones specific for the infectious agent, some of which crossreact onto self peptides. Despite its limitations the overall benefits of the immune system are clear from the fate of patients with combined immunodeficiency disease who rapidly succumb unless placed in a germ free environment.

Human Class I HLA Proteins

The antibodies found in the blood of multiparous women were first used to define the class I antigens, HLA-A,B,C. These molecules are found on the surface of most cell types, but the levels vary and can be regulated by γ interferon and other mediators of the inflammatory response.

The HLA-A,B,C proteins are formed by a membrane anchored heavy chain, encoded by either the HLA-A,B, or C gene, associated with a common light chain, β_2-microglobulin (β_2-m). Crystallographic structures of HLA-A and B molecules (Bjorkman, Saper, Samraoui, Bennett, Strominger, and Wiley, 1987*a;* Madden, Gorga, Strominger, and Wiley, 1992) show they consist of a pair of immunoglobulin domains supporting a peptide binding site which is speculated to have similarity with the substrate-binding domain of heat shock proteins in the hsc70 family (Rippmann, Taylor, Rothbard, and Green, 1991; Flajnik, Canel, Kramer, and Kasahara, 1991).

HLA-A,B,C proteins bind short peptides in the endoplasmic reticulum of human cells and transport them to the plasma membrane where the protein-peptide complex forms a potential ligand for the antigen receptors of circulating CD8 positive T lymphocytes. In healthy cells, the bound peptides are the products of normal turnover of cellular proteins "self" against which the T cells are unreactive. When a cell is virally infected, peptides of viral origin get bound and presentation of such foreign antigens stimulates a cytotoxic T cell response which kills the infected cells, thereby preventing viral replication and further infection of healthy tissue. HLA-A,B, and C heavy chains are homologous proteins and their codominant expression is believed to increase the number of foreign peptides presented during viral infection and thus to the vigor of the T cell response.

Like other components contributing to the specificity of the immune system, antigen presenting molecules have only been found in the vertebrates and thus appear to be of recent evolutionary origin. Sequences for class I heavy chains from species in the major classes of vertebrates (fish, reptiles, amphibians, birds and mammals) have been determined. Their comparison reveals only seven invariant positions of a total of 182 residues in the peptide binding domains (α_1 and α_2) and eight invariant positions of the 92 residues in the α_3 domain. These conserved residues are exclusively in positions that are important for maintaining either the structure of individual domains or interdomain contacts. In the α_3 domain the conserved residues are common to immunoglobulin domains and those in the α_1 and α_2 domains provide interactions between α_1 and α_2 and their supporting Ig-like domains (Grossberger and Parham, 1992). The majority of positions within the class I structure can tolerate substitution and even insertion or deletion without losing antigen presenting activity, a view finding support in the experience of investigators performing site-directed mutagenesis.

Class I HLA Gene Structure

HLA-A,B,C genes consist of eight exons which correspond to distinctive protein domains. The genes are closely linked to each other and also to ~17 homologous pseudogenes, gene fragments and intact genes of unknown function (Geraghty, Koller, Hansen, and Orr, 1992*a;* Geraghty, Koller, Pei, and Hansen, 1992*b*). The number of class I genes varies greatly between species and it is only in the most closely related species, e.g., human and chimpanzee, that orthologous genes can be identified. Such comparison shows that class I genes frequently undergo gene duplication, deletion and changes in functional status.

Mice are the favorites of immunologists and much is known about their MHC (the H-2 region) and the function of their antigen presenting molecules. The H-2 region contains more class I genes than the human HLA region. Orthologies between the human and mouse class I genes have not been found (reviewed in Lawlor, Zemmour, Ennis, and Parham, 1990). When phylogenetic trees are constructed they show that the mouse genes form one cluster and the human genes another. Within the common laboratory strains of mice the numbers of class I genes differ as a result of large duplication and deletion events. This propensity of class I genes has been taken to an extraordinary degree in one African rodent, which has over 1,500 class I genes (Delarbre, Jaulin, Kourilsky, and Gachelin, 1992).

Class I HLA Gene Polymorphism

Further variation is seen at the level of individual class I genes. The most characteristic feature of the HLA-A,B and C genes is their genetic polymorphism. Unlike most other genes, for HLA-A,B, or C, the term wild type has no meaning: various alleles are evenly distributed in human populations. The total number of alleles is unknown. Their identification and structural characterization is still in progress, but some ninety HLA-B alleles, sixty HLA-A alleles, and forty HLA-C alleles have been defined so far.

Polymorphism is correlated with function. Whereas HLA-A,B, and C have many alleles, the pseudogenes, gene fragments and genes of unknown function have few.

HLA-A,B,C alleles can differ by 1–77 nucleotide substitutions, a disproportionate number of which change amino acid residues within the peptide binding site and the specificity of peptide binding and T cell recognition (Bjorkman, Saper, Samraoui, Bennett, Strominger, and Wiley, 1987*b;* Parham, Lomen, Lawlor, Ways, Holmes, Coppin, Salter, Wan, and Ennis, 1988). Overall, the data are consistent with the polymorphism being the product of natural selection for antigen presenting function.

The polymorphism of HLA genes influences medical practice, most obviously in transplantation. The number of alleles is such that most unrelated individuals express a different combination of class I and class II molecules. The sequence differences between the HLA molecules are themselves capable of stimulating strong T cell responses and it is precisely this type of response that causes rejection of tissues transplanted between HLA mismatched individuals.

Two approaches have been made to deal with the problem of rejection. One uses immunosuppressive drugs to ablate the response, the other attempts to match recipients and donors through serological typing for HLA polymorphism. Molecular and immunological analysis has revealed that single serotypes often embrace multiple alleles, for example, 12 in the case of HLA-A2, and that their differences provoke T cell responses capable of rejecting transplanted tissue (López de Castro, 1989; Fleischhauer, Kernan, O'Reilly, Dupont and Yang, 1990). Such studies indicate that the polymorphism is greater than has been supposed previously, prompting the search for more accurate, DNA-based, methods of tissue typing. These findings also increase the potential use for HLA polymorphisms as a tool in the study of human genetics and evolution.

Mechanisms of Diversification

Comparison of HLA-A,B,C allelic sequences reveals a patchwork pattern in which an individual allele comprises a unique combination of sequence motifs, each of which is shared with other alleles. This pattern suggests that recombination plays a key role in the generation of alleles, an impression supported by pairwise comparison of the sequences. The most closely related pairs of alleles usually differ by localized clusters of substitutions for which both sequence motifs can be found in other alleles. This pattern implicates interallelic conversion or double recombination as the diversifying mechanism. Although the vast majority of such events appear to involve recombination between allele of the same locus, in three cases (HLA-B46,B54 and B76), we have good evidence for intergenic conversion.

The role of recombination in generating polymorphism is illustrated by the HLA-A,B,C alleles of American Indian populations. These people derive from small founding populations that crossed the Behring strait during the last ice age. Only a fraction of the HLA allelic lineages found in Europe and Asia are represented in American Indians (Kostyu and Amos, 1981). However, the founding alleles have recombined to produce many new alleles not found in other populations. This is particularly true for HLA-B in South American Indian populations (Belich, Madrigal, Hildebrand, Williams, Luz, Petzl-Erler, and Parham, 1992; Watkins, McAdam, Liu, Strang, Milford, Levine, Garber, Dogon, Lord, Ghim, Troup, Hughes, and Letvin, 1992).

In comparison to the many pairs of alleles that differ by localized clusters of substitutions, few pairs differ by point substitutions and of these only a handful differ by a substitution that has not been found in another allele. Thus, it appears that the rate at which point mutations create new alleles is slower than the rate at which new mutations are subsequently recombined with existing mutations.

Evolution of An HLA-B Lineage

As an example of the patterns that emerge let us consider the allele B*4002. This allele can be found throughout Europe and Asia but is particularly frequent in the Japanese population. Also present in the Japanese is a closely related allele, B*4006, which differs by a cluster of substitutions that change the amino acid residues at positions 95 and 97 from L → W and R → T respectively (Kawaguchi, Kato, Kashiwase, Karaki, Kohsaka, Akaza, Kano, and Takiguchi, 1993). The crystallographic structures of HLA-A2,A68 and B27 show that residues 95 and 97 interact directly with bound peptide and affect the sequences of peptide bound (Bjorkman et al., 1987*a;* Garrett, Saper, Bjorkman, Strominger, and Wiley, 1989; Madden et al., 1992). The W95 T97 motif that distinguishes B*4006 from B*4002, is common to other HLA-B alleles. Therefore, B*4006 and B*4002 are probably related in evolution either by gene conversion or double recombination. As B*4002 has a widespread distribution whereas B*4006 has only been found in Japanese, it seems likely that B*4002 is the older allele and B*4006 the derivative. If correct then a corollary is that the conversion event would have occurred either in a Japanese individual or an individual from one of the populations from which the modern Japanese population is derived. Current serological methods of tissue typing do not distinguish the protein products of the B*4002 and B*4006 alleles (both are typed as the B61 antigen) and we cannot be sure that the B*4006 allele is not present in other East Asian populations, for example, Koreans. Application of DNA typing can resolve such questions as it should be trivial to distinguish B*4002 and B*4006 at the DNA level.

In North America, another product of the B*4002 allele is encountered, however, it exists in serological disguise. For many years it was appreciated that the Pima and certain other Indian populations in the South Western part of the United States had an antigen with serological crossreactivity with the B21 antigen. This specificity became known as BN21 (Williams and McAuley, 1992). In itself, this was a surprising observation as antigens in the HLA-B21 group were absent in Amerindians from other regions of either North, South or Central America (Kostyu and Amos, 1981).

Sequencing the allele encoding BN21 revealed a structure that was highly divergent from the two known alleles in the B21 group (B*4901 and B*5001), but closely related to B*4002. The BN21 encoding allele (now called B*4005) differs from B*4002 by substitutions which change residues 152 from V → E and 163 from T → L. These are both substitutions known to affect antigen presentation. The E152, L163 motif is found in other HLA-B molecules and we therefore hypothesize that B*4002 and B*4005 are related in evolution by a gene conversion (Hildebrand, Madrigal, Belich, Zemmour, Ward, Williams, and Parham, 1992). Again, because B*4002 is widespread and B*4005 of highly restricted distribution, it seems likely that B*4005 was derived from B*4002. Consistent with this interpretation is the presence in the Piman Indians of both B*4002, the acceptor gene, and B alleles with the postulated donor sequence.

In the case of B*4005, we are confident that it is not present in other populations because it can be distinguished both from B21 and B61 (B*4002) by serological typing. Pertinent at this point is to consider why the B*4002/B*4005 difference is distinguished by serology whereas the B*4002/B*4006 is not; after all, they both involve two amino acid differences.

The substitutions between B*4002 and B*4005 are positioned on the helix of the α_2 domain which flanks the peptide binding groove. These residues are relatively exposed to solvent and can therefore be contacted directly by antibodies. The B*4005 residues at these positions are identical to those found in B21 molecules explaining the serological cross-reactivities between B21 and B*4005. In contrast the differences distinguishing B*4002 and B*4006 are located at the bottom of the peptide binding groove. They are not available for direct contact with antibodies and are serologically silent.

In South America, we have found two alleles in the Guarani tribe of southern Brazil, B*4003 and B*4004, that are also related to B*4002 (Belich et al., 1992). They differ from B*4002 by substitutions at residues 114 and 116 (B*4003) and 94, 95, and 103 (B*4004). As seen for B*4005 and B*4006, the differences affect residues within the antigen binding site of the class I molecule and both clusters of differences are found in other HLA-B alleles.

We now see how B*4002 has spawned a series of daughter alleles, B*4003–B*4006, through the mechanism of gene conversion. Each of the daughter alleles is found in a different modern population and in some populations, the ancestral B*4002 allele has been retained (Japanese and Piman) and in another, the Guarani, it has not. I have used B*4002 as an example of a more general phenomenon. Similar allelic radiations of the other B lineages: B15, B35, B51, and B48, in Amerindian groups are being described, particularly in South America.

Other Polymorphisms Affecting Antigen Presentation By Class I Molecules

Although most of the polymorphism in class I molecules is found in and around the peptide binding groove there are examples of polymorphism in α_3, the immunoglobulin-like domain, which affect binding to the CD8 coreceptor of T cells. The molecules HLA-A*6801, A*6802 and B*4801 all have substitution at position 245 of α_3 that reduces affinity for CD8 (Salter, Benjamin, Wesley, Buxton, Garrett, Clayberger, Krensky, Norment, Littman, and Parham, 1990). Other proteins involved in loading class I MHC molecules with peptide also exhibit polymorphism, for example two subunits of the protease and the two subunit of the peptide transporter, all of which are encoded in the MHC. In rats, there are two divergent alleles of the TAP2 gene of the transporter which affect the nature of peptides reaching class I MHC molecules (Powis, Deverson, Coadwell, Ciruela, Huskisson, Smith, Butcher and Howard, 1992).

The Role of Selection in Generating Class I HLA Polymorphism

During the ~500 generations that humans have lived in the Americas, there has been a considerable evolution of HLA-B alleles. Of current interest is whether this evolution is the result of selection, genetic drift or some mixture of the two. The impression so far is that the generation of new alleles has not led to significant increase in the number of alleles within a particular tribe. The generation of new alleles appears accompanied by the replacement of old alleles rather than expansion of the overall number of alleles.

Most serological analysis of HLA-A,B,C antigens has focused on the urban populations in which clinical transplantation is practiced. In such populations the number of alleles as defined by serotype is 4–5 times that found in tribes of American Indians (Williams and McAuley, 1992). Consideration of the data from urban populations has suggested that balancing selection acts to maintain such large numbers of alleles in human populations. If this were indeed the case then one might expect that populations starting with small numbers of alleles, e.g., Amerindian, might incorporate new alleles without loss of the older ones, which does not appear to be the case. This issue is of course confounded by the fact that Amerindian populations underwent massive attrition due to the diseases brought by Europeans and Africans during the last 500 years.

Disease Specific Selection

The idea that individual HLA class I alleles might be selected by particular infectious agents has seemed plausible for years, but only recently has evidence in its support have been obtained. In West Africa, endemic malaria is the cause of significant infant mortality. HLA-B53 (B*5301) protects infants against severe malaria and as a consequence of this property is at high frequency in the West African population whereas in other parts of the world it is rare or absent (Hill, Allsopp, Kwiatkowski, Anstey, Twumasi, Rowe, Bennett, McMichael, and Greenwood, 1991). HLA-B53 presents a peptide from the liver form of the malarial parasite and the protective effect is speculated to stem from the cytotoxic T cell response made to that peptide (Hill, Elvin, Willis, Aidoo, Allsopp, Gotch, Gao, Takiguchi, Greenwood, Townsend, McMichael, and Whittle, 1992).

The selection for B53 is seen in a situation where a single disease has been a major cause of death for many generations. Furthermore, death from malaria commonly occurs before reproductive age and the epitope presented by B53 appears to be one that cannot be mutated without impairing viability of the pathogen. This may not be a common scenario. Survival in most environments requires effective immune responses against an array of microorganisms, viruses and antigenic compounds. That the alleles now seen in the population have survived suggest that on average a random set of two HLA-A alleles, two HLA-B alleles and two HLA-C alleles is sufficient to permit survival and that, again on average, any one set of six alleles is as good as another.

Overdominant Selection

From their analysis of class I and II allelic sequences, Hughes and Nei (Hughes and Nei, 1988) have concluded that overdominance (heterozygote advantage) is a major component of the selection upon MHC molecules. This can be understood readily in terms of their biological function. Possession of 2 HLA-A,B or C alleles, with distinct peptide binding specificities, should expand the range of peptide sequences that can be presented by a cell and by inference the strength of the T cell response should that cell become infected with a virus. In this model, the selective advantage of additional alleles will tend to decrease as the number of alleles increases and the number of homozygotes decrease. That Native American populations have limited numbers of alleles may therefore be explained by this effect. Urban populations have larger numbers of alleles but are the result of recent admixture drawing on people from wide catchment areas, the non-Native American population epitomizing this phenomenon. In such populations the large numbers of HLA class I alleles are likely to result from admixture, not balancing selection.

Frequency Dependent Selection

Within the general scheme of heterozygote advantage one can see how new alleles might be favored over older ones. Everything else being equal, an individual with a new allele will produce a greater proportion of heterozygotes than other members of the population. The increase being due to the fact that germ cells carrying the new allele cannot give rise to a homozygote. This advantage will act to increase the frequency of the new allele at the expense of the older ones. As its frequency increases the selective advantage of the new allele will decrease until it reaches a level comparable to that of the older alleles. In the absence of disease-specific selection targeted to particular alleles this effect will tend to bring new alleles into the population, as seems to be the case for HLA-B in South American Indians. Conversely, the frequency of older alleles will decrease and the chances of them being lost altogether goes up. Both the selective advantage of low frequency alleles and the probability of them being lost through genetic drift should be dependent upon population size and will be accentuated in smaller populations. For the American Indian groups the rates of gain and loss appear to be balanced as the total number of alleles in different populations is similar.

That the class I alleles of human populations are continually turning over is supported by comparison of HLA-A,B with their homologues in chimpanzees and gorillas (Lawlor, Warren, Taylor, and Parham, 1991). Although there are similarities in the alleles of these three closely related species, no allele is shared. Moreover, those alleles which are most similar exhibit amino acid changes in the peptide binding site. This contrasts with many other proteins which exhibit identical sequences in the three species. Thus, it is possible that no class I allele present in the ancestral population common to humans, chimpanzees and gorilla has survived unmodified to the present day.

References

Belich, M. P., J. A. Madrigal, W. H. Hildebrand, J. Zemmour, R. C. Williams, R. Luz, M. L. Petzl-Erler, and P. Parham. 1992. Unusual HLA-B alleles in two tribes of Brazilian Indians. *Nature.* 357:326–329.

Bjorkman, P. J., M. A. Saper, B. Samraoui, W. S. Bennett, J. L. Strominger, and D. C. Wiley. 1987*a*. Structure of the human class I histocompatibility antigen, HLA-A2. *Nature.* 329:506–512.

Bjorkman, P. J., M. A. Saper, B. Samraoui, W. S. Bennett, J. L. Strominger, and D. C. Wiley. 1987*b*. The foreign antigen binding site and T cell recognition regions of class I histocompatibility antigens. *Nature.* 329:512–518.

Delarbre, C., C. Jaulin, P. Kourilsky, and G. Gachelin. 1992. Evolution of the major histocompatibility complex: a hundred-fold amplification of MHC class I genes in the African pigmy mouse Nanomys selulosus. *Immunogenetics.* 37:29–38.

Flajnik, M. F., C. Canel, J. Kramer, and M. Kasahara. 1991. Evolution of the major histocompatibility complex: molecular cloning of major histocompatibility complex class I from the amphibian *Xenopus. Proceedings of the National Academy of Sciences, USA.* 88:537–541.

Fleischhauer, K., N. A. Kernan, R. J. O'Reilly, B. Dupont, and S. Y. Yang. 1990. Bone marrow allograft rejection by T lymphocytes recognizing a single amino acid in HLA-B44. *New England Journal of Medicine.* 323:1818–1822.

Garrett, T. P. J., M. A. Saper, P. J. Bjorkman, J. L. Strominger, and D. C. Wiley. 1989. Specificity pockets for the side chains of peptide antigens in HLA-Aw68. *Nature.* 342:692–696.

Geraghty, D. E., B. H. Koller, J. A. Hansen, and H. T. Orr. 1992*a*. The HLA class I gene family includes at least six genes and twelve pseudogenes and gene fragments. *Journal of Immunology.* 149:1934–1946.

Geraghty, D. E., B. H. Koller, J. Pei, and J. A. Hansen. 1992*b*. Examination of four HLA class I pseudogenes. Common events in the evolution of HLA genes and pseudogenes. *Journal of Immunology.* 149:1947–1956.

Grossberger, D., and P. Parham. 1992. Reptilian class I Major Histocompatibility complex genes reveal conserved elements in class I structure. *Immunogenetics.* 36:166–174.

Hildebrand, W. H., J. A. Madrigal, M. P. Belich, J. Zemmour, F. W. Ward, R. C. Williams, and P. Parham. 1992. Serologic crossreactivities poorly reflect allelic relationships in the HLA-B12 and HLA-B21 groups: dominant epitopes of the α_2 helix. *Journal of Immunology.* 149:3563–3568.

Hill, A. V. S., C. E. M. Allsopp, D. Kwiatkowski, N. M. Anstey, P. Twumasi, P. A. Rowe, S. Bennett, D. Brewster, A. J. McMichael, and B. M. Greenwood. 1991. Common West African HLA antigens are associated with protection from severe malaria. *Nature.* 352:595–600.

Hill, A. V. S., J. Elvin, A. C. Willis, M. Aidoo, C. E. M. Allsopp, F. M. Gotch, X. Ming Gao, M. Takiguchi, B. M. Greenwood, A. R. M. Townsend, A. J. McMichael, and H. C. Whittle. 1992. Molecular analysis of the association of HLA-B53 and resistance to severe malaria. *Nature.* 360:434–439.

Hughes, A. L., and M. Nei. 1988. Pattern of nucleotide substitution at major histocompatibility complex class I loci reveals overdominant selection. *Nature.* 335:167–170.

Kawaguchi, G., N. Kato, K. Kashiwase, S. Karaki, T. Kohsaka, T. Akaza, K. Kano, and M. Takiguchi. 1993. Structural analysis of HLA-B40 epitopes. *Human Immunology.* 36:193–198.

Kostyu, D. D., and D. B. Amos. 1981. Mysteries of the Amerindians. *Tissue Antigens.* 17:111–123.

Lawlor, D. A., E. Warren, P. Taylor, and P. Parham. 1991. Gorilla class I MHC alleles: comparison to human and chimpanzee class I. *Journal of Experimental Medicine.* 174:1491–1509.

Lawlor, D. A., J. Zemmour, P. D. Ennis, and P. Parham. 1990. Evolution of class I major histocompatibility complex genes and proteins: from natural selection to thymic selection. *Annual Review of Immunology.* 8:23–63.

López de Castro, J. A. 1989. HLA-B27 and HLA-A2 subtypes: structure, evolution and function. *Immunology Today.* 10:239–246.

Madden, D. R., J. C. Gorga, J. L. Strominger, and D. C. Wiley. 1992. The three-dimensional structure of HLA-B27 at 2.1A resolution suggests a general mechanism for tight peptide binding to MHC. *Cell.* 70:1035–1048.

Parham, P., C. E. Lomen, D. A. Lawlor, J. P. Ways, N. Holmes, H. L. Coppin, R. D. Salter, A. M. Wan, and P. D. Ennis. 1988. The Nature of Polymorphism in HLA-A,B,C Molecules. *Proceedings of the National Academy of Sciences, USA.* 85:4005–4009.

Powis, S. J., E. V. Deverson, W. J. Coadwell, A. Ciruela, N. S. Huskisson, H. Smith, G. W.

Butcher, and J. C. Howard. 1992. Effect of polymorphism of an MHC-linked transporter on the peptides assembled in a class I molecule. *Nature.* 357:211–215.

Rippmann, F., W. R. Taylor, J. B. Rothbard, and N. M. Green. 1991. A hypothetical model for the peptide binding domain of hsp70 based on the peptide binding domain of HLA. *Embo Journal.* 10:1053–1059.

Salter, R. D., R. J. Benjamin, P. K. Wesley, S. E. Buxton, T. P. J. Garrett, C. Clayberger, A. M. Krensky, A. M. Norment, D. R. Littman, and P. Parham. 1990. A binding site for the T-cell co-receptor CD8 on the α_3 domain of HLA-A2. *Nature.* 345:41–46.

Watkins, D. I., S. N. McAdam, X. Liu, C. R. Strang, E. L. Milford, C. G. Levine, T. L. Gerber, A. L. Dogon, C. I. Lord, S. H. Ghim, G. M. Troup, A. L. Hughes, and N. L. Letvin. 1992. New recombinant HLA-B alleles in a tribe of South American Amerindians indicate rapid evolution of MHC class I loci. *Nature.* 357:329–333.

Williams, R. C., and J. E. McAuley. 1992. HLA class I variation controlled for genetic admixture in the Gila River Indian community of Arizona: a model for the Paleo-Indians. *Human Immunology.* 33:39–46.

The Generation of Protein Isoform Diversity by Alternative RNA Splicing

David M. Helfman

Department of Molecular Genetics, Cold Spring Harbor Laboratory, Cold Spring Harbor, New York 11724

The generation of protein isoform diversity by alternative RNA splicing is a fundamental mechanism in eukaryotes, which contributes to tissue-specific and developmental patterns of gene expression. A growing number of cellular and viral genes have been characterized that encode multiple protein isoforms via the use of alternatively spliced exons. One such example is the tropomyosin (TM) gene family. Tropomyosins are a diverse group of actin filament-binding proteins found in all eukaryotic cells, with distinct isoforms found in muscle (skeletal, cardiac, and smooth), brain and various nonmuscle cells. Many animals including nematodes, flies, frogs, birds, and mammals possess multiple TM isoforms. This isoform diversity is generated by a combination of multiple genes, some of which contain alternative promoters and some of which exhibit alternative RNA splicing of primary RNA transcripts. The TM gene family appears to have arisen through the duplication of an ancestral gene. There are four different TM genes that have been characterized in vertebrates. The alternatively spliced exons among different isoforms correspond to functional domains of the proteins. The expression of a diverse group of isoforms in a cell-type specific manner strongly suggests that the various isoforms will be required for specific cellular functions. The implications of this isoform diversity with respect to cellular functions are discussed.

A General Overview of RNA Splicing for the Generation of Protein Diversity

The formation of a mature mRNA from a primary transcript generally requires excision of intervening sequences (introns) with the subsequent joining together (splicing) of exons. The general splicing reaction of metazoan introns has been well characterized and involves a two-step process (reviewed in Green, 1991; Rio, 1992). In the first step, a 5′ exon is cleaved and two intermediates are formed: a 5′ exon with a 3′ hydroxyl end and an RNA species containing the downstream intron sequence and 3′ exon in which the 5′ terminal guanosine of the intron is covalently linked to a residue, usually an adenosine, located 18–40 nucleotides upstream of the 3′ splice site via a 2′–5′ phosphodiester bond. In the second step, cleavage at the 3′ splice site results in the release of a lariat intron and concomitantly the two exons are ligated together.

Despite great advances in our understanding of the general splicing reaction, a number of important problems remain unresolved. How splicing is carried out with such fidelity is still not understood, both at the level of ligation at a specific 5′ or 3′ splice site or at the level of joining exons together without inadvertently failing to include an exon, i.e., "exon skipping." Most genes contain multiple exons, e.g., collagen has more than 50 exons, and the dystrophin gene has more than 70 exons. Clearly if the splicing machinery were to make an error at the level of a single nucleotide, or miss including an exon, the result would likely be a mRNA that encoded a nonfunctional protein. In addition, some introns are very large and how the individual 5′ and 3′ splice sites are brought together is not known. Take for example, the dystrophin gene which is over two million base pairs long and contains a number of introns which are larger than 200,000 nucleotides. Whether the 5′ and 3′ splice sites of these introns are brought together by free diffusion or require an active mechanism remains to be determined. Although consensus sequences are known for 5′ and 3′ splice sites, it is still not possible to predict with 100% accuracy the position of authentic splice sites based solely on sequence analysis. Finally, the nature of the cellular factors that mediate pre-mRNA splicing are just beginning to emerge and include a group of small nuclear ribonuclear particles (snRNPs) and other proteins (reviewed in Green, 1991; Rio, 1992). In the case of alternative splicing, two additional questions need to be considered including (*a*) the nature of the splicing signals in alternatively spliced exons that allow them to be regulated in a tissue and cell type-specific manner, and (*b*) what are the cellular factors that are responsible for alternative splicing.

It is well established that many proteins expressed in eukaryotic cells exist in various forms or isoforms. In many cases, the different isoforms exhibit cell-type specific patterns of expression which are subject to developmental control. For example, different forms of globin are known to be expressed during different developmental stages. The generation of protein isoforms is most often associated with the expression of different members of multigene families, but a large number of genes have been identified that can encode more than a single protein by alternative RNA processing. Alternative RNA processing has been identified in a wide range of systems including the genomes of various DNA and RNA viruses, invertebrates and vertebrates. In the case of viruses, the use of alternative splicing may facilitate the expression of a larger number of proteins within a relatively small genome. The organization of genes that are subject to alternative processing is quite varied, and can be due to use of alternative promoters and alternative exons located at internal as well as at the 3′ end of genes (reviewed in Breitbart, Andreadis, and Nadal-Ginard, 1987).

A number of cellular genes and processes have been characterized that involve alternative splicing in the control of developmental and various physiological processes. For example, sex determination in *Drosophila* is dependent on a cascade of regulatory genes, each of which is subject to alternative RNA splicing. Whether a *Drosophila* will become a male or female is largely dependent on the pattern of alternatively spliced exons (reviewed in Baker, 1989). During the development of humoral immunity, the switch from the membrane-bound form of antibody to the secreted form is dependent on alternative RNA processing at the 3′ end of the primary transcript where distinct carboxy-terminal coding exons are used. The membrane bound immunoglobulin contains the membrane anchor sequence whereas the secreted form does not (reviewed in Breitbart et al., 1987). There are countless other examples of alternative RNA splicing for the generation of protein isoform diversity including cell surface receptors, adhesion molecules, neuropeptides, transcription factors and contractile proteins (see below).

The Tropomyosin System

One classic physiological response is that of the excitation/contraction response of skeletal and cardiac muscles. The protein components that comprise the actomyosin/ATPase responsible for muscle contraction have been well characterized and are known to involve not only actin and myosin, but tropomyosin and the troponin complex (troponin-I, -T, and -C). The function of TM in skeletal and cardiac muscle is, in association with the troponin complex (troponin-I, -T, and -C), to regulate the calcium-sensitive interaction of actin and myosin (Smillie, 1979). Although most eukaryotic cells contain an actin-based contactile system, the regulation of contractility in different cell types is achieved by different mechanisms. Nonmuscle and smooth muscle cells are devoid of a troponin complex, and the calcium-sensitive regulatory mechanism controlling the interaction of actin and myosin in these cells involves at least two Ca^{2+} calmodulin regulated proteins; myosin light chain kinase and caldesmon. These differences in the regulation of the contractile apparatus in various cell types appears to require structurally and functionally distinct forms of TM. In contrast to muscle cells, microfilaments in nonmuscle cells do not appear to function primarily in contraction, but play a more significant role in intracellular transport, secretion, motility and cell structure. Nonmuscle cells express a larger number of tropomyosin isoforms than cells of muscle origin. This likely reflects the greater diversity of cellular functions in which they participate rather than functional redundancy. Most nonmuscle cells, such as fibroblasts, express a greater number of TM isoforms (five to eight isoforms) than do skeletal or cardiac muscle (one to two isoforms), and smooth muscle (two isoforms). Multiple forms of TM have been detected in cultures of different nonmuscle cells, including rat, mouse, human, and chick fibroblasts (reviewed in Lees-Miller and Helfman, 1991). For example, rat fibroblasts contain seven isoforms of tropomyosin: three TMs of major abundance termed TMs 1, 2, and 4 (apparent M_r = 40,000, 36,500 and 32,400, respectively), and four TMs relatively minor in abundance termed TMs 3, 5, 5a, and 5b (apparent M_r = 35,000, 32,000, 32,000, and 32,000, respectively) (Matsumura, Lin, Yamashiro, Thomas, and Topp, 1983*a,b;* Goodwin et al., 1991, and Guo and Helfman, unpublished results). Thus, nonmuscle cells may have adapted TM isoforms to serve several different functions. Expression of the different fibroblast-type isoforms are not restricted to fibroblasts but are found in different combinations in various nonmuscle cells and tissues, e.g., intestinal epithelium, blood cells, liver, and kidney. In addition, brain expresses three unique isoforms, two of which are found only in neurons (Lees-Miller et al., 1990; Weinberger, Henke, Tolhurst, Jeffrey, and Gunning, 1993; Stamm, Caspar, Lees-Miller, and Helfman, 1993).

Like many structural proteins, tropomyosin (TM) isoform expression involves the use of multiple genes, but diversity is also generated by alternative RNA processing (reviewed in Lees-Miller and Helfman, 1991). Alternative RNA processing for the generation of TM isoforms has been found in invertebrates (*C. elegans* and *Drosophila*) and vertebrates. For example, four different TM genes have been characterized in mammals (Fig. 1). Each of these genes has been named after a protein they encode. The α and β genes are named after striated muscle α- and β-TM, respectively. The TM-4 and hTMnm genes are named after the rat fibroblast TM-4 and human fibroblast TM30 nm isoforms, respectively. The skeletal muscle isoform of the α gene corresponds to the α fast-twitch isoform (fast-twitch fibers), whereas the skeletal muscle isoform of the hTMnm gene corresponds to the α slow-twitch isoform (slow-twitch fibers). The isoforms known to be encoded by each gene are shown schematically in Fig. 2. The α gene is known to encode at least nine different isoforms, the β and hTMnm genes each encode at least two distinct isoforms, while the TM-4 gene encodes a single isoform. One interesting point is the intron-exon organization of the mammalian α TM gene is almost identical to that of a *Drosophila* or *C. elegan* TM gene, indicating that alternative splicing for the generation of TM isoform diversity arose relatively early in metazoan evolution (reviewed in Lees-Miller and Helfman, 1991). In addition to the isoforms depicted in Fig. 2, other isoforms are known to be expressed from the chick β-TM gene (Forry-Schaudies, Maihle, and Hughes, 1990) and mouse β-TM gene (Wang and Rubenstein, 1992). Given the complexity due to alternative RNA processing, it will not be surprising if additional isoforms will be identified in the future. At the protein level TMs are elongated proteins that possess a simple dimeric α-helical coiled-coil structure along their entire length (Smillie, 1979). The coiled-coil structure is based on a repeated pattern of seven amino acids with hydrophobic residues at the first and fourth

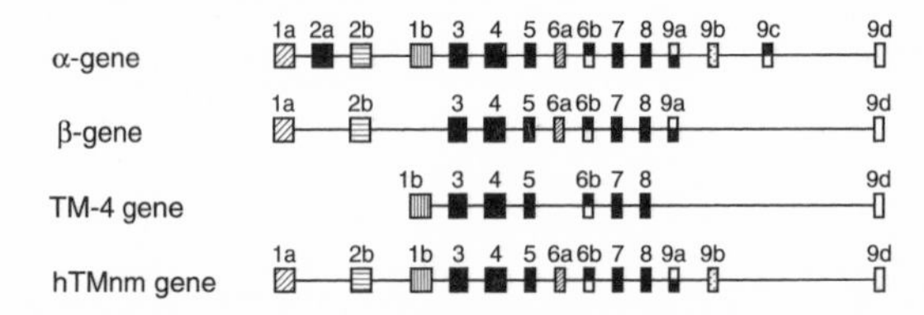

Figure 1. The intron-exon organization of the mammalian TM genes. Exons are represented by boxes and introns by horizontal lines. The exons are numbered from 1a through 9d in order to facilitate comparison between TM genes. The depicted α gene structure is that of rat (Ruiz-Opazo and Nadal Ginard, 1987; Wieczorek et al., 1988; Lees-Miller et al., 1990; Goodwin et al., 1991). The β gene shown is that of rat (Helfman et al., 1986). The hTMnm gene shown is that of human (Clayton et al., 1988), and the TM-4 gene is that of rat (Lees-Miller et al., 1990). Alternatively spliced exons are indicated as 1a/1b; 2a/2b; 6a/6b; and 9a/9b/9c/9d. Exons common to all mRNAs are indicated by exons 3, 4, 5, 7, and 8. The isoforms known to be encoded by each gene are shown schematically in Fig. 2.

positions and is highly conserved in all TM isoforms found in eukaryotic organisms from yeast to man. The TMs bind to themselves in a head-to-tail manner, and lie in the grooves of F-actin, with each molecule interacting with six or seven actin monomers. As indicated in Fig. 2, vertebrate cells express a number of different isoforms that are expressed in a cell type-specific manner. A major unresolved question is the functional significance of the different isoforms. The expression of a diverse group of isoforms in a highly tissue-specific manner suggests that each isoform is required to carry out specific functions in the actin-based filaments of various muscle and nonmuscle cell types.

The alternatively spliced exons appear to correspond to different functional domains including actin binding regions, head-to-tail overlap domains, and sequences involved in interaction to TM-binding proteins. For example, troponin-T interacts with TM in two regions of the TM molecule, the regions represented by exon 6b (amino acids 188–213) and exon 9a (amino acids 254–284) in the α, β, and hTMnm genes (Figs. 1 and 2). In addition, sequences at the amino- and carboxy-terminal ends of the protein are important for head-to-tail binding, which is important for cooperative binding to actin filaments. Thus, the inclusion of various exons influences the ability of the different isoforms to interact with F-actin, bind to each other and interact with different TM-binding proteins, e.g., troponin-T or caldesmon (see below).

As mentioned above, nonmuscle cells express a larger number of TM isoforms compared to muscle (skeletal, cardiac, and smooth). What is the role of each TM isoform in nonmuscle cells? Immunofluorescence studies of nonmuscle cells, such as

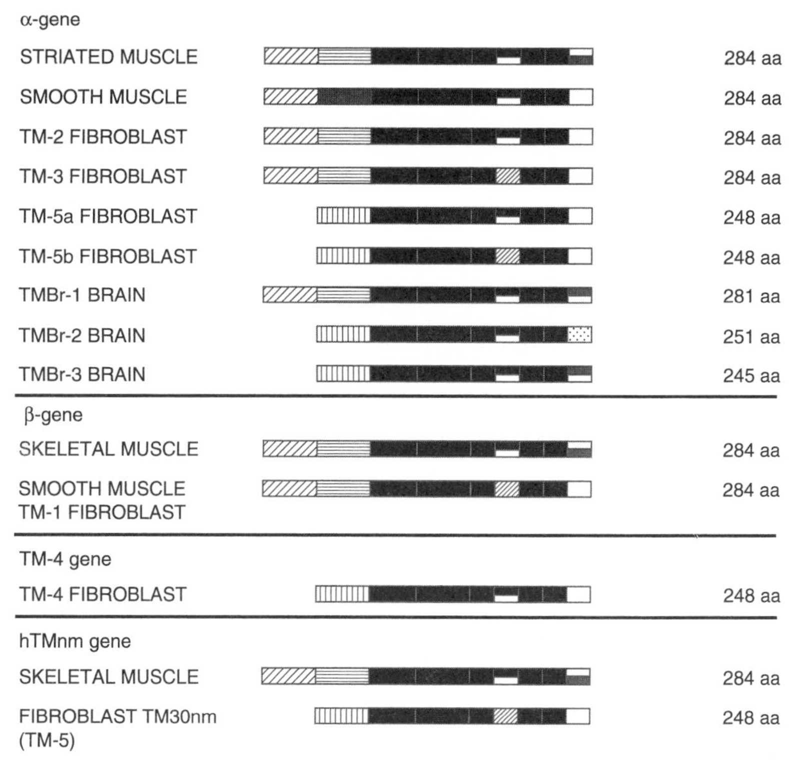

Figure 2. Schematic representation of the TM isoforms expressed from the mammalian TM genes. The α gene striated muscle isoform is frequently referred to as the α-fast twitch isoform, while the hTMnm gene encodes a skeletal muscle isoform which corresponds to the α-slow twitch isoform. Fibroblast TM-4 corresponds to the human platelet TM30 isoform. The human hTMnm isoform has also been identified in mouse (Takenaga et al., 1990) and rat (Guo and Helfman, unpublished data) as is also referred to as TM-5. The length of each protein in amino acids is indicated on the right.

fibroblasts, reveal that tropomyosin is associated with the actin-containing microfilaments of the cytoskeleton. Microfilaments of fibroblasts represent dynamic structures that can exist in different supramolecular forms, such as microfilament bundles (stress fibers), microfilament meshworks, polygonal networks, and contractile rings. The particular TM isoforms involved in these higher-order actin structures are not known. The multiplicity of TM isoforms raises the possibility that specific associations of given isoforms are required for these distinct actin structures. Differences in the localization of various isoforms have been demonstrated with antibodies recognizing only the high molecular weight TMs versus those recognizing only the low molecular weight isoforms (Lin, Hegmann, and Lin, 1988). The low molecular weight forms were found in both ruffles and stress fibers whereas the high molecular weight forms were found only in stress fibers. These differences in the distribution of different TM isoforms suggest distinct, although unidentified, roles for the various isoforms.

Different biochemical properties have been reported for various nonmuscle isoforms. For example, studies of rat fibroblast TMs demonstrated that a mixture of the low molecular weight TM isoforms (termed TM-4 and TM-5) bound weakly to F-actin whereas the higher molecular weight isoforms (TM-1, TM-2, and TM-3) bound more strongly (Matsumura and Yamashiro-Matsumura, 1985; Yamashiro-Matsumura and Matsumra, 1988). Similarly, it was shown that the low molecular weight TM isoform of equine platelets bound actin with low affinity (Cote', Lewis, and Smillie, 1978) as does the low molecular weight TM isoform of brain (Broschat and Burgess, 1986). In contrast, however, the low molecular weight TMs isolated from human erythrocytes (Fowler and Bennett, 1984) and chicken intestinal epithelial cells (Broschat and Burgess, 1986) have been shown to bind with much greater affinity, similar to high molecular weight isoforms. These studies clearly indicate that different biochemical properties are a characteristic of specific isoforms, suggesting the possibility of distinct functions in vivo. For example, one possibility is that TM may stabilize actin filaments in nonmuscle cells. This function has been inferred from in vitro experiments which have also shown that TM protects actin filaments against the severing action of gelsolin (Fattoum, Hartwig, and Stossel, 1983; Ishikawa, Yamashiro, and Matsumura, 1989) or the activities of brain depolymerizing factor (Bernstein and Bamburg, 1982). Other experiments have shown that the nonmuscle TMs prevent villin induced severing or bundling of F-actin (Burgess, Broschat, and Haydon, 1987), but they were unable to prevent bundling caused by 55-kD protein (Matsumura and Yamashiro-Matsumura, 1986). These studies used mixtures of individual isoforms, but indicate that different isoforms can have different biochemical properties.

At present, few studies have addressed the function of TM isoforms in living cells. Alteration of intracellular granule movement resulted after microinjection of antibodies to TM into chick fibroblasts, suggesting an involvement of TM in intracellular motility (Hegmann, Lin, and Lin, 1989). In yeast, gene disruption studies suggest a role for TM in actin filament assembly, secretion and cell division (Liu and Bretscher, 1989, 1992; Balasubrammanian, Helfman, and Hemmingsen, 1992).

Transformation of cells in tissue culture result in a variety of cellular changes including alterations in cell growth, adhesiveness, motility, morphology, and organization of the cytoskeleton. Morphological changes are perhaps the most readily apparent feature of transformed cells. These changes in cell shape are clearly associated with the cytoskeleton. Immunofluorescence and electron microscopic studies have shown that transformed cells exhibit profound changes in the organization of microfilaments. Such studies reveal that microfilament bundles exist in a more dispersed state and are reduced in size and number in transformed cells. In addition, these changes in microfilament structure are highly related to both anchorage-independent growth and cellular tumorigenicity (Shin, Freedman, Risser, and

Pollack, 1975), suggesting a role for microfilament alteration in oncogenic transformation. The molecular, biochemical and structural bases for these alterations of microfilament cables in transformed cells are poorly understood. The altered expression of a number of individual cytoskeletal proteins have been implicated in the transformed phenotype including actin, tropomyosin, gelsolin, caldesmon, and myosin light chain. A number of groups have analyzed the expression of the protein components of microfilaments before and after transformation. Interestingly, comparative two-dimensional protein gel analysis of normal and transformed cells has revealed that of the numerous cytoskeletal proteins, tropomyosin expression is selectively altered in some transformed cells (Hendricks and Weintraub, 1981, 1984; Leonardi, Warren, and Rubin, 1982; Matsumura et al., 1983*a,b;* Cooper, Feuerstein, Noda, and Bassin, 1985; Lin, Helfman, Hughes, and Chou, 1985; Leavitt, Latter, Lutomski, Goldstein, and Burbeck, 1986; Fujita et al., 1990). In general, these studies demonstrate that in several transformed cells one or more of the major TM isoforms of higher molecular weight (284 amino acids) are decreased or missing, whereas the levels of one or more of the lower molecular weight (248 amino acids) TM isoforms are increased. These alterations in TM expression appear to correlate well with the rearrangement of microfilament bundles and morphological alterations observed in transformed cells. Recently it was reported that forced expression of fibroblast TM-1 in transformed cells led to reversion of some features associated with the transformed phenotype including loss of anchorage-independent growth and the ability to form tumors in athymic mice (Prasad, Fuldner, and Cooper, 1993). Using a Kirsten virus transformed rat kidney cell line (NRK1569) in which TM-1 expression is reduced about 50% and which expresses no detectable TM-2 or TM-3 (Matsumura et al., 1983*a*), we found that stable expression of TM-2 or TM-3 leads to a partial restoration of microfilament bundles and cell spreading (Kazzaz, Pittenger, and Helfman, unpublished observations). Collectively, such studies indicate that alterations in TM-1, TM-2, and TM-3 play a role in the transformed phenotype. One hypothesis to emerge from these studies is that formation of stable microfilament bundles in fibroblasts will require the presence of a high molecular weight TM isoform (284 amino acids) from both the α gene (TM-2 and/or TM-3) and β gene (TM-1).

While the function of TM in the regulation of skeletal and cardiac muscle contraction is fairly well understood, the role that TM plays in the regulation of myosin motor proteins in other cell types is not known. At present, three major types of motor proteins have been identified: myosins, dyneins, and kinesins. These molecules move along either actin filaments (for myosin) or microtubules (for dynein and kinesin). TM can influence the interaction of myosin and actin. As already discussed, in skeletal muscle TM in association with the troponin complex regulates the interaction of myosin heads with thin filament actin in a Ca^{2+}-dependent manner. The TM-troponin complex is inhibitory to contraction when the intracellular calcium concentration is low, working either by a steric blocking mechanism or through allosteric interactions. In smooth muscle and nonmuscle cells TM plays a different role because there is no troponin complex. Biochemical studies have shown that under conditions that would produce inhibition of the actin activated myosin MgATPase by skeletal TM, smooth muscle and nonmuscle TMs have a stimulatory effect. For example, when smooth muscle or brain tropomyosins (Sobiezek and Small, 1981) were substituted for skeletal TM in an actin activated MgATPase assay using heavy meromyosin, these TMs were found to be stimulatory while the skeletal TM was inhibitory. In nonmuscle cells, exactly which myosins take part in such cellular functions as locomotion, extension of filapodia or lamellapodia, mitosis and cytokinesis is an area of active investigation. Cells contain both single headed (myosin I's) and two-headed myosins (myosin II's). A specific isoform of myosin II is usually found in a particular cell type (striated muscle, smooth muscle or nonmuscle cell) while several types of myosin I are expressed in a given cell. The nonmuscle myosin Is are by far the larger class and appear to be involved in vesicle movement and membrane association. The motor activities of these myosins may be regulated by cellular calcium levels, distinct kinases or TM isoforms. With the identification of at least four subclasses of myosin I's (Cheney, Riley, and Mooseker, 1993), the effects of different TM isoforms on myosin ATPase activity and motor functions will need to be carefully investigated, and despite great interest in these actin based motor proteins only a few studies have addressed these questions.

Evolutionary Considerations of Alternative RNA Processing and Isoform Diversity

Tropomyosins represent just one family of proteins that exhibit extensive isoform diversity due to alternative RNA splicing. The different isoforms appear to be required for the physiological needs of various cells and tissues, although the function of each isoform is just beginning to emerge. The expression of a diverse group of isoforms in a highly tissue-specific manner strongly suggests that each isoform will be required to carry out specific cellular functions. In addition to the obvious questions regarding the functional significance of protein isoform diversity, the generation of protein isoforms via alternative RNA splicing raises a number of general issues. For one, some gene families exhibit a large number of family members without exhibiting alternative RNA processing for the generation of protein isoform diversity e.g., intermediate filament proteins, actins, and G-proteins. In contrast, other gene families exhibit complex patterns of alternative splicing for the generation of diversity, e.g., tropomyosins, troponins, and N-CAMs (neural adhesion molecules). What factors played a role during evolution that led to an increase in complexity due to alternative splicing, as opposed to expansion of the number of members in a gene family, i.e., gene duplication, are not known. The explanations for multigene families include a number of hypotheses including a need for functional redundancy, a requirement for distinct cis-elements to regulate the level of gene expression, i.e., transcription, and isoform-specific functions in different cell-types. While examples can be found in nature to support each hypothesis, in the case of alternatively spliced genes it would appear that the requirement for isoform-specific functions is a major factor for such protein diversity. Another question is the evolution of RNA splicing (reviewed in Gestland and Atkins, 1993), and for alternative RNA splicing, how the *cis*-acting elements and cellular factors involved in the regulation of alternatively spliced exons "co-evolved." Clearly, to regulate a particular splice choice used in different tissues and cell types, regulatory molecules must also be expressed in a cell type-specific manner. Understanding the evolution of RNA processing and alternative RNA splicing will no doubt give greater insights into the complex questions of molecular evolution.

Acknowledgements

Special thanks to Jeff Kazzaz for help with art work.

D. M. Helfman is the recipient of grants GM43049 and CA58607 from the NIH and is an Established Investigator of the American Heart Association.

References

Baker, B. S. 1989. Sex in flies: the splice of life. *Nature.* 340:521–524.

Balasubrammanian, M. K., D. M. Helfman, and S. M. Hemmingsen. 1992. A new tropomyosin essential for cytokinesis in the fission yeast *S. pombe. Nature.* 360:84–87.

Bernstein, B. W., and J. R. Bamburg. 1982. Tropomyosin binding to F-actin protects the F-ctin from dissaembly by brain actin-depolymerizing factor (ADF). *Cell Motility.* 2:1–8.

Breitbart, R. E., A. Andreadis, and B. Nadal-Ginard. 1987. Alternative splicing: A ubiquitous mechanism for the generation of multiple protein isoforms from single genes. *Annual Review Biochemistry.* 56:467–495.

Broschat, K. O., and D. R. Burgess. 1986. Low Mr tropomyosin isoforms from chicken brain and intestinal epithelium have distinct actin-binding properties. *The Journal of Biological Chemistry.* 261:13350–13359.

Burgess, D. R., K. O. Broschat, and J. M. Hayden. 1987. Tropomyosin distinguishes between two actin-binding sites of villin and affects actin-binding properties of other brush border proteins. *The Journal of Cell Biology.* 104:29–40.

Cheney, R. E., M. A. Riley, and M. S. Mooseker. 1993. Phylogenetic analysis of the myosin superfamily. *Cell Motility and Cytoskeleton.* 24:215–223.

Clayton, L., F. C. Reinach, G. M. Chumbley, and A. R. MacLeod. 1988. Organization of the hTMnm gene: implications for the evolution of muscle and non-muscle tropomyosins. *The Journal of Molecular Biology.* 201:507–515.

Cooper, H. L., N. Feuerstein, M. Noda, and R. H. Bassin. 1985. Suppression of tropomyosin synthesis, a common biochemical feature of oncogenesis by structurally diverse retroviral oncogenes. *Molecular and Cellular Biology.* 5:972–83.

Cote', G., W. G. Lewis, and L. B. Smillie. 1978. Non-polymerizability of platelet tropomyosin and its NH_2- and COOH-terminal sequences. *FEBS Letters.* 91:237–241.

Fattoum, A., J. H. Hartwig, and T. P. Stossel. 1983. Isolation and some structural and functional properties of macrophage tropomyosin. *Biochemistry.* 22:1187–1193.

Forry-Schaudies, S., N. J. Maihle, and S. H. Hughes. 1990. Generation of skeletal, smooth, and low molecular weight non-muscle tropomyosin isoforms from the chick tropomyosin 1 gene. *The Journal of Molecular Biology.* 211:321–330.

Fowler, V. M., and V. Bennet. 1984. Erhthrocyte membrane tropomyosin: purification and proerties. *The Journal of Biological Biochemistry.* 259:5978–5989.

Fujita, H., H. Suzuki, N. Kuzumaki, L. Mullauer, Y. Ogiso, A. Oda, K. Ebisawa, T. Sakurai, Y. Nonomura, and S. Kijimoto-Ochiai. 1990. A specific protein, p92, detected in flat revertants derived from NIH/3T3 transformed by human activated c-Ha-*ras* oncogene. *Experimental Cell Research.* 186:115–121.

Gestland, R. F., and J. F. Atkins, editors. 1993. The RNA World. Cold Spring Harbor Labortory Press, Cold Spring Harbor, NY.

Goodwin, L. O., J. P. Lees-Miller, S. Cheley, M. Leonard, and D. M. Helfman. 1991. Four rat fibroblast tropomyosin isoforms are expressed from a single gene via alternative RNA splicing and utilization of two promoters. *The Journal of Biological Chemistry.* 266:8408–8415.

Green, M. R. 1991. Biochemical mechanisms of constitutive and regulated pre-mRNA splicing. *Annual Review of Genetics.* 7:559–599.

Hegmann, T. E., J. L.-C. Lin, and J. J.-C. Lin. 1989. Probing the role of nonmuscle tropomyosin isoforms in intracellular granule movement by microinjection of monoclonal antibodies. *The Journal of Cell Biology.* 109:1141–1152.

Helfman, D. M., S. Cheley, E. Kuismanen, L. A. Finn, and Y. Yamawaki-Kataoka. 1986. Nonmuscle and muscle tropomyosin isoforms are expressed from a single gene by alternative RNA splicing and polyadenylation. *Molecular and Cellular Biology.* 6:3582–3595.

Hendricks, M., and H. Weintraub. 1981. Tropomyosin is decreased in transformed cells. *Proceedings of the National Academy of Sciences, USA.* 78:5633–5637.

Hendricks, M., and H. Weintraub. 1984. Multiple tropomyosin polypeptides in chicken embryo fibroblasts: differential repression of transcription by Rous sarcoma virus transformation. *Molecular and Cellular Biology.* 4:1823–1833.

Ishikawa, R., S. Yamashiro, and F. Matsumura. 1989. Differential modulation of actin-severing activity of gelsolin by multiple isoforms of cultured rat cell tropomyosin. *The Journal of Biological Chemistry.* 264:7490–7497.

Leavitt, I., G. Latter, L. Lutomski, D. Goldstein, and S. Burbeck. 1986. Tropomyosin isoform switching in tumorigenic human fibroblasts. *Molecular and Cellular Biology.* 6:2721–2726.

Lees-Miller, J. P., and D. M. Helfman. 1991. The molecular basis for tropomyosin isoform diversity. *BioEssays.* 13:429–437.

Lees-Miller, J. P., L. O. Goodwin, and D. M. Helfman. 1990. Three novel brain tropomyosin isoforms are expressed from the rat α-tropomyosin gene through the use of alternative promoters and alternative RNA processing. *Molecular and Cellular Biology.* 10:1729–1742.

Lees-Miller, J. P., A. Yan, and D. M. Helfman. 1990. Structure and complete nucleotide sequence of the gene encoding rat fibroblast tropomyosin 4. *Journal of Molecular Biology.* 213:399–405.

Leonardi, C. L., R. H. Warren, and R. W. Rubin. 1982. Lack of tropomyosin correlates with the absence of stress fibers in transformed rat kidney cells. *Biochimica et Biophysica Acta.* 720:154–162.

Lin, J. J.-C., T. E. Hegmann, and J. L.-C. Lin. 1988. Differential localization of tropomyosin isoforms in cultured nonmuscle cells. *The Journal of Cell Biology.* 107:563–572.

Lin, J. J.-C., D. M. Helfman, S. H. Hughes, and C. S. Chou. 1985. Tropomyosin isoforms in chicken embryo fibroblasts: purification, characterization and changes in Rous sarcoma virus-transformed cells. *The Journal of Cell Biology.* 100:692–703.

Liu, H., and A. Bretscher. 1989. Disruption of the single tropomyosin gene in yeast results in the disappearance of actin cables from the cytoskeleton. *Cell.* 57:233–242.

Liu, H., and A. Bretscher. 1992. Characterization of TPM1 disrupted yeast cells indicates and involvement of tropomyosin in directed vesicle transport. *The Journal of Cell Biology.* 118:285–300.

Matsumura, F., J. J.-C. Lin, S. Yamashiro-Matsumura, G. P. Thomas, and W. C. Topp. 1983*a*. Differential expression of tropomyosin forms in the microfilaments isolated from normal and transformed rat cultured cells. *The Journal of Biological Chemistry.* 258:13954–13964.

Matsumura, F., S. Yamashiro-Matsumura, and J. J.-C. Lin. 1983*b*. Isolation and characterization of tropomyosin-containing microfilaments from cultured cells. *The Journal of Biological Chemistry*. 258:6636–6644.

Matsumura, F., and S. Yamashiro-Matsumura. 1985. Purification and characterization of multiple isoforms of tropomyosin from rat cultured cells. *The Journal of Biological Chemistry*. 260:13851–13859.

Matsumura, F., and S. Yamashiro-Matsumura. 1986. Modulation of Actin-bundling activity of 55-kDa Protein by multiple isoforms of tropomyosin. *The Journal of Biological Chemistry*. 261:4655–4659.

Prasad, G. L., R. A. Fuldner, and H. L. Cooper. 1993. Expression of transduced tropomyosin 1 cDNA suppresses neoplastic growth of cells transformed by the ras oncogene. *Proceedings of the National Academy of Sciences, USA*. 90:7039–7043.

Rio, D. C. 1992. RNA processing. *Current Opinion in Cell Biology*. 4:444–452.

Ruiz-Opazo, N., and B. Nadal-Ginard. 1987. α-Tropomyosin gene organization. *The Journal of Biological Chemistry*. 262:4755–4765.

Shin, S., V. H. Freedman, R. Risser, and R. Pollack. 1975. Tumorigenicity of virus-transformed cells in nude mice is correlated specifically with anchorage independent growth in vitro. *Proceedings of the National Academy of Sciences, USA*. 72:4435–4439.

Smillie, L. B. 1979. Structure and functions of tropomyosins from muscle and non-muscle sources. *Trends in Biochemical Science*. 4:151–155.

Sobieszek, A., and J. V. Small. 1981. Effect of muscle and nonmuscle tropomyosins in reconstituted skeletal muscle actomyosin. *European Journal of Biochemistry*. 118:533–539.

Stamm, S., D. Casper, J. P. Lees-Miller, and D. M. Helfman. 1993. Brain-specific tropomyosins TMBr-1 and TMBr-3 have distinct patterns of expression during development and in adult brain. *Proceedings of the National Academy of Sciences, USA*. 90:9857–9861.

Takenaga, K., Y. Nakamura, H. Kageyama, and S. Sakiyama. 1990. Nucleotide sequence of cDNA for nonmuscle tropomyosin 5 of mouse fibroblast. *Biochimica et Biophysica Acta*. 1087:101–103.

Wang, Y. C., and P. A. Rubenstein. 1992. Splicing of two alternative exon pairs in β-tropomyosin pre-mRNA is independently controlled during myogenesis. *The Journal of Biological Chemistry*. 267:12004–12010.

Weinberger, R. P., R. C. Henke, O. Tolhurst, P. L. Jeffrey, and P. Gunning. 1993. Induction of neuron specific tropomyosin mRNAs by nerve growth factor is dependent on morphological differentiation. *The Journal of Cell Biology*. 120:205–215.

Wieczorek, D. F., C. W. J. Smith, and B. Nadal-Ginard. 1988. The rat α-tropomyosin gene generates a minimum of six different mRNAs coding for striated, smooth, and nonmuscle isoforms by alternative splicing. *Molecular and Cellular Biology*. 8:679–694.

Yamashiro-Matsumura, S., and F. Matsumura. 1988. Characterization of 83-kilodalton nonmuscle caldesmon from cultured rat cells: stimulation of actin binding of nonmuscle tropomyosin and periodic localization along microfilaments like tropomyosin. *The Journal of Cell Biology*. 106:1973–1983.

Chapter 3

Evolution of the Actin and Myosin-related Families of Cytoskeletal Proteins

Profilins, Ancient Actin Binding Proteins with Highly Divergent Primary Structures

Thomas D. Pollard* and Stephen Quirk*‡

*Departments of *Cell Biology and Anatomy and ‡Biophysics and Biophysical Chemistry, The Johns Hopkins University Medical School, Baltimore, Maryland 21205*

The actin filament-based cytoskeleton is a hallmark of eukaryotic cells where it may comprise over 25% of total cellular protein. None of the many proteins which comprise the actin cytoskeleton are known to exist in eubacteria or archaebacteria, hence it has been speculated that this system is one of the features that made eukaryotic life possible on earth and differentiates the prokaryotic and eukaryotic divergence. The argument is that the ability of this system to resist and transmit mechanical forces and to produce internal forces for cellular motility alleviated the need for a protective cell wall and allowed early eukaryotic cells the freedom to move toward prey and away from danger.

We know about more than 50 different distinct actin binding proteins, many of which are found in all branches of the eukaryotic phylogenetic tree (Pollard, 1993). Thus, although exceptions may exist, most of these proteins were present in the primitive eukaryotic progenitor of animals, plants and fungi and are therefore as evolutionarily ancient as actin itself. The actin binding proteins fall into several classes including actin monomer binding proteins, actin filament capping proteins, actin filament cross-linking proteins and myosin motors that move along actin filaments.

Profilin Structure and Function

We will concentrate on a small actin monomer binding protein called profilin. Profilins are remarkable in an evolutionary context because, in spite of the fact that they bind several different ligands and have the same three dimensional structure in amoebas and mammals, the sequences of the known profilins are extremely divergent. In addition, the available evidence suggests that the isoforms present in contemporary species have arisen after the divergence of the major branches of the phylogenetic tree, a feature that differs from the actin-based motor proteins, the myosins (Goodson and Spudich, 1993; Cheney, Riley, and Mooseker, 1993). For these reasons, profilins may be useful in studying the evolution of the components of the actin cytoskeleton.

Profilins are small proteins, consisting of 125 to 153 amino acid residues, present in very high concentrations in the cytoplasm of vertebrates, plants, echinoderms, slime moulds, ciliates, and amoebae (Machesky and Pollard, 1993). Even vaccinia virus has a profilin gene that is expressed at high levels late in infection (Blasco, Cole, and Moss, 1991). Of all of the proteins in nonmuscle cells, only actin (100–250 μM) usually exceeds profilin (20–100 μM) in concentration. Profilin is essential for normal growth of yeast cells which are temperature sensitive and severely disabled without it (Haarer, Lillie, Adams, Magdolen, Bandlow, and Brown, 1990). Plant profilins are now known to be major human allergens (Valenta, Duchene, Pettenburger, Sillaber, Valent, Bettelheim, Breitenbach, Rumpold, Kraft, and Scheiner, 1991).

The atomic structures of *Acanthamoeba* profilin-I (Vinson, Archer, Lattman, Pollard, and Torchia, 1993; Fedorov, Magnus, Graupe, Lattman, Pollard, and Almo, 1994) and bovine profilin (Schutt, Myslik, Rozychi, Goonesekere, and Lindberg, 1993) share most important features. The amoeba profilin structure (Fig. 1) was determined independently by multidimensional NMR (Vinson et al., 1993) and by x-ray crystallography (Fedorov et al., 1994). The bovine profilin structure was determined by x-ray crystallography as a complex with actin (Schutt et al., 1993). Both profilins are built around a six-stranded, antiparallel beta sheet flanked on one side by two long helices comprised of the NH_2- and COOH-terminal sequences of the polypeptide. The other side of the sheet is covered by short helices and a pair of beta

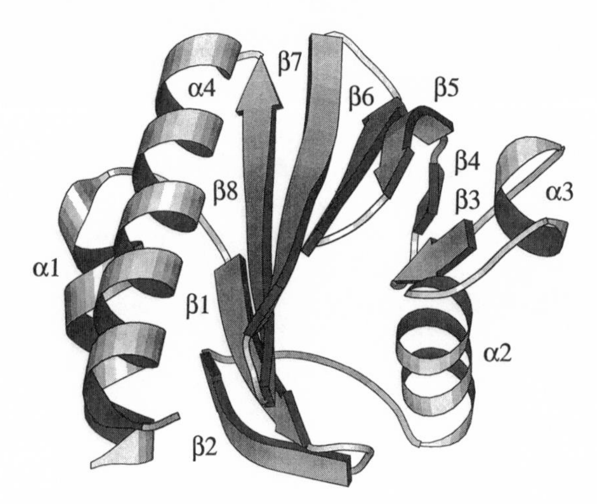

Figure 1. Crystal structure of *Acanthamoeba* profilin-I. A ribbon diagram of the fold of the protein based on the x-ray structure of Fedorov et al., (1994; Vinson et al., 1993). This figure was kindly provided by Dr. Valda Vinson.

strands. The architecture of the bovine profilin is similar, except that the bovine profilin has somewhat longer surface loops connecting beta strands 1 to 2, 5 to 6, and 6 to 7. The similarity of tertiary structure is remarkable given the many differences in primary structure discussed below.

Profilins bind at least four different ligands (Fig. 2). Profilin was originally discovered as actin monomer binding protein (Carlsson, Nystrom, Sundkvist, Markey, and Lindberg, 1976). From the crystal structure of the complex we know that the long COOH-terminal helix of profilin along with adjacent residues from helix-3, beta strands 4, 5, 6, and 7 contact the actin molecule. Earlier chemical cross-linking experiments (Vandekerckhove, Kaiser, and Pollard, 1989) had established that COOH-terminal helix of profilin contacts the COOH-terminal part of actin monomers, strongly supporting the conclusion that the contacts observed in the crystal are used in solution.

Binding profilin alters the behavior of the actin molecule in two ways. First, profilin strongly accelerates the dissociation of the adenine nucleotide bound to actin (Mockrin and Korn, 1980). Because profilin binds and dissociates on a subsecond time scale, it can move rapidly form one actin molecule to the next and promote nucleotide exchange in a catalytic fashion (Goldschmidt-Clermont, Machesky, Doberstein, and Pollard, 1991*b*). Second, profilin bound to actin monomers inhibits the nucleation and elongation of actin filaments, especially at their slow growing end (Pollard and Cooper, 1985; Lal and Korn, 1985; Goldschmidt-Clermont et al., 1991*b*).

Profilins also bind poly-L-proline (Tanaka and Shibata, 1985). This interaction involves aromatic and other hydrophobic residues between the NH_2- and COOH-terminal helices (Archer, Vinson, Pollard, and Torchia, 1994) and mutation of these residues can prevent binding (Bjorkegren, Rozycki, Schutt, Lindberg, and Karlsson, 1993). The physiological significance of this activity remains to be established, although a number of cytoplasmic proteins have sequences enriched in proline.

Many profilins also bind to polyphosphoinositides (Lassing and Lindberg, 1988; Machesky, Goldschmidt-Claremont, and Pollard, 1990), a minor class of phospholipids found the cellular membranes. One of these lipids, phosphatidylinositol bisphos-

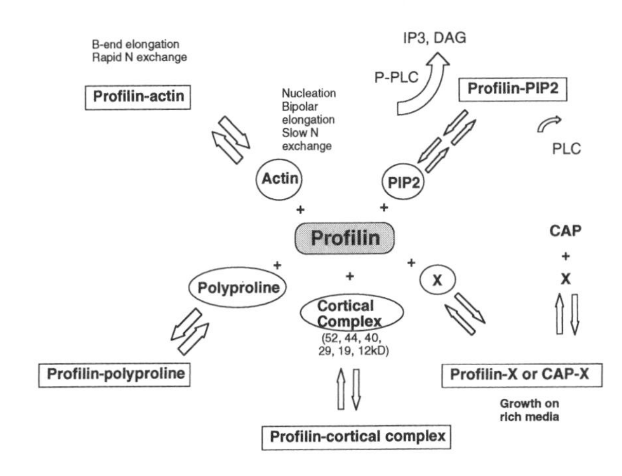

Figure 2. Diagram of the interaction profilins with their various ligands. The size of the arrows are roughly proportional to the values of the rate constants for the reactions.

phate (PIP_2), is the substrate for the production of the important second messengers inositol trisphosphate and diacylglycerol. The polyphosphoinositides compete with actin for binding profilin (Lassing and Lindberg, 1985). Candidate polyphosphoinositide binding sites have been identified by comparing the sequences (Machesky et al., 1990) and atomic structures (Fedorov et al., 1994) of two isoforms of *Acanthamoeba* profilin. Profilin-II binds PIP_2 10 times more strongly than profilin-I and has a surface pocket a much higher positive charge than profilin-I. This site is the right size to bind a small cluster of negatively charged PIP_2 head groups exposed on the surface of a lipid bilayer. This interaction of profilin with phosphatidylinositol-bisphosphate may regulate the production of the second messengers because profilin inhibits the access of phospholipase Cγ1 to its substrate, unless this enzyme is phosphorylated by activated growth factor receptor tyrosine kinases (Goldschmidt-Clermont, Kim, Machesky, Rhee, and Pollard, 1991*a*). In this way, profilin is an essential negative regulator required for the receptors to turn on the activity of the phospholipase.

Amoeba profilins also bind to a complex of seven polypeptides that are localized in the cortex of the cell (Machesky, Cole, Moss, and Pollard, 1994). The complex includes two unconventional actins, the most likely candidates in the complex for binding profilin. In addition, over expression of profilin can suppress defects in yeast cyclase associated protein called CAP (Vojtek, Haaer, Field, Gerst, Pollard, Brown, and Wigler, 1991). The vertebrate homologues of CAP are actin binding proteins (Gieselmann and Mann, 1992), and the yeast CAP is also involved with signaling from activated ras to adenyl cyclase.

An interesting but untested hypothesis is that profilin may act as a second messenger in cells. By virtue of its multiple binding sites, it may be able to carry information from growth factor signaling pathways to the cytoskeleton or other cellular systems.

In following sections, we examine the phylogenetic relationships of profilin in view of its proposed multiple functions and the great variability of its primary structure.

TABLE I
Organism and Sequence References

Organism	Code	Accession number	Reference
Homo sapiens	HSAP	Genbank:J03191	Kwiatkowski and Bruns (1988)
Mus musculus	MMUS	Genbank:X14425	Sri Widada et al. (1989)
Bos primigenius taurus	BTAU	SwissProt: P02584	Nystrom et al. (1979)
Vaccinia virus	VACC	PIR:B40897	Blasco et al. (1991)
Drosophila melanogaster	DMEL	Genbank:M84529	Cooley et al. (1992)
Betula verrucosa	BVER	EMBL:M65179	Valenta et al. (1991)
Clypeaster japonicus	CJAP	PIR:S13197	Takagi et al. (1991)
Anthocidaris crassispina	ACRA	PIR:S13198	Takagi et al. (1991)
Saccharomyces cerevisiae	SCER	Genbank:Y00463	Oechsner et al. (1987)
Dictyostelium discoideum	DDI1	Genbank:X61581	Haugwitz et al. (1991)
Dictyostelium discoideum	DDI2	Genbank:X61580	Haugwitz et al. (1991)
Physarum polycephalum	PPO1	Genbank:M38037	Binette et al. (1990)
Physarum polycephalum	PPO2	Genbank:M38038	Binette et al. (1990)
Acanthamoeba castellanii	ACC1	PIR:A22163	Vandekerckhove et al. (1989)
Acanthamoeba castellanii	ACC2	PIR:S00282	Ampe et al. (1988)
Tetrahymena pyriformis	TPYR	Genbank:D00813	Edamatsu et al. (1991)

Genbank v. 71, EMBL v. 30, PIR v. 31, SwissProt v. 21.

Profilin sequences were obtained for the 16 organisms from the sources and databases as noted. The codes are used for clarity and to label the phylogenetic trees.

Methods for Sequence Analysis

The 16 profilin sequences used in this study were collected in mid 1993 from the on line sequence resources listed in Table I. Additional sequences are rapidly being added to the data base, so that this analysis can be extended significantly in the near future. Amino acid sequences were used for the phylogenetic analyses to avoid codon bias errors. All computer programs were run on a Macintosh IICi. The sequences were aligned by the progressive pairwise alignment method of Feng and Doolittle (1987, 1990) including prealignment using the programs PREALIGN and ALIGN. For a comparison, the alignment program CLUSTAL V (Higgins, Bleasby, and Fuchs, 1992) was employed. The ALIGN generated alignment was used in subsequent analyses. Two separate phylogenetic methods were utilized, maximum parsimony and distance matrix, as implemented in the PHYLIP package of programs (Felsenstein, 1989, 1993).

For the maximum parsimony analysis, the alignment of the 16 profilin sequences (including gaps) served as input to the PHYLIP program SEQBOOT. This program constructed a multiple sequence data set comprised of 50 random resamplings of the original input alignment and included five rounds of randomized order entry to avoid load order bias. This bootstrapped data set (Felsenstein, 1985) was used as input to the PHYLIP parsimony program PROTPARS. This program searched for the most parsimonious tree from each data set and was configured for five rounds of randomized order entry per data set. The output file was used as input to the PHYLIP program CONSENSE which calculates a strict majority rule consensus tree. For this part of the analysis TPYR was used as the outgroup root, but the resulting tree is still considered unrooted.

In the distance matrix method, the ALIGN alignment file was input into the PHYLIP program PROTDIST which calculates a distance matrix using the Dayhoff PAM 001 matrix, (Dayhoff, 1979). The resulting distance matrix was input into the PHYLIP program FITCH which implements the Fitch-Margoliash (1967) method of fitting trees to distance data. The best tree minimizes:

$$\sum_{i}\sum_{j}\frac{n_{ij}(D_{ij}-d_{ij})^2}{PDij}$$

where D_{ij} and d_{ij} are the observed and expected distances between organisms i and j, n_{ij} is the number of times a distance has been replicated (here $n = 1$), and P is the power ($P = 2$ for the Fitch-Margoliash method [Fitch-Margoliash, 1967]). FITCH was run with nonnegative branch lengths, the TPYR sequence as the outgroup root, ten rounds of randomized order entry, and global branch swapping. This last option reconsiders the position of every organism in the tree, after the tree is calculated, by removing and replacing every member, hence the final tree is improved. Our distance matrix analysis concerns only tree topology and not time of divergence.

Sequence Comparisons and Evolutionary Relationships of Profilins

In spite of major differences in primary structures and polymer length, the full-length sequences of all 16 profilins can be aligned in a significant fashion by means of the Feng and Doolittle (1987, 1990) algorithm of progressive pairwise alignment including the prealignment of highly homologous sequences (Fig. 2). This alignment required a number of insertions. Relating the positions of these gaps to the secondary structure deduced from the atomic models of profilins, most, but not all, of these insertions are in surface loops connecting elements of secondary structure (Compare Fig. 1 with 3). Within the groupings of mammalian, fungal, echinoderm and slime mold profilins, the sequences are similar throughout their entire lengths. Simple pairwise alignment, as implemented in the program CLUSTAL V (Higgins et al., 1992), gave a similar alignment and similar distance scores (data not shown) indicating the robustness of the observed relationships.

Sequence similarity ranges from 11 to 96%, with corresponding evolutionary distances of 213 to 2. *Tetrahymena* profilin (TPYR) is clearly distinct from the other 15 profilins in that it contains two amino acid insertions that force gaps in the alignment. The sequence YNNYQIDVEGQ is inserted in the connection between beta strand-2 and helix-2. The sequence QQNGV is inserted just before the

COOH-terminal helix (see Figs. 1 and 3). The sea urchin (ACRA and CJAP) and birch (BVER) profilin sequences each contain a single insertion after the NH_2-terminal helix. The six residue sequence (DLRTKS) in the human profilin (HSAP) forms most of the beta-6 strand, but does not align with the beta-6 strand of the amoeba profilin. This region will require further detailed analysis, because it

```
           *  *    *              * *         **    *
HSAP    MAGWNAYID.NLM......ADGTCQDAAIVGYKDSPSVWAAVPGKTFV...........N
MMUS    ---------.S--......---------------------------...........S
BTAU    .--------.---......-N-------------------------...........-
VACC    --E-HKI-E.DIS......KNNKFE-----D--TTKN-L--I-NR--A...........K
ACC1    ..T-QS-V-T--V......GT-AVTQ---L-LDG..NT--SFA-..-............A
ACC2    ..T-QS-V-T--V......GT-AVTQ---I-HDG..NT--TSA-..-............A
PPO1    .MS-Q--V-DQ-V......GT-HVIG---I-HDG..N---SKN....L...........S
PPO2    .MS-QT-VHEQ-V......GT-QLDG-I-I-LDG..NS--SKN....L...........T
DDI1    .MS-QQ-V-EQ-T......GA-LS-G-ILGANDG..G---KSS-...I...........-
DDI2    .MT-Q--V-N--L......GA-FASA-LLGAADG..----HSA-................
BVER    .MS-QT-V-EH--.CDIDGQASNSLAS----HDG..----QSSS..-P...........Q
SCER    .MS-Q................-KVDK-V-YSRAG DA---TSG-...L...........S
DMEL    .MS-QD-V-NQ-L......-SQCVTK-C-A-HDG..NI--QSS-...F...........E
ACRA    ..S-DS---NLIAQTKDASGT-HSDK-C-I-IDG.GAP-TTAGHANAL...........K
CJAP    ..S-DS---NLVAQTKDASGTAHSDR-C-I-LDG.GAP-TTAGHANAL...........K
TPYR    -S--DQ-VQ.Y-T......-NQQVEYGL-L-KT-.GTI--SNV-L-TLYNNYQIDVEGQK

            *                   *    *                         *
HSAP    ITPAEVGVLVG..KDR..SSFYVNGLTLGGQKCSVIRDSLLQDGEFSMDLRTKSTGGAPT
MMUS    -----------..---..---F-----------------------T------------
BTAU    -------I---..---..---F--------E---------------T------------
VACC    -N-G-IIP-I..TNR..N..ILKPL.IGQKYCIVYTNSLMDENTYAMEL...LTGYAPV.
ACC1    V---QGTT-A-AFNNA..DAIRAS-FD-A-VHYVTL-AD..DRSIYGK......K-A-..
ACC2    VS--NGAA-ANAF--A..TAIRS--FE-A-TRYVT--AD..DRSVYGK......K-S-..
PPO1    LKAG-GAKI-NGF--S..A-VLSG-IFVD---YLT-KAD..DKSIYGK......K-AG..
PPO2    LKAG-GQAIAALF-TP..ANVFAS-I-IN-I-YMG-KGD..SRSIYGK......K-AT..
DDI1    --KP-GDGIAALF-NP..AEVFAK-ALI--V-YMG-KGD..PQSIYGK......K-AT..
DDI2    FNV--GKAITALFQKD..GAAFAT-IHVA-K-YMA-KSD..TRSAYGK......L-AG..
BVER    FK-Q-ITGIMKDFEEP..GHLAPT--H---I-YM--QGE.AGAVIRGK......K-SG..
SCER    LQ-N-I-EI-QGFDNP..AGLQS---HIQ---FMLL-AD..DRSIYGR......HDAE..
DMEL    V-KE-LSK-ISGFDQQ..DGLTS--V--A--RYIYLSGT..DRVVRAK......L-RS..
ACRA    LEGQ-GPNIARCF-SKDFTP-MSS-IVAD-T-YQFL-EE.DGKLVLAK.....KK-QG..
CJAP    LQGT-GANIAKCF-SKDF-A-MAG-VHAE-L-YQFL-EE.DAKLVLAK.....KK-EG..
TPYR    ANVN-TAN-LAAMNNNGVPTDPLC-IRIMN--YYTVKYDADSQVWYLK......KDHG..

               *                       *            *
HSAP    FNVTVTKTDKTLVLL...MGKE.....GVHGGLINKKCYEMASHLRRSQY
MMUS    ------M-A------...----.....-----------------------
BTAU    --I---M-A------...---Q.....-----M------Q---------
VACC    SPIVIAR-HTA-IF-...---P.....TTSRRDVYRT-RDH-TRV-ATGN
ACC1    .G-ITV--S-AILVG...VYN-.....KIQP-TAANVVEKL-DY-IGQGF
ACC2    .G-ITV--S-AILIG...VYN-.....KIQP-TAANVVEKL-DY-IGQGF
PPO1    .G-VLV--GQSVLIG...HYN-.....TIQP-QATTVVEKL-DY--ENG-
PPO2    .G-ATVI-GQCILIG...YYN-.....KQQP-NAALVVEKL-DY-IENG-
DDI1    .GCVLVR-GQAIIVG...IYDD.....K-QP-SAALIVEKLGDY--DNG-
DDI2    .G-VCV--LTCIIVA...VYDD.....KLQP-AAANIAEKL-DY-IDNNC
BVER    .GI-IK--GQA--FG...IYE-.....P-TP-QC-MVVERLGDY-IDQGL
SCER    .G-VCVR-KQ-VIIA...HYPP.....T-QA-EAT-IVEQL-DY-IGV--
DMEL    .G-HCM--TQAVIVS...IYED.....P-QPQQAASVVEKLGDY-ITCG-
ACRA    .AL-LQSSKTAI-IG...HAP-.....-GQQ-NT--GVAVI-EY-ESLGM
CJAP    .AI-LQASKTAI-IA...HCP-.....-GQQ-NT--GVSVI-EY-ESLGM
TPYR    .GACIAI-NQA--IGTFDIT-KQQNGVAQNP-QV--VVESL-AT-KQAG-
```

Figure 3. The alignment of the 16 profilin amino acid sequences based on progressive pairwise alignment. The uppercase letters refer to the standard single letter amino acid code and asterisks denote individual residues in the alignment that are common to at least 75% of the organisms. Dots within the sequence are gaps placed by the alignment program for optimality and dashes represent identical residues in comparison to HSAP. Sequence code names are as in Table I.

participates in binding actin (Schutt et al., 1993). In addition to these major sequences, many of the profilins have scattered double or single amino acid insertions.

In spite of the highly conserved tertiary structure of profilin, only 15 amino acid residues are shared by 75% of the profilin sequences that we have surveyed and only the tryptophan near the NH_2-terminus is invariant! We expect that this sequence diversity will grow with the size of the data base. The majority of amino acid substitutions are conservative. The profilins range in length from 117 amino acids (SCER) to 153 amino acids (TPYR). As expected, closely related profilins have similar numbers of residues, but there is no real trend in length and evolutionary distance.

Profilin isoforms range in identity from 83% for *Acanthamoeba* ACC1 and ACC2, to 66% for *Physarum* PPO1 and PPO2 to 55% for *Dictyostelium* DDI1 and DDI2. Profilin isoforms from single organisms generally differ in isoelectric points, with an acidic and basic isoform. In one example, the isoforms have different activities. In *Acanthamoeba,* the basic isoform, profilin-II, binds PIP_2 much more strongly than the acidic isoform, profilin-I (Machesky et al., 1990).

The isoforms from each species are more closely related to each other than to any other profilins, suggesting that they have all arisen by gene duplication after the

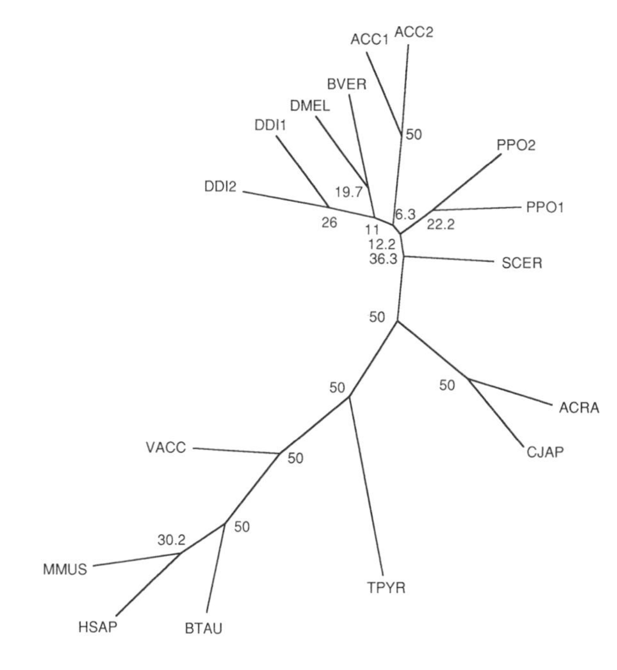

Figure 4. Majority-rule strict consensus phylogenetic tree of the profilins based on a maximum parsimony analysis of the aligned amino acid sequences in Fig. 1. Branch lengths are not proportional to evolutionary distance. The numbers at each branch indicate the bootstrap value, which is the number of parsimony trees calculated out of a total of 50 which indicate the division of organisms defined by that branch. Fractional values arise when multiple trees of equal parsimony are found for any one of the 50 samples.

species diverged from each other. The evolution of myosin isoforms differs dramatically, because the main isoforms appear to have been present in the progenitors of fungi, insects and animals (Goodson and Spudich, 1993; Cheney, Riley, and Mooseker, 1993).

Maximum parsimony analysis (Fig. 4) indicates that profilin proteins cluster into five main groups: the mammal and the animal virus (HSAP, MMUS, BTAU, VACC), *Tetrahymena* alone, the echinoderms (CJAP and ACRA), the *Physarum-Saccharomyces* group (PPO1, PPO2, and SCER) and the remaining sequences which comprise a rather diverse group including plants, flies, amoebas, and a cellular slime mold (BVER, DMEL, ACC1, ACC2, DDI1, DDI2). Each isoform from a single organism clusters together indicating intra-species gene duplication. Bootstrap values of 100% are identified for branch points separating the urchin sequences from each other, as well as from the ancestral node; and for the BTAU, VACC, and ACC nodes. Lower confidence bootstrapping (from 12.6 to 72.6%) is found for the remaining nodes. The lowest confident node of 12.6%, which marks the *Acanthamoeba* branch from the main trunk, is however followed by the 100% bootstrap level for the ACC1-ACC2 subgroup. The lower bootstrap confidence in the invertebrate group is reflected by an average lower divergence value calculated by distance methods and is reflected in the phylogenetic tree generated from the distance matrix.

Distance measurements (Table II) and the resulting distance matrix tree (Fig. 5) show that the profilins cluster into three main groups: the mammalian profilins (including VACC, although that sequence is diverging rapidly from the group), *Tetrahymena,* which as in the maximum parsimony tree, is a possible outgroup root, and all the remaining profilins (although the sea urchin sequences are diverging from this grouping). Generally speaking, the results of this analysis are in accord with accepted evolutionary taxonomy, that is the invertebrate proteins diverge from the ancestral node before the vertebrate profilins. The first invertebrate branch sepa-

TABLE II
Percent Identity after Progressive Pairwise Alignment/Evolutionary Distance (×100)

	HSAP	MMUS	BTAU	VACC	ACC1	ACC2	PPO1	PPO2	DDI1	DDI2	BVER	SCER	DMEL	ACRA	CJAP	TPYR
HSAP		95.7	92.1	32.3	27.1	28.7	27.1	19.7	22.8	21.5	28.0	25.2	22.0	20.3	20.3	21.5
MMUS	1.8		93.5	32.3	25.4	27.1	27.1	20.5	22.8	20.7	27.2	26.1	21.1	20.3	20.3	21.5
BTAU	6.9	5.6		33.3	24.6	26.2	23.8	18.9	23.6	20.7	27.2	24.4	19.5	19.5	20.3	20.5
VACC	79.3	81.0	85.9		15.3	14.4	15.3	11.0	16.0	12.0	17.4	18.9	13.5	10.5	10.5	19.1
ACC1	107.4	110.4	113.8	163.0		83.2	51.2	52.9	44.4	47.2	36.0	37.7	42.7	28.8	28.8	29.0
ACC2	104.8	107.4	108.2	166.1	8.8		56.1	52.0	41.1	47.2	37.6	38.6	44.4	33.6	31.2	27.4
PPO1	109.2	110.9	111.6	165.1	44.6	36.9		65.6	51.2	44.7	40.8	49.6	41.6	37.1	34.7	33.1
PPO2	125.2	126.0	124.3	185.8	44.8	43.1	28.5		60.8	48.8	40.8	42.6	36.8	34.7	33.1	33.1
DDI1	114.2	117.2	113.9	175.2	54.0	50.1	39.8	34.0		54.8	44.4	42.2	40.5	27.2	26.4	28.8
DDI2	128.2	129.7	132.5	200.1	56.0	53.0	53.5	47.9	44.8		37.1	39.5	37.1	23.6	25.2	23.6
BVER	124.8	124.5	124.5	186.5	59.9	55.9	57.3	73.2	67.8	63.9		32.8	41.3	26.5	26.5	27.6
SCER	118.5	122.9	121.6	162.7	69.3	63.5	60.2	68.8	65.2	77.6	64.9		39.7	29.3	29.3	24.8
DMEL	129.8	128.5	130.1	171.3	69.4	65.6	59.5	72.6	76.5	74.8	69.2	87.5		24.0	23.2	27.2
ACRA	138.6	141.5	142.8	208.2	103.3	96.5	86.0	95.7	109.5	124.9	110.5	103.5	106.7		83.5	22.9
CJAP	138.5	140.1	142.6	212.8	100.2	98.4	86.2	95.9	113.1	120.8	110.4	101.9	103.3	9.1		22.1
TPYR	137.7	134.9	140.9	176.5	124.7	124.3	111.2	118.7	127.1	133.1	132.5	133.1	133.5	140.4	140.3	

The set of numbers above the matrix diagonal refer to the percent identity between profilins after progressive pairwise alignment. Values below the diagonal are evolutionary distances.

rates the *Acanthamoeba* and echinoderm sequences (ACC1, ACC2, ACRA, and CJAP) from the remaining invertebrates which then diverge into two main groups: the *Dictyostelium*/*Physarum* branch and the yeast/birch/*Drosophila* branch.

The general topological features of the distance matrix tree mirror those of the consensus majority rule maximum parsimony tree. Both methods illustrate the slight taxonomic independence of subgroups within the invertebrate out grouping. The major difference in the two methods is the branch position of PPO1 and PPO2. The distance matrix tree does not place the two *Physarum* isoforms on an iso-branch point, although they are separated by a distance value of 28.5. Both methods clearly show that the mammalian profilins (HSAP, BTAU, MMUS, and VACC) along with *Tetrahymena* profilin (TPYR) have diverged most from the ancestral profilin.

Vaccinia profilin (VACC) is an interesting outlier. Vaccinia virus specifically infects bovine species, producing cowpox, so the viral profilin gene must have originated in the cow genome (BTAU) and later was incorporated into the genome of the virus. Phylogenetic analysis (distance matrix or maximum parsimony) still group VACC with or close to the three mammalian profilins (HSAP, MMUS,

BTAU) supporting the bovine origin of vaccinia profilin. Since entering the viral genome, the VACC sequence has diverged more than any other profilin in the analysis (Table II). This is reflected in its unique properties, having much lower affinity for actin and poly-L-proline than any other characterized profilin. The divergence of the sequence is demonstrated by the greater distances between VACC and the other profilins compared to any of the other inter sequence distances. In addition, vaccinia profilin shows lowest average sequence identity to the other profilins. Most striking is the observation that in both tree types VACC splits before, and is distinct from, the vertebrate subgroup. The physiological role of profilin in the virus life cycle is unknown, so we do not understand the selective pressures that have fostered the divergence of the sequence. Null mutations of the profilin gene do not prevent growth in culture, but virologists believe that all of the viral genes provides some selective advantage.

An unexpected result of this analysis is that *Drosophila* profilin (DMEL) is closely related to the birch profilin (BVER), a conclusion supported by both phylogenetic methods. There is no a priori reason to suspect that tree and insects would have evolutionarily related profilin. More examples of insect and plant profilins are needed to elucidate the phylogenetic relationships between the DMEL/BVER node specifically, and in general, to establish the profilin phylogeny of species that are intermediate to the vertebrates and lower eukaryotes in this study.

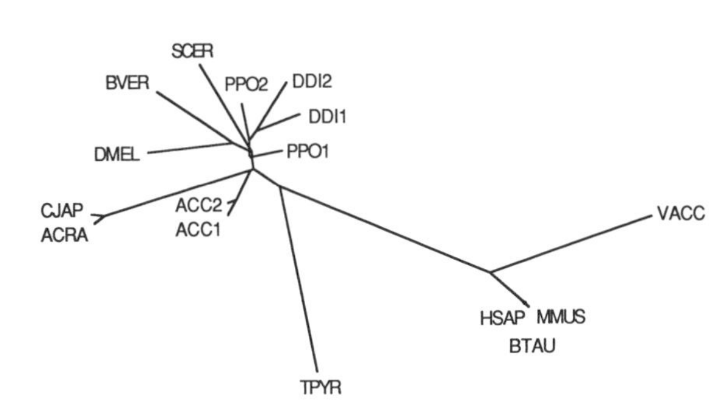

Figure 5. Phylogenetic tree based on analysis of the distance matrix in Table II of the aligned sequences in Fig. 1. The tree branch lengths are proportional to distance and the tree has an average standard deviation of 8.7 percent. The tree is unrooted and is calculated using nonnegative branch lengths and global optimization as implemented in FITCH.

The profilin family may be a good example of the inherent randomness of amino acid sequences (White, 1994), illustrating how a common tertiary structure can be maintained in the face of great variability in primary structure. An interesting possibility is that the abundance and variability of their primary structures may well contribute to making plant profilins major human allergens (Valenta et al., 1991). Clearly, many different amino acid sequences are compatible with the profilin fold, but variations of specific surface residues determine the ability of each to interact with the various ligands. For example, vaccinia profilin cannot interact with actin, because bulky side chains (tyrosine 79, arginine 114, and tyrosine 117) all interfere sterically with the contact with actin. Similarly, the viral profilin cannot bind poly-L-proline (Machesky, Ampe, Vandekerckhove, and Pollard, 1994), because several of hydrophobic residues contributing to the binding site in *Acanthamoeba* profilin (Archer et al., 1994) are missing. Characterization of additional profilins may reveal how the ligand binding sites have evolved and may clarify, for example, whether lipid binding is a relatively new function that has been acquired more than once by convergent evolution of ancient profilins that originally bound to actin and/or poly-L-proline. In addition, the profilin family of actin binding proteins provides an excellent opportunity for the evolutionary study of the actin based cytoskeleton.

Acknowledgements

Our research was supported by an NSRA Postdoctoral Fellowship to S. Quirk and NIH Research Grants GM-26338 to T. D. Pollard and GM-35171 to E. E. Lattman.

References

Ampe, C., M. Sato, T. D. Pollard, and J. Vandekerckhove. 1988. The primary structure of the basic isoform of *Acanthamoeba* profilin. *European Journal of Biochemistry.* 170:597–601.

Archer, S., V. Vinson, T. D. Pollard, and D. Torchia. 1994. Identification of the poly-L-proline binding site in Acanthamoeba profilin I by NMR spectroscopy. *FEBS Letters.* 337:145–151.

Binette, F., M. Benard, A. Laroche, G. Pierron, G. Lemieux, and D. Pallotta. 1990. Cell-specific expression of a profilin gene family. *DNA Cell Biology.* 9:323–334.

Bjorkegren, C., M. Rozycki, C. E. Schutt, U. Lindberg, and R. Karlsson. 1993. Mutagenesis of human profilin locates its poly(L-proline) binding site to a hydrophobic patch of aromatic amino acids. *FEBS Letters.* 333:123–126.

Blasco, R., N. B. Cole, and B. Moss. 1991. Sequence analysis, expression, and deletion of a vaccinia virus gene encoding a homolog of profilin, a eukaryotic actin-binding protein. *Journal of Virology.* 65:4598–4608.

Carlsson, L., L. E. Nystrom, L. Sundkvist, F. Markey, and U. Lindberg. 1977. Actin polymerizability is influenced by profilin, a low molecular weight protein in non-muscle cells. *Journal of Molecular Biology.* 115:465–483.

Cheney, R., M. Riley, and M. S. Mooseker. 1993. Phylogenetic analysis of the myosin superfamily. *Cell Motility of the Cytoskeleton* 24:215–223.

Cooley, L., E. Verheyen, and K. Ayers. 1992. Chickadee encodes a profilin required for intercellular cytoplasmic transport during Drosophila oogenesis. *Cell.* 69:173–184.

Dayhoff, M. O. 1979. Atlas of protein sequence and structure. Vol. 5. Supplement 3, 1978. National Biomedical Research Foundation, Washington, DC.

Edmatsu, M., M. Hirono, and T. Takemasa. 1991. The primary structure of Tetrahymena profilin. *Biochemical Biophysical Research Communications.* 175:543–550.

Fedorov, A. A., K. A. Magnus, H. Graupe, E. E. Lattman, T. D. Pollard, and S. C. Almo. 1994. X-ray structures of isoforms of the actin binding protein profilin that differ in their affinity for polyphosphoinositides. *Proceedings of the National Academy of Sciences, USA.* In press.

Felsenstein, J. 1985. Confidence limits on phylogenies: an approach using the bootstrap. *Evolution.* 39:783–791.

Felsenstein, J. 1989. PHYLIP-Phylogeny Inference Package. Version 3.2. *Cladistics.* 5:164–166.

Felsenstein, J. 1993. PHYLIP-Phylogeny Inference Package. Version 3.5c. Distributed by the author. Department of Genetics. University of Washington, Seattle.

Feng, D. F., and R. F. Doolittle. 1987. Progressive sequence alignment as a prerequisite to correct phylogenetic trees. *Journal of Molecular Evolution.* 25:351–360.

Feng, D. F., and R. F. Doolittle. 1990. In Molecular Evolution: Computer Analysis of Protein and Nucleic Acid Sequences. R. F. Doolittle, editor. *Methods in Enzymology.* 183:375–387.

Fitch, W. M., and R. L. Margoliash. 1967. Construction of phylogenetic trees. *Science.* 155:279–284.

Gieselmann, R., and K. Mann. 1992. ASP-56, a new actin sequestering protein from pig platelets with homology to CAP, an adenylate cyclase associated protein from yeast. *FEBS Letters.* 298:149–153.

Goldschmidt-Clermont, P. J., J. W. Kim, L. M. Machesky, S. G. Rhee, and T. D. Pollard. 1991*a*. Regulation of phospholipase C-gamma 1 by profilin and tyrosine phosphorylation. *Science.* 251:1231–1233.

Goldschmidt-Clermont, P. J., L. M. Machesky, S. K. Doberstein, and T. D. Pollard. 1991*b*. Mechanism of interaction of human platelet profilin with actin. *Journal of Cell Biology.* 113:1081–1089.

Goodson, H., and J. A. Spudich. 1993. Molecular evolution of the myosin family: relationships derived from comparisons of amino acid sequences. *Proceedings of the National Academy of Sciences, USA.* 90:659–663.

Haarer, B. K., S. H. Lillie, A. E. M. Adams, V. Magdolen, W. Bandlow, and S. S. Brown. 1990. Purification of profilin from *Saccharomyces cerevisiae* and analysis of profilin-deficient cells. *Journal of Cell Biology.* 110:105–114.

Haugwitz, M., A. A. Noegel, D. Rieger, F. Lottspeich, and M. Schleicher. 1991. *Dictyostelium* contains two profilin isoforms that differ in structure and function. *Journal of Cell Science* 100:481–489.

Higgins, D. G., A. J. Bleasby, and R. Fuchs. 1992. Clustal V: improved software fro multiple sequence alignment. *CABIOS.* 8:189–191.

Kwiatkowski, D. J., and G. A. Bruns. 1988. Human profilin: molecular cloning sequence comparison, and chromosomal analysis. *Journal of Biological Chemistry.* 263:5910–5915.

Lal, A. A., and E. D. Korn. 1985. Reinvestigation of the inhibition of actin polymerization by profilin. *Journal of Biological Chemistry.* 260:10132–10138.

Lassing, I., and U. Lindberg. 1985. Specific interaction between phosphatidylinositol 4,5 bisphosphate and profilactin. *Nature.* 314:472–474.

Lassing, I., and U. Lindberg. 1988. Specificity of the interaction between PIP2 and the profilin:actin complex. *Journal of Cellular Biochemistry.* 37:255–268.

Machesky, L. M., P. J. Goldschmidt-Clermont, and T. D. Pollard. 1990. The affinities of human platelet and *Acanthamoeba* profilin isoforms for polyphosphoinositides account for their relative abilities to inhibit phospholipase *Cell Regulation.* 1:937–950.

Machesky, L. M., N. B. Cole, B. Moss, and T. D. Pollard. 1994. Vaccinia virus expresses a novel profilin with a higher affinity for polyphosphoinositides than actin. *Biochemistry.* In press.

Machesky, L. M. and T. D. Pollard. 1993. Profilin as a potential mediator of membrane-cytoskeletal communication. *Trends in Cell Biology.* 3:381–385.

Machesky, L. M., C. Ampe, S. Atkinson, J. Vandekerckhove, and T. D. Pollard. 1994. A cortical complex of seven *Acanthamoeba* polypeptides including two unconventional actin binds to profilin. *Journal of Cell Biology.* In press.

Mockrin, S. C., and E. D. Korn. 1980. *Acanthamoeba* profilin interacts with G-actin to increase the exchange of actin bound adenosine 5′-triphosphate. *Biochemistry.* 19:5359–5362.

Nystrom, L. E., U. Lindberg, J. Kendrick-Jones, and R. Jakes. 1979. *FEBS Letters.* 101:161–165.

Oechsner, U., V. Magdolen, and W. Bandlow. 1987. The cDNA and deduced amino acid sequence of profilin from Saccharomyces cerevisiae. *Nucleic Acids Research.* 15:9078–9078.

Pollard, T. D., and J. A. Cooper. 1984. Quantitative analysis of the effect of *Acanthamoeba* profilin on actin filament nucleation and elongation. *Biochemistry.* 23:6631–6641.

Pollard, T. D. 1993. Actin and actin-binding proteins. *In* Guidebook to the Cytoskeletal and Motor Proteins. T. Kreis and R. Vale, editor. Oxford University Press, Oxford, UK. 3–11.

Schutt, C., J. C. Myslik, M. D. Rozychi, N. C. W. Goonesekere, and U. Lindberg. 1993. *Nature.* 365:810–816.

Sri Widada, J., C. Ferraz, and J. P. Liautard. 1989. Total coding sequence of profilin cDNA from Mus musculus macrophage *Nucleic Acids Research.* 17:2855–2855.

Takagi, T., I. Mabuchi, H. Hosoya, K. Furahashi, and S. Hatano. 1991. Primary structure of profilins from two species of *Echinoidea* and *Physarum polycephalum European Journal of Biochemistry.* 192:777–781. (Published erratum *European Journal of Biochemistry.* 1991. 197: 819.)

Tanaka, M. and H. Shibata. 1985. *European Journal of Biochemistry.* 151:291–297.

Valenta, R., M. Duchene, K. Pettenburger, C. Sillaber, P. Valent, P. Bettelheim, M. Breitenbach, H. Rumpold, D. Kraft, and O. Scheiner. 1991. Identification of profilin as a novel pollen allergen; IgE autoreactivity in sensitized individuals. *Science.* 253:557–560.

Vandekerckhove, J. S., D. A. Kaiser, and T. D. Pollard. 1989. *Acanthamoeba* actin and profilin can be cross-linked between glutamic acid 364 of actin and lysine 115 of profilin. *Journal of Cell Biology.* 109:619–626.

Vinson, V., S. Archer, E. E. Lattman, T. D. Pollard, and D. Torchia. 1993. Three dimensional solution structure of *Acanthamoeba* profilin-I. *Journal of Cell Biology.* 122:1277–1283.

Vojtek, A., B. Haarer, J. Field, J. Gerst, T. D. Pollard, S. Brown, and M. Wigler. 1991. Evidence for a functional link between profilin and CAP in the yeast *S. cerevisiae. Cell.* 66:497–505.

White, S. H. 1994. Global statistics of protein sequences: implications for the origin, evolution and prediction of structure. *Annual Review of Biophysics Biomolecular Structure.* In press.

The Evolution of the Chicken Sarcomeric Myosin Heavy Chain Multigene Family

Everett Bandman, Laurie A. Moore, Maria Jesús Arrizubieta, William E. Tidyman, Lauri Herman, and Macdonald Wick

Department of Food Science and Technology, University of California at Davis, Davis, California 95616

Summary

This manuscript describes the chicken sarcomeric myosin heavy chain (MyHC) multigene family and how it differs from the sarcomeric MyHC multigene families of other vertebrates. Data is discussed that suggests the chicken fast MyHC multigene family has undergone recent expansion subsequent to the divergence of avians and mammals, and has been subjected to multiple gene conversion-like events. Similar to human and rodent MyHC multigene families, the chicken multigene family contains sarcomeric MyHC genes that are differentially regulated in developing embryonic, fetal, and neonatal muscles. However, unlike mammalian genes, chicken fast MyHC genes expressed in developing muscles are also expressed in mature muscle fibers as well. The potential significance of conserved and divergent sequences with the MyHC rod domain of five fast chicken isoforms that have been cloned and sequenced is also discussed.

Sarcomeric MyHCs Are Encoded in a Differentially Regulated Multigene Family

Sarcomeric MyHC isoform diversity is a common theme in all eukaryotes with striated muscles that have been studied (Bandman, 1985). With the single exception of *Drosophila melanogaster* in which a single MyHC gene produces different transcripts through alternative splicing events (Rozek and Davidson, 1983), the development of large and complex multigene families has been responsible for generating and maintaining MyHC diversity. However, alternative splicing of some nonmuscle and smooth muscle MyHC genes has been observed (Hamada, Yanagisawa, Katsurgawa, Doleman, Nagata, Matsuda, and Masaki, 1990) suggesting that at least some subclasses of the myosin superfamily in birds and mammals may utilize differential splicing to generate diversity.

Sarcomeric MyHC gene structure has been highly conserved. The organization of the rat embryonic (Strehler, Strehler-Page, Perriard, Periasamy, and Nadal-Ginard, 1986) and chicken embryonic (Molina, Kropp, Gulick, and Robbins, 1987) MyHC genes are very similar. Both genes contain two exons which encode the 5′-untranslated region, the initiating codons of both genes are located in the third exon, and both encode proteins that are 1940 amino acids. Although intron sequences and sizes are highly variable, all intron positions are exactly conserved with the single exception that the rat exon equivalent to the 40th exon of the chicken gene is divided into two exons 24 bp upstream from the stop codon. Five of the eight intron positions of a nematode MyHC gene (McLachlin, 1983) have been preserved in rat and chicken genes. This striking conservation of intron positions among vertebrate and invertebrate sarcomeric myosin genes suggests that the ancestral MyHC gene contained at least the common introns. However, it is unclear whether this pre-vertebrate MyHC gene was more highly split with introns removed during the evolution of lower organisms, or whether the additional introns found in present day vertebrate MyHC genes result from intron insertion events (Strehler et al., 1986).

In vertebrates, isoform diversity at the protein level is accomplished by switching among the members of the MyHC multigene family. Differential gene expression during development and in different physiological fiber types, results in distinct MyHC isoforms in embryonic, neonatal, and adult muscles (Bandman, Matsuda, and Strohman, 1982). It is widely accepted that muscle fibers contain actin-activated myosin ATPase activities proportional to their contractile properties (Barany, 1967). Thus, rapidly contracting fibers are associated with fast MyHC isoforms while slow twitch or tonic fibers have slow MyHCs (Reiser, Moss, Giulian, and Greaser, 1985). This functional separation into two classes of MyHCs, fast and slow isoforms, is reflected in the genomes of species in which sarcomeric MyHCs have been cloned and sequenced. Fast and slow MyHC genes are found on separate chromosomes in rodent and human genomes (Leinwand, Saez, McNally, and Nadal-Ginard, 1983). The evolutionary divergence of fast and slow MyHC genes is also illustrated by the observation that fast MyHCs from phylogenetically diverse organisms such as fish, amphibia, birds, and mammals (Stedman, Eller, Julian, Fertels, and Sarkar, 1990) are much more homologous to each other than to slow MyHC genes. The same is true of the slow MyHC genes that have been characterized (Stedman et al., 1990). Both slow and fast MyHC gene subsets have been shown to exhibit gene switching during muscle development (Mahdavi, Izumo, and Nadal-Ginard, 1987).

Although initially it appeared as though similar MyHC transitions occurred in chicken and mammalian muscles (Whalen, Sell, Butler-Browne, Schwartz, Bouveret, and Pinset-Harstrom, 1981; Bandman et al., 1982; Winkelmann, Lowey, and Press, 1983), more recent studies indicate otherwise. Specifically, it appears that MyHC isoforms present in developing muscles of the chicken (embryonic and neonatal isoforms) are also expressed in many adult muscle fibers (Crow and Stockdale, 1986; Bandman and Bennett, 1988) and that there appear to be distinct programs of fast MyHC expression in different fast muscle fibers (Bandman and Bennett, 1988). With the exception of the extraocular and masseter (jaw) muscles of humans (Wieczorek, Periasamy, Butler-Browne, Whalen, and Nadal-Ginard, 1985; Soussi-Yanicostas, Barbet, Laurent-Winter, Barton, and Butler-Browne, 1990), adult mammalian skeletal muscles contain no embryonic or perinatal myosins and all fast fibers appear to have a similar program of MyHC expression during their development (Mahdavi et al., 1987).

While the contribution of fast and slow MyHC isoforms to the contractile properties of adult muscles is well established (Barany, 1967; Reiser et al., 1985), the functional significance of MyHC switching during muscle development is unclear. In addition to its enzymatic properties, myosin is a structural protein and plays a role in myofibrillogenesis. In nematode muscles, different sarcomeric MyHCs are found in the center and periphery of the thick filament (Miller, Ortiz, Berliner, and Epstein,

1983). In vertebrate muscles, the distribution of MyHC isoforms also may not be uniform (Taylor and Bandman, 1989; Gauthier, 1990), but this does not appear to be an intrinsic property of the MyHC sequence (Taylor and Bandman, 1989). Because many fast isoforms appear to have similar enzymatic properties, it is possible that the multigene family encodes many functionally equivalent MyHCs, and switching occurs because different regulatory elements are required to ensure expression during specific developmental periods and in physiologically different muscle fiber types.

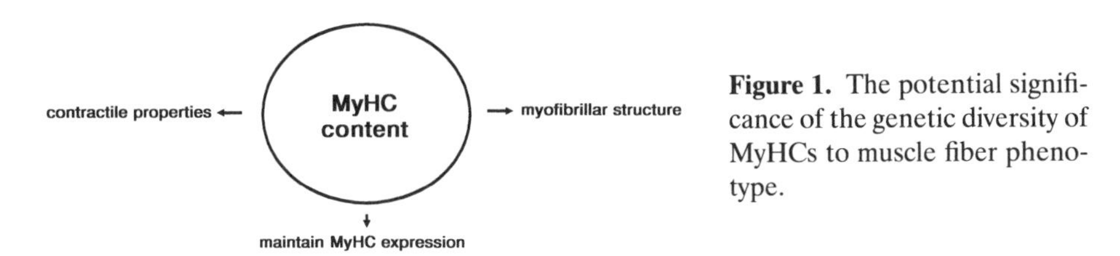

Figure 1. The potential significance of the genetic diversity of MyHCs to muscle fiber phenotype.

Complexity and Organization of Mammalian and Avian MyHC Multigene Families

It is likely that the multigene families of sarcomeric MyHCs have arisen from gene duplications and subsequent divergence of each of the members of the gene family. Two linkage groups (fast and slow/cardiac) comprise the mammalian sarcomeric MyHC multigene families. The human fast skeletal MyHCs are clustered on chromosome 11, while the mouse locus is found on chromosome 17 (Leinwand, Fournier, Nadal-Ginard, and Shows, 1983). The α and β cardiac MyHC genes are arranged in tandem in both human and rat genomes at a separate chromosomal locus from that of the fast skeletal MyHC genes (Mahdavi, Chambers, and Nadal-Ginard, 1984). Nine yeast artificial chromosomes containing human MyHC genes have been used to construct a contiguous 650-kb segment encompassing six fast MyHC genes (Yoon, Sellar, Kucherlapati, and Leinwand, 1992). Unlike the globin and homeobox gene clusters where the arrangement of the genes reflects their developmental pattern of expression (Efstratiadis, Posakony, Maniatis, Lawn, O'Connell, Spritz, DeRiel, Gorget, Weissman, Slightom et al., 1980; Hart, Fainsod, and Ruddler, 1987), the order of the fast MyHC genes on chromosome 17 does not correspond to the developmental pattern of expression of their expression.

The genomic organization of the chicken sarcomeric MyHC gene family differs from that of mammals. At least three linkage groups have been partially characterized by pulsed-field gel electrophoresis, a fast MyHC gene cluster, a slow MyHC gene cluster, and a cardiac gene cluster (Chen, Q., L. A. Moore, and E. Bandman, manuscript in preparation). At least five genes are found at the fast locus, two at the slow locus, and at least two genes encoding cardiac MyHCs represent a third MyHC cluster. The arrangement of the genes within each cluster has not been determined. It is known however that at the fast locus, the C_{emb1} MyHC gene is linked within 7.5 kb of the C_{neo} MyHC gene (Gulick, Kropp, and Robbins, 1987).

The precise number of sarcomeric MyHC genes in the chicken is unclear. It has been estimated that 31 MyHC-like genes are present in chicken genome (Robbins, Horan, Gulick, and Kropp, 1986). This is considerably greater than the estimated 7–10 genes observed in mouse, rat, or human genomes (Leinwand et al., 1983;

Wydro, Nguyen, Gubits, and Nadal-Ginard, 1983). However, based on northern blot analysis with 5′ probes only seven of the 31 potential MyHC genes appear to be related to fast MyHC genes (Kropp, Gulick, and Robbins, 1987). Our laboratory using 3′-UTR probes has shown that five fast MyHC genes (Moore, Tidyman, Arrizubieta, and Bandman, 1992) are expressed in developing chicken fast muscles, and two slow MyHC genes are expressed in developing slow muscles. These two MyHCs correspond to the SM1 and SM2 MyHC proteins that have been characterized by SDS-PAGE and peptide mapping (Hoh, 1979; Matsuda et al., 1983). The number of cardiac MyHC genes in the chicken is also unclear. At least two genes, one ventricular (Stewart, Camoretti-Mercado, Perlman, Gupta, Jakovcic, and Zak, 1991; Bisaha and Bader, 1991) and a second atrial (D. Bader, personal communication) have been cloned and sequenced. Because the ventricular MyHC gene is also transiently expressed in developing chicken muscles (Bourke, Wylie, Wick, and Bandman, 1991), it has been referred to as a primordial myosin (Sweeny, Kennedy,

CHICKEN SARCOMERIC MYOSIN HEAVY CHAINS

"FAST"	"SLOW"	"CARDIAC"
CEMB1	CSM1	CVMHC1
CEMB2	CSM2	CATR
CEMB3		
CNEO		
CAD		

MAMMALIAN SARCOMERIC MYOSIN HEAVY CHAINS

"FAST"	"SLOW/CARDIAC"
EMB	β
PERI	α
IIA	
IIB	
IID/IIX	
EXTRAOCULAR	
SUPERFAST/MASTICATORY	

Figure 2. A comparison of mammalian and chicken MyHC gene families. Chicken isoforms are grouped into fast, slow, and cardiac classes. Mammalian MyHCs have been grouped into two classes because the β-MyHC gene is expressed in both adult slow muscle fibers and adult cardiac muscle cells. GenBank accession numbers for cDNA sequences can be found in Moore et al. (1993). The mammalian superfast/masticatory MyHC has only been characterized at the protein level.

Zak, Kokjohn, and Kelly, 1989). A newly identified atrial cDNA (D. Bader, personal communication) expressed in very early heart development, may correspond to an atrial isoform that is transiently expressed during development of slow muscle fibers (Page, Miller, DiMario, Hager, and Stockdale, 1992). A chicken genomic DNA digest separated by pulsed-field electrophoresis and hybridized with a ventricular cDNA specific probe identified a single 200-kb fragment (L. A. Moore, unpublished observation). Although the sequence of the atrial cDNA is more analogous to the ventricular form than either the fast or slow MyHC genes, it is not known whether the atrial and ventricular genes are linked. Fig. 2 shows a comparison of mammalian and chicken MyHC gene families.

Gene Conversion Among MyHC Genes

Data has been presented for gene conversion-like events among MyHC genes (Moore et al., 1992, 1993). The most striking evidence comes from a comparison of the C_{emb1} and C_{neo} sequences within the LMM domain. A stretch of 309 nucleotides is identical in the two cDNA sequences and a second stretch of 564 nucleotides contains only a single silent substitution (Moore et al., 1992, 1993). While the five fast MyHCs of the chicken are quite similar (nucleotide homology ranges from 91–95%), on average there is one substitution per 15 nucleotides. These regions devoid of silent site substitutions suggest recent gene conversion events have occurred. Although codon bias may result in a low level of synonymous substitutions in G + C-rich sequences (Ticher and Graur, 1989), further evidence in support of gene conversion comes from an analysis of the genomic sequences and organization of the C_{emb1} and C_{neo} genes. The 564 nucleotide homologous coding sequence in these genes spans two introns which are also completely identical in both genes (Moore et al., 1992) and the C_{emb1} and C_{neo} genes are known to be linked within 7.5 kb in chicken genome (Gulick et al., 1987). Thus, the proximity of the genes coupled with the selective pressure to maintain the α-helical coiled-coil structure of the myosin rod would provide favorable conditions for the mechanism by which gene conversions are believed to occur. Mammalian α- and β-MyHC genes, but not fast MyHC genes, also exhibit similar stretches of nucleotide identity suggesting recent gene conversion-like events have occurred (Moore et al., 1993). Like the C_{emb1} and C_{neo} genes, α- and β-cardiac MyHC genes are also closely linked in the mammalian genome. However, while proximity may help promote gene conversions, it is likely that the degree of sequence similarity is also important. While the overall homology of the human α- and β-MyHC pair is similar to that of the C_{emb1} and C_{neo} genes pair, the homology of the human embryonic and neonatal cDNA sequences is significantly less, and these two genes show no evidence of recent gene conversion (Moore et al., 1993).

The Significance of Conserved and Diverging MyHC Domains in the Myosin Rod

Analysis of the protein sequences of MyHCs has identified domains that are highly conserved and other regions which appear to be less conserved. All sarcomeric MyHC rod sequences show highly conserved seven residue and 28 residue repeating periodicities (McLachlin, 1983). All vertebrate sarcomeric MyHC rods contain 39.5 28 residue repeats. Four aa, referred to as skip residues, are added to repeats 13, 20, 27, and 35 (McLachlin, 1983). While the position of the skip residues has been conserved in all sarcomeric MyHCs, removal of these residues by site-directed mutagenesis has no clear effect on myosin rod properties (Atkinson and Stewart, 1991).

Fig. 3 is a diagram of the myosin rod indicating sequences which have been conserved in chicken MyHCs. These regions include all of repeats 20, 34, and 38. Repeats 20 and 34 show a high degree of sequence conservation among all sequenced vertebrate MyHCs (Stedman et al., 1990). Their position within the light meromyosin domain suggests these residues may play a key role in myosin filament formation. The diagram also indicates the position of divergent sequences (two consecutive heptads with four or more aa substitutions). The two heptads at the COOH-terminus in repeat 40 show a great deal of diversity (seven substitutions), as do the two heptads at the light meromyosin—hinge junction in repeats 16 and 17 (seven substitutions). Similar diversity in these two regions is found in mammalian MyHCs. While it is likely that highly conserved sequences encode domains that are crucial to myosin functionality, the importance and role of sequences that are less conserved is less clear because divergent sequences can indicate specialization of isoform specific functions, or simply that the sequence is not important for functionality. Since different chicken fast MyHC rods cannot form heterodimers (Lowey, Waller, and Bandman, 1991; Kerwin and Bandman, 1991) some of the aa substitutions likely destabilize the α-helical coiled-coil comprised of different isoforms. However, because sarcomeric MyHCs in mammals can be heterodimeric (Dechesne, Bouvagnet, Walzthony, and Leger, 1987), the importance of limiting chicken MyHCs to forming homodimeric molecules is unclear.

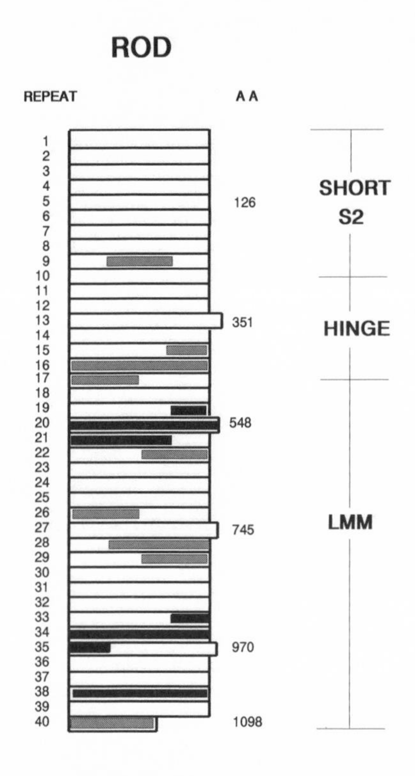

Figure 3. The position of highly conserved and divergent sequences in chicken MyHC rods. The figure shows a schematic diagram to illustrate the 39.5 28 aa repeats that comprise vertebrate sarcomeric MyHC rods. Repeats 13, 20, 27, and 34 contain 29 aa because of the presence of the skip residue (McLachlin, 1983). The repeats that make up the short S2, the hinge, and the light meromyosin domain are also indicated. The solid bars are aa sequences that are identical in all chicken fast MyHCs that have been sequenced, whereas the striped bars represent sequences that exhibit two or more consecutive heptads with four or more aa differences between the five fast MyHCs.

The Evolutionary Relationship of Chicken and Mammalian MyHC Genes

Analysis of the five fast MyHCs in the chicken indicate that the chicken isoforms are more homologous to one another than are the mammalian embryonic, neonatal, and adult fast MyHCs (Moore, Arrizubieta, Tidyman, Herman, and Bandman, 1992). These observations were the basis for subsequent studies that suggest that the chicken fast MyHC family differs from the mammalian fast MyHC family and that at least some of the chicken genes are the result of gene duplications that have occurred subsequent to separation of birds and mammals from their common ancestor (Moore et al., 1993). The method of Perler, Efstratiadis, Lomedico, Gilbert, Kolodner, and Dodgson (1980) was used to quantitate the divergence of the five fast chicken MyHCs and many mammalian MyHC sequences that can be found in GenBank. Although the results are complicated by the evidence of gene conversion-like events, the embryonic, neonatal, and adult fast MyHCs in the chicken have apparently evolved independently of the mammalian embryonic, neonatal, and adult isoforms (Moore et al., 1993). Fig. 4 shows a similar evolutionary analysis using the multisequence alignment program PILEUP (Devereaux, 1989). It appears that the chicken MyHC gene family expanded after the mammalian-avian split approximately 270 MY ago (Shapiro, 1991) and that the most likely ancestral gene(s) is the mammalian proto-adult or proto-neonatal/adult genes (Moore et al., 1993). If this hypothesis is correct, then it is likely that the developmental regulation of chicken and mammalian MyHC gene families has evolved independently. This would also

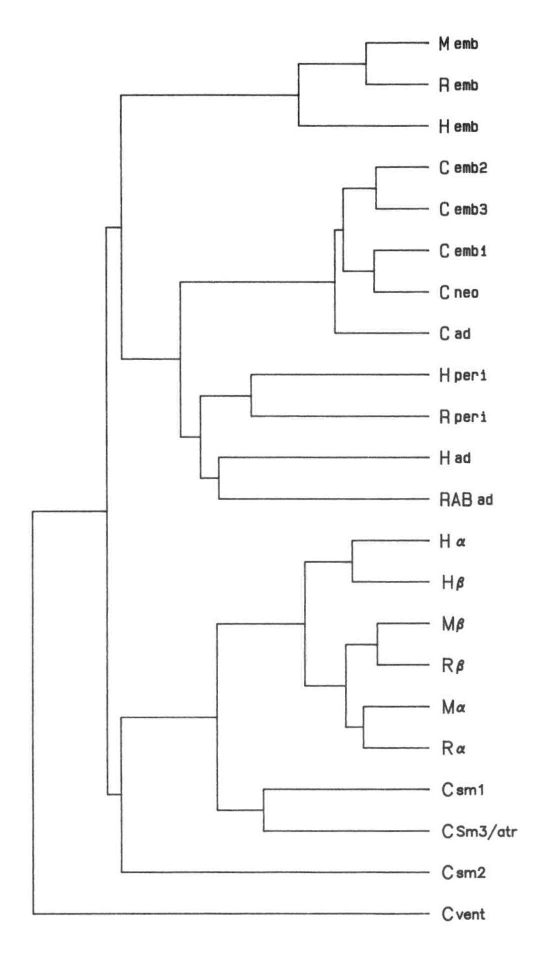

Figure 4. A dendrogram to illustrate the relationship of chicken and mammalian MyHC sequences. The multisequence alignment program PILEUP was used to analyze vertebrate MyHC sequences contained in GenBank (Moore et al., 1993). C_{sm1} and C_{sm2} cDNA sequences were kindly provided by D. Essig (unpublished observations) and C_{atr} was provided by F. Stockdale (unpublished observations). Only isoforms in which a majority of the LMM sequence has been determined were included in the analysis. The following abbreviations are used: *C,* chicken; *H,* human; *M,* mouse; *R,* rat; *R*ab, rabbit; *emb,* embryonic; *peri,* perinatal; *neo,* neonatal; *ad,* adult; *vent,* ventricular; *atr,* atrial; *sm,* slow myosin.

explain the different patterns of MyHC gene expression seen in birds and mammals described above.

If our hypothesis that the chicken MyHC multigene family has undergone recent expansion is correct, this would explain the greater number of gene conversions seen in chicken fast MyHC genes and the lack of distinct functional properties among chicken fast MyHCs. Why has the chicken MyHC multigene family recently expanded while mammalian MyHC gene families, such as rodent and primate, have been fixed in number? It has been suggested that when functional diversity is needed, concerted evolution results in gene families with variable members, whereas homogeneously redundant genes under the control of different promoter and/or enhancer elements have been acquired when large amounts of uniform gene product is required (Ohta, 1989). If positive natural selection is driving the expansion of the chicken MyHC family, diversity of contractile function in muscle fibers of birds may be less dependent on myosin isoform composition than in mammals. Accordingly, classification of MyHCs by their pattern of expression in adult muscle fiber types is more appropriate for mammalian MyHC gene families (Bandman, Moore, Arrizubieta, and Tidyman, 1992).

References

Atkinson, S. J., and M. Stewart. 1992. Molecular interactions in myosin assembly. Role of the 28-residue charge repeat in the rod. *Journal of Molecular Biology.* 226:7–13.

Bandman, E. 1985. Myosin isozyme transitions during muscle development, maturation, and disease. *International Review of Cytology.* 97:97–131.

Bandman, E., L. A. Moore, M. J. Arrizubieta, and W. E. Tidyman. 1992. Chicken fast myosin heavy chains. *In* Gene Expression in Neuromuscular Development. A. M. Kelly and H. Blau, editor. Raven Press, New York. 173–181.

Bandman, E., R. Matsuda, and R. C. Strohman. 1982. Developmental appearance of myosin heavy and light chain isoforms in vivo and in vitro in chicken skeletal muscle. *Developmental Biology.* 93:508–518.

Bandman, E., and T. Bennett. 1988. Diversity of fast myosin heavy chain expression during development of gastrocnemius, bicep brachii, and posterior latissimus dorsi muscles in normal and dystrophic chickens. *Developmental Biology.* 130:220–231.

Barany, M. 1967. ATPase activity of myosin correlated with speed of muscle shortening. *Journal of General Physiology.* 50:197–218.

Bisaha, J. G., and D. Bader. 1991. Identification and characterization of a ventricular-specific avian myosin heavy chain, VMHC1: expression in differentiating cardiac and skeletal muscle. *Developmental Biology.* 148:355–364.

Bourke, D. L., S. R. Wylie, M. Wick, and E. Bandman. 1991. Differentiating skeletal muscle cells initially express a ventricular myosin heavy chain. *Basic and Applied Myology.* 1:16–26.

Crow, M. T., and F. E. Stockdale. 1986. The developmental program of fast myosin heavy chain expression in avian skeletal muscle fibers. *Developmental Biology.* 118:333–342.

Dechesne, S. A., P. Bouvagnet, D. Walzthony, and L. L. Leger. 1987. Visualization of cardiac ventricular myosin heavy chain homodimers and heterodimers by monoclonal antibody epitope mapping. *Journal of Cell Biology.* 105:3031–3037.

Devereaux, J. 1989. Sequence analysis software package of the genetics computer group, version 7.0. University of Wisconsin Biotechnology Center, Madison, WI.

Efstratiadis, A., J. W. Posakony, T. Maniatis, R. M. Lawn, C. O'Connell, R. A. Spritz, J. K. DeRiel, B. G. Gorget, S. M. Weissman, J. L. Slightom, A. E. Blechl, O. Smithies, F. E. Baralle, C. C. Shoulders, and N. J. Proudfoot. 1980. The structure and evolution of the human beta-globin gene family. *Cell.* 21:653–668.

Gauthier, G. F. 1990. Differential distribution of myosin isoforms among the myofibrils of individual developing muscle fibers. *Journal of Cell Biology.* 119:693–701.

Gulick, J., K. Kropp, and J. Robbins. 1987. The developmentally regulated expression of two linked myosin heavy-chain genes. *European Journal of Biochemistry.* 169:79–84.

Hamada, Y., M. Yanagisawa, Y. Katsuragawa, J. R. Doleman, S. Nagata, G. Matsuda, and T. Masaki. 1990. Distinct vascular and intestinal smooth muscle myosin heavy chain mRNAs are encoded by a single-copy gene in the chicken. *Biochemical Biophysical Research Communications.* 170:53–58.

Hart, C. P., A. Fainsod, and F. H. Ruddle. 1987. Sequence analysis of the murine Hox-2.2, -2.3, and -2.4 homeoboxes: evolutionary and structural comparisons. *Genomics.* 1:182–195.

Hoh, J. F. Y. 1979. Developmental changes in chicken skeletal muscle isozymes. *FEBS Letters.* 98:267–270.

Kerwin, B., and E. Bandman. 1991. Assembly of avian skeletal muscle myosins: Evidence that homodimers of the heavy chain subunit are the thermodynamically stable form. *Journal of Cell Biology.* 113:311–320.

Kropp, K., J. Gulick, and J. Robbins. 1987. Structural and transcriptional analysis of a chicken myosin heavy chain gene subset. *Journal of Biological Chemistry.* 262:16536–16545.

Leinwand, L. A., L. Saez, E. McNally, and B. Nadal-Ginard. 1983. Isolation and characterization of human myosin heavy chain genes. *Proceedings of the National Academy of Sciences, USA.* 80:3716–3720.

Leinwand, L. A., R. E. Fournier, B. Nadal-Ginard, and T. B. Shows. 1983. Multigene family for sarcomeric myosin heavy chain in mouse and human DNA: localization on a single chromosome. *Science.* 221:766–799.

Lowey, S., G. S. Waller, and E. Bandman. 1991. Neonatal and adult myosin heavy chains form homodimers during avian skeletal muscle development. *Journal of Cell Biology.* 113:303–310.

Mahdavi, V., A. P. Chambers, and B. Nadal-Ginard. 1984. Cardiac α- and β-myosin heavy chain genes are organized in tandem. *Proceedings of the National Academy of Sciences, USA.* 81:2626–2630.

Mahdavi, V., S. Izumo, and B. Nadal-Ginard. 1987. Developmental and hormonal regulation of sarcomeric myosin heavy chain gene family. *Circulation Research.* 60:804–814.

Matsuda, R., E. Bandman, and R. C. Strohman. 1983. The two myosin isozymes of chicken anterior latissimus dorsi muscle contain different myosin heavy chains encoded by separate mRNAs. *Differentiation.* 23:36–42.

McLachlan, A. D. 1983. Analysis of gene duplication repeats in the myosin rod. *Journal of Molecular Biology.* 169:15–30.

Miller, III, D. M., I. Ortiz, G. C. Berliner, and H. F. Epstein. 1983. Differential localization of two myosins within nematode thick filaments. *Cell.* 34:477–490.

Molina, M. I., K. E. Kropp, J. Gulick, and J. Robbins. 1987. The sequence of an embryonic myosin heavy chain gene and isolation of its corresponding cDNA. *Journal of Biological Chemistry.* 262:6478–6488.

Moore, L. A., W. E. Tidyman, M. J. Arrizubieta, and E. Bandman. 1992. Gene conversions within the skeletal myosin multigene family. *Journal of Molecular Biology.* 223:383–387.

Moore, L. A., W. E. Tidyman, M. J. Arrizubieta, and E. Bandman. 1993. The evolutionary relationship of avian and mammalian myosin heavy-chain genes. *Journal of Molecular Evolution.* 36:21–30.

Moore, L. A., M. J. Arrizubieta, W. E. Tidyman, L. A. Herman, and E. Bandman. 1992. Analysis of the chicken fast myosin heavy chain family: localization of isoform-specific antibody epitopes and regions of divergence. *Journal of Molecular Biology.* 223:383–387.

Ohta, T. 1989. Role of gene duplication in evolution. *Genome.* 31:304–310.

Page, S., J. B. Miller, J. DiMario, E. J. Hager, and F. E. Stockdale. 1992. Developmentally regulated expression of 3 slow myosin isoforms of myosin heavy chain: diversity among the 1st fibers to form in avian muscle. *Developmental Biology.* 154:118–128.

Perler, F., A. Efstratiadis, P. Lomedico, W. Gilbert, R. Kolodner, and J. Dodgson. 1980. The evolution of genes: the chicken preproinsulin gene. *Cell.* 20:555–566.

Reiser, P. J., R. L. Moss, G. C. Giulian, and M. L. Greaser. 1985. Shortening velocity in single fibers from adult rabbit soleus muscles is correlated with myosin heavy chain composition. *Journal of Biological Chemistry.* 260:9077–9080.

Robbins, J., T. Horan, J. Gulick, and K. Kropp. 1986. The chicken myosin heavy chain family. *Journal of Biological Chemistry.* 261:6606–6612.

Rozek, C. E., and N. Davidson. 1983. *Drosophila* has one myosin heavy chain gene with three developmentally regulated transcripts. *Cell.* 32:23–34.

Shapiro, S. 1991. Uniformity in the nonsynonymous substitution rates of embryonic β-globin genes of several vertebrate species. *Journal of Molecular Evolution.* 32:122–127.

Soussi-Yanicostas, N., J. P. Barbet, C. Laurent-Winter, P. Barton, and G. S. Butler-Browne. 1990. Transition of myosin isozymes during development of human masseter muscle. Persistance of developmental isoforms during postnatal stage. *Development.* 108:239–249.

Stedman, H. H., M. Eller, E. H. Julian, H. Fertels, and S. Sarkar. 1990. The human embryonic myosin heavy chain. *Journal of Biological Chemistry.* 265:3568–3576.

Stewart, A. F. R., B. Camoretti-Mercado, D. Perlman, M. Gupta, S. Jakovcic, and R. Zak. 1991. Structural and phylogenetic analysis of the chicken ventricular myosin heavy chain rod. *Journal of Molecular Evolution.* 33:357–366.

Strehler, E. E., M. A. Strehler-Page, J. C. Perriard, M. Periasamy, and B. Nadal-Ginard. 1986. Complete nucleotide and encoded amino acid sequence of a mammalian myosin heavy chain gene. *Journal of Molecular Biology.* 190:291–317.

Sweeney, L. J., J. M. Kennedy, R. Zak, K. Kokjohn, and S. W. Kelly. 1989. Evidence for expression of a common myosin heavy chain phenotype in future fast and slow skeletal muscle during initial stages of embryogenesis. *Developmental Biology.* 133:361–374.

Taylor, L. D., and E. Bandman. 1989. Distribution of fast myosin heavy chain isoforms in thick filaments of developing chicken pectoral muscle. *Journal of Cell Biology.* 108:533–542.

Ticher, A., and D. Graur. 1989. Nucleic acid composition, codon usage, and the rate of sequences. *Journal of Molecular Biology.* 210:665–671.

Whalen, R. G., S. M. Sell, G. S. Butler-Browne, K. Schwartz, P. Bouveret, and I. Pinset-Harstrom. 1981. Three myosin heavy chain isoenzymes appear sequentially in rat muscle development. *Nature.* 292:805–809.

Wieczorek, D. F., M. Periasamy, G. S. Butler-Browne, R. G. Whalen, and B. Nadal-Ginard. 1985. Co-expression of multiple myosin heavy chain genes, in addition to a tissue-specific one, in extraocular muscle. *Journal of Cell Biology.* 101:618–629.

Winkelmann, D. A., S. A. Lowey, and J. L. Press. 1983. Monoclonal antibodies localize changes in myosin heavy chain isozymes during avian myogenesis. *Cell.* 34:295–306.

Wydro, R. M., H. T. Nguyen, R. M. Gubits, and B. Nadal-Ginard. 1983. Characterization of sarcomeric myosin heavy chain genes. *Journal of Biological Chemistry.* 258:670–678.

Yoon, S. J., S. H. Sellar, R. Kucherlapati, and L. A. Leinwand. 1992. Organization of the human skeletal myosin heavy chain gene cluster. *Proceedings of the National Academy of Sciences, USA.* 89:12078–12082.

Molecular Evolution of the Myosin Superfamily: Application of Phylogenetic Techniques to Cell Biological Questions

Holly V. Goodson

Department of Biochemistry, Beckman Center for Molecular and Genetic Medicine, Stanford University Medical School, Stanford, California 94305

We have used distance matrix and maximum parsimony methods to study the evolutionary relationships between members of the myosin superfamily of molecular motors. Amino acid sequences of the conserved core of the motor region were used in the analysis. Our results show that myosins can be divided into at least three main classes, with two types of unconventional myosin being no more related to each other than they are to conventional myosin. Myosins have traditionally been classified as conventional or unconventional, with many of the unconventional myosin proteins thought to be distributed in a narrow range of organisms. We find that members of all three of these main classes are likely to be present in most (or all) eukaryotes. Three proteins do not cluster within the three main groups and may each represent additional classes. The structure of the trees suggests that these ungrouped proteins and some of the subclasses of the main classes are also likely to be widely distributed, implying that most eukaryotic cells contain many different myosin proteins. The groupings derived from phylogenetic analysis of myosin head sequences agree strongly with those based on tail structure, developmental expression, and (where available) enzymology, suggesting that specific head sequences have been tightly coupled to specific tail sequences throughout evolution. Analysis of the relationships within each class has interesting implications. For example, smooth muscle myosin and striated muscle myosin seem to have independently evolved from nonmuscle myosin. Furthermore, brush border myosin I, a type of protein initially thought to be specific to specialized metazoan tissues, probably has relatives that are much more broadly distributed.

Introduction

Myosin was originally identified as the molecular motor in muscle cells that converts chemical energy of ATP into force. It has been found in every eukaryotic organism thus far examined. It powers not only the muscles of metazoan organisms, but also generates the force for the contractile ring of dividing cells (Mabuchi and Okuno, 1977; De Lozanne and Spudich, 1987; for review see Emerson and Bernstein, 1987; Warrick and Spudich, 1987). This muscle and contractile ring protein, now called "conventional" myosin or myosin II, has a motor domain that is connected to a filament-forming, coiled-coil tail (Fig. 1). Metazoan organisms contain many different isoforms (paralogues) of conventional myosin that are specialized for acting in different muscle and nonmuscle tissues. The evolutionary relationships between these proteins have been the subject of a number of studies (Stedman, Eller, Jullian, Fertels, Sarkar, Sylvester, Kelly, and Rubinstein, 1990; Stewart, Camoretti, Perlman, Gupta, Jakovcic, and Zak, 1991; Moore, Tidyman, Arrizubieta, and Bandman, 1992, 1993). Within the last five years, researchers have identified a whole series of proteins that have motor domains or "heads" clearly related to those of conventional myosins but have very different tail regions. These proteins are called "unconventional" myosins (see Fig. 1) (for review see Pollard, Doberstein, and Zot, 1991 and Mooseker, 1993). Here, we will define an unconventional myosin as any protein with a recognizable myosin motor domain and a non–filament-forming tail. Unconventional myosins have sometimes been referred to as "myosin I," although this name is now reserved for a particular type of unconventional myosin (see below).

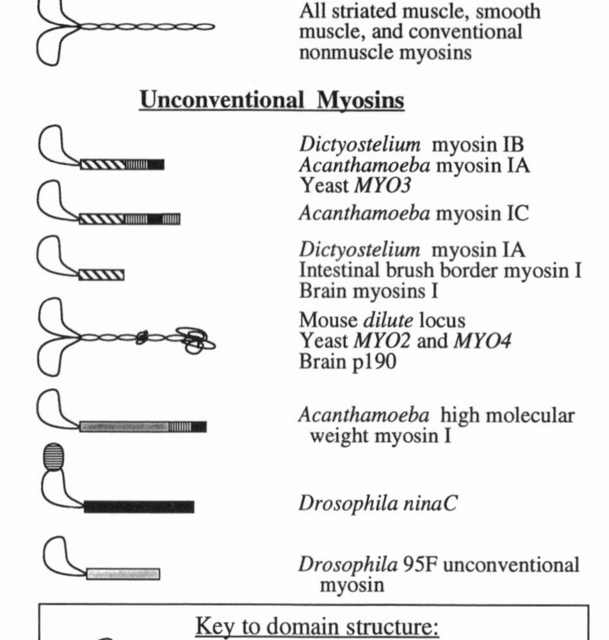

Figure 1. Myosin motor superfamily: Classification by tail domains. Adapted from Goodson and Spudich (1993). Lengths are not to scale.

The existence of these many very different myosin proteins suggests four major questions:

(*a*) How many types of myosin are there? Are unconventional myosins more related to each other than to myosin II, or are there several completely different types of myosin?

(*b*) How may types of myosin is a given cell or organism likely to have? Are most of these myosins organism-specific proteins with highly specialized functions, or are they ubiquitous proteins that are likely to play an important role in basic cell biology?

(*c*) Have myosin heads and tails been exchanged over evolutionary time frames? A common model of myosin function suggests that myosin heads are interchangeable units, with the specific function of the protein being determined completely by the tail. One way to test this idea is to see whether myosin heads and tails have been exchanged across evolution.

(*d*) Which myosin was first? Although this question cannot be answered in any definitive way, hypotheses of which type(s) of myosin existed in the most primitive eukaryotic cells could be useful for considering which actin-based activities are the most fundamental and what cytoskeletal functions could be performed by these early cells.

We have attempted to address these questions by molecular phylogenetic analysis of myosin motor domain amino acid sequences. Much of this work was presented in an earlier paper (Goodson and Spudich, 1993); the reader is referred to this paper for additional discussion and information on the methods. Our discussion here will be divided into two sections: first, a review of the conclusions drawn from study of the family as a whole, and second, a discussion of the relationships within subfamilies. Included in this section is analysis of new sequences not included in our previous work.

Methods

Two tree-building methods that are based on different principles and make somewhat different assumptions were used for this analysis: a distance matrix method (the "Tree" package of programs by Feng and Doolittle [1990]) and a maximum parsimony method (the PROTPARS program from the PHYLIP package by J. Felsenstein [1989]). Bootstrap resampling (100 trials) was used to judge the robustness of the nodes in Fig. 3 (the PROTPARS tree). An estimation of the error in branch length and position in Fig. 2 (the "Tree" tree) can be obtained from the percent standard deviation, as found in Fig. 2. However, since bootstrapping is a better test of the strength of the data supporting a particular relationship, we have used the PROTPARS bootstrap values as our main indicator of the robustness of a given node. Though use of other methods (e.g., maximum likelihood) or other versions of the methods used here (i.e., bootstrapped distance matrix) might allow a more sophisticated analysis of our data set and better resolution of some relationships that are indeterminate in our trees, the combination of bootstrapped maximum parsimony and distance matrix allows determination of the basic topology of the tree and identification of robust relationships.

Distance matrix methods can artifactually shorten long branches when no correction is made for multiple substitutions (Felsenstein, 1988). The "Tree" programs do not explicitly correct for multiple substitutions (Feng and Doolittle, 1990), but since we are primarily interested in branching order and are not trying to correlate branch length with time, artifacts of this type are not expected to affect the main body of results. However, we have based a number of inferences on the apparent consistency of the molecular clock within and between most myosin proteins as judged from the branch lengths of the distance matrix tree. Artificial suppression of large branch lengths could suggest that the molecular clock has been more consistent than it actually has. We have tested the consequences of failing to explicitly correct for multiple substitutions by analyzing the data set with the ClustalV tree-building program (Higgins, Bleasby, and Sharp, 1992) and including Kimura's correction for multiple substitution (Kimura, 1983). The relative branch lengths in the ClustalV tree are very similar to those shown in Fig. 2, and the "Tree" and ClustalV tree topologies differ only at nodes not well supported by bootstrapping (data not shown). "Tree" distances are calculated as a logarithmic function of the similarity score (Feng and Doolittle, 1990), while ClustalV branch lengths (after Kimura's correction) are calculated as a logarithmic function of the percent identity

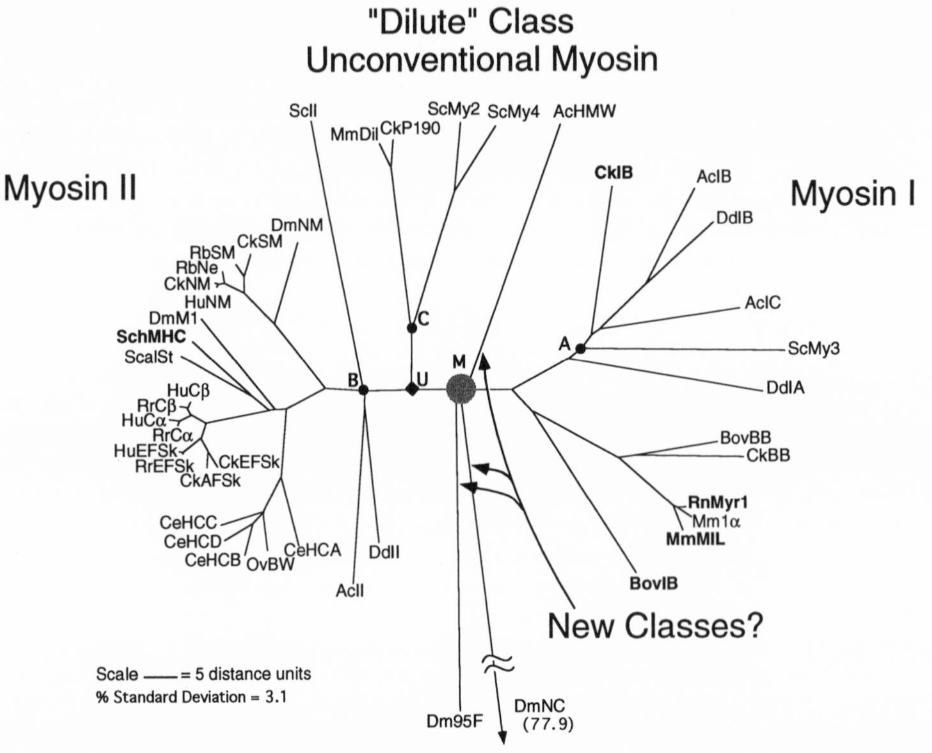

Figure 2

(Kimura, 1983; Higgins et al., 1992). Thus, errors in the "Tree" branch lengths caused by multiple substitutions would be expected to be small, at least as compared to percent identity–based distances corrected by Kimura's method. We have chosen to present the results from the "Tree" program as our distance matrix method because "Tree" distances are based on similarity and are thus expected to give a more sensitive measure of branch length than ClustalV's percent identity measurements.

The alignment input to the "Tree" program was created by the progressive alignment functions of the "Tree" program package (Feng and Doolittle, 1990); the alignment used by PROTPARS was created by ClustalV (Higgins et al., 1992).

Alignments are available upon request. Only motor domain sequences were used so that relationships between proteins with unrelated tail structures could be examined. The alignments were truncated to begin and end at universally conserved residues. Amino acid sequences were used to avoid artifacts caused by codon bias and include sequences determined by protein sequencing. The PROTPARS tree is a consensus tree constructed from evaluation of 100 bootstrap resamplings of the entire data

---

Figure 2. Phylogenetic tree obtained from distance matrix analysis of myosin motor domain amino acid sequences. Branch lengths are drawn to scale indicated in units of distance as calculated in the distance matrix. The percent standard deviation (*lower left corner*) gives an estimate of the error in branch length and position. This tree is drawn unrooted (without definition of the position of the "trunk"). However, biological considerations suggest that the root attaches somewhere internal to points *A–C* (justification of this suggestion and significance of points *A–C* and *U* are as explained in the text). Area M marks the approximate midpoint of the tree and represents the hypothesized point of the root connection under the assumption of a consistent molecular clock. Sequences in bold are those not included in the earlier analysis (Goodson and Spudich, 1993). Note: the branch lengths of MuDil and CkP190 were in error in our earlier analysis and are corrected here. Published protein sequences were obtained from the Protein Identification Resource data bank (PIR), version 32, or the University of Geneva protein sequence data bank (Swiss-Prot), version 22, unless otherwise noted. Accession numbers and abbreviations are as follows: *Acanthamoeba* myosin II (*AcII*): Swiss-Prot P05659; *Acanthamoeba* high molecular weight myosin I (*AcHMW*): PIR A23662; *Acanthamoeba* myosin IB (*AcIB*): Swiss-Prot P19706; *Acanthamoeba* myosin IC (*AcIC*): Swiss-Prot P10569; bovine brush border myosin I (*BovBB*): Swiss-Prot P10568; Bovine myosin IB (*BovIB*): GenBank Z22852; *Caenorhabditis elegans* myosin heavy chain A (*CeHCA*): Swiss-Prot P12844; *Caenorhabditis elegans* myosin heavy chain B (*CeHCB*): Swiss-Prot P02566; *Caenorhabditis elegans* myosin heavy chain C (*CeHCC*): Swiss-Prot P12845; *Caenorhabditis elegans* myosin heavy chain D (*CeHCD*): Swiss-Prot P02567; chicken brain P190 (*CkP190*): PIR S19188; chicken myosin IB (*CkIB*): EMBL X70400; chicken embryonic fast skeletal muscle myosin (*CkEFSk*): Swiss-Prot P02565; Chicken adult skeletal muscle myosin (*CkAFSk*): Swiss-Prot P13538; chicken gizzard smooth muscle myosin (*CkSm*): Swiss-Prot P10587; chicken nonmuscle myosin (*CkNM*): Swiss-Prot P14105; chicken brush border myosin I (*CkBB*): PIR A33620; *Dictyostelium* myosin II (*DdII*): PIR A26655; *Dictyostelium* myosin IB (*DdIB*): PIR A33284; *Dictyostelium* myosin IA (*DdIA*): Swiss-Prot P22467; *Drosophila nina*C (*DmNC*): PIR A29813; *Drosophila* muscle myosin (*DmM1*) was spliced by hand from sequence PIR A32491 and splice junction information for cDNA cD301 in George, Ober and Emerson (1989); *Drosophila* nonmuscle (*DmNM*): PIR A36014; *Drosophila* 95F unconventional myosin (*Dm95F*): EMBL X67077; human cardiac alpha isoform (*HuCα*) was entered from Matsuoka, Beisel, Furutani, Arai and Takao (1991); human cardiac beta isoform (*HuCβ*): Swiss-Prot P12883; human nonmuscle myosin (*HuNM*): GenBank M81105; human embryonic fast skeletal (*HuEFS*): Swiss-Prot P11055; mouse brain myosin I alpha (*MuIα*): GenBank L00923; mouse myosin IL (*MuIL*): GenBank X69987; mouse dilute locus (*Mudil*): PIR S13652; *Onchocerca volvulus* body wall myosin (*OvBW*): GenBank M74066; rabbit smooth muscle (*RbSM*): GenBank M77812; rabbit neuronal myosin (*RbNe*): EMBL X62659; rat myosin MYR1 (*RnMyr1*): GenBank (X68199); rat cardiac alpha (*RrCα*): PIR S06005; rat cardiac beta (*RrCβ*): Swiss-Prot P02564; rat embryonic fast skeletal (*RrEFS*): Swiss-Prot P12847; scallop striated (*ScalSt*) PIR S13557; *Schistosoma mansoni* myosin heavy chain (*SchMHC*): GenBank L01634; *Saccharomyces cerevisiae* MYO1 (*ScII*): PIR S12323; *S. cerevisiae* MYO2 (*ScMY2*): PIR A38454; *S. cerevisiae* MYO3 (*ScMY3*) unpublished results; *S. cerevisiae* MYO4 (*ScMY4*): GenBank M90057.

set, not by the splicing of overlapping consensus subtrees, as was done for the PROTPARS tree in Goodson and Spudich (1993). The position of the approximate midpoint (*M* in Fig. 2) was estimated by inspection of the tree without consideration of the DmNC branch. DmNC was not included because of the possibility that this branch is abnormally long as a result of an increase in the rate of the DmNC molecular clock. If DmNC is included in determination of the midpoint, the midpoint lies along the DmNC branch. Other methods, controls, and procedures are as in Goodson and Spudich (1993).

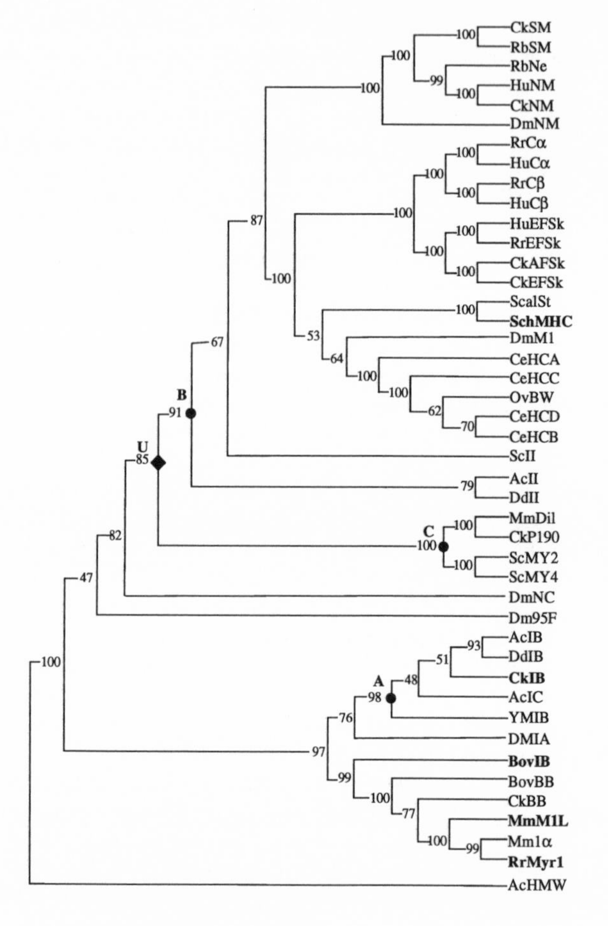

Figure 3. Phylogenetic tree obtained from maximum parsimony analysis of myosin head protein sequences. This tree has been arbitrarily rooted at AcHMW for the purposes of more easily viewing topology. Branch lengths have no meaning. Numbers beside nodes are the percentage of bootstrapping trials in which an identical node was produced (see Methods), and thus are a measure of the robustness of the data generating that particular node. Sequences in bold are those not included in our original analysis. The significance of nodes *A–C* and *U* is explained in the text. References and abbreviations are the same as in Fig. 1.

Results and Discussion

Figs. 2 and 3 show the trees obtained from the distance matrix and maximum parsimony analyses, respectively, of myosin motor domain sequences in the databases as of September, 1993. Nodes (*branch points*) on these trees can represent either protein divergence (*gene duplication*) or species divergence, so all interpretations of these trees must accommodate this ambiguity. Both trees should be regarded as unrooted (without definition of where the ancestral myosin attaches to the trees), although the maximum parsimony tree has been arbitrarily rooted at AcHMW for purposes of better viewing the topology (abbreviations are given in Fig. 2; see below for further discussion of rooting).

Examination of Figs. 2 and 3 leads to three immediate conclusions:

(*a*) Myosin motor domain amino acid sequence data is conducive to phylogenetic analysis and allows clear definition of many relationships. Both methods give very similar tree topologies and differ only at nodes that are not well supported by bootstrap analysis (nodes with low bootstrap percentages are not well supported by the data and should be regarded as ambiguous). Much of the tree topology is well supported (found in $>90\%$ of bootstrap trials). There is a high degree of symmetry about the approximate midpoint of the distance matrix tree (marked *M* in Fig. 2; this midpoint is a possible position for the attachment of the root or ancestral myosin as discussed below), and distances between analogous points are in most (though clearly not all) cases similar. The symmetry about the midpoint and similarities in branch lengths imply that the molecular clock has been fairly consistent within and between branches, which suggests that tree-building artifacts resulting from large changes in the rate of the molecular clock are not likely to be a problem in most cases. The sum of these observations suggests that the relationships strongly supported by these analyses are likely to reflect the actual paths of myosin evolution.

(*b*) There are at least three distinct classes of myosin: myosin II (conventional myosin), myosin I, and "dilute-type" unconventional myosin. There may be many more. The myosin family has traditionally been divided into conventional and unconventional myosins. Though all of the conventional myosins clearly cluster together in both trees, both examination of the maximum parsimony bootstrap values and measurement of distance matrix branch lengths show that the two major branches of unconventional myosin are not detectably more related to each other than they are to myosin II. AcHM, Dm95F, and DmNC (see the legend to Fig. 2 for sequence abbreviations) fail to group with these three major clusters and may each represent additional classes, although these sequences are the only published representatives of these potential classes as yet. In addition, a partial *Drosophila* sequence (unconventional myosin 35B,C) that also fails to group with the major clusters has recently been identified, suggesting that there may be as many as seven classes of myosin in the sequences now known (Kiehart, D., personal communication; data not shown).

(*c*) The groupings that result from phylogenetic analysis of myosin motor domain sequences agree with those based on tail structure and (where known) enzymology (Fig. 4). For example, mouse dilute and yeast MYO2, which have similar tail structures, both group together in these motor domain–based trees. On a much smaller scale, human and rat beta cardiac myosins, which have closely related tail sequences, enzymatic activity, and developmental expression patterns, group together in Figs. 2 and 3, apart from the other cardiac and skeletal myosin sequences. This agreement between nonmotor structural classification and the relationships derived from phylogenetic analysis of motor sequences provides external support for the relationships drawn by the phylogenetic analysis. It also supports the practice of hypothesizing activities and functions for uncharacterized myosins on the basis of sequence similarity. In addition, it suggests that phylogenetic analysis might be a useful way of categorizing or assigning relevant names to myosins. Clearly, the classification of myosin into "conventional" and "unconventional" groups is no longer sufficient. In our original classification (Goodson and Spudich, 1993), we tried to work within the old framework. However, Mooseker and co-workers proposed a new myosin nomenclature based on their own phylogenetic analysis of the myosin family (Cheney, R. E., M. A. Riley, and M. S. Mooseker, 1993). As this nomenclature has come into general use, it is presented in Fig. 4 and will be used through the rest of this discussion.

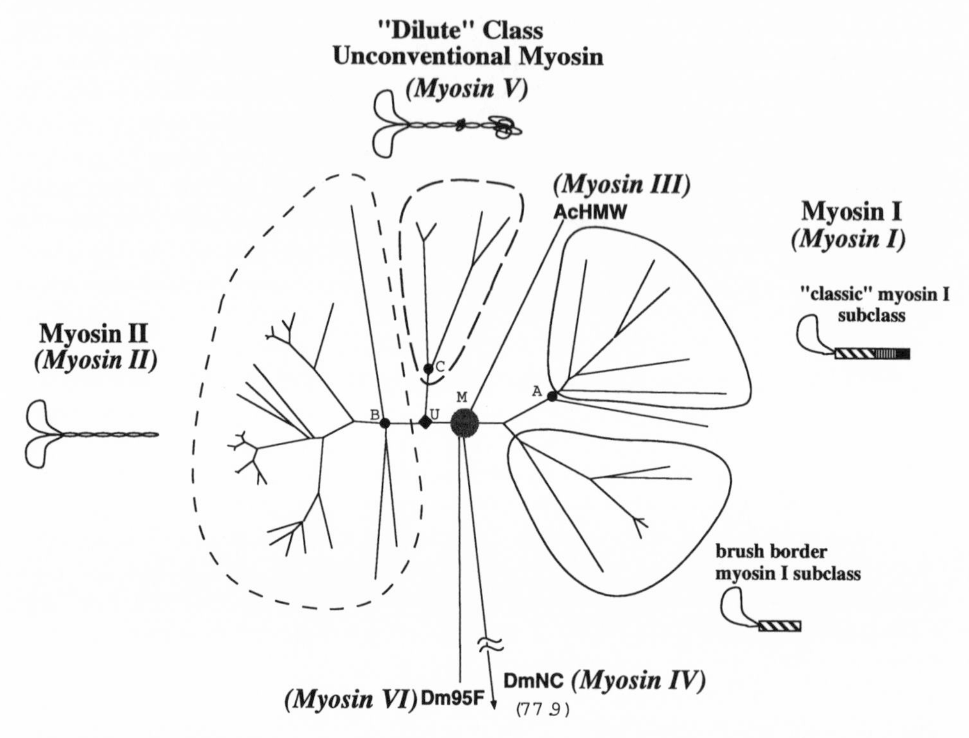

Figure 4. Agreement between groupings derived from phylogenetic analysis of motor domain sequences and those derived from tail structures: the distance matrix tree. Names in itallics give the new nomenclature proposed by Cheney (1993). Though clearly a myosin I by both tail and motor similarities, DdIA has not been grouped with either the classic or brush border subclasses because of the intermediate nature of its sequence. Points *A–C, U,* and M, as well as the abbreviations are as explained in Fig. 2; myosin structure key is as in Fig. 1.

How Many of These Classes of Myosin Are Likely To Be Present in Each Organism?

Examination of Fig. 2 and 3 show that all three of the major classes have representative proteins from widely divergent organisms. This distribution implies that proteins belonging to the three major classes existed in the common ancestor of these organisms. If these proteins were present in the ancestor of all (or most) eukaryotes and had functions fundamental enough to cell function that they have been retained in divergent modern organisms with divergent lifestyles, then it seems likely that members of these three classes still exist in most (or all) eukaryotic organisms.

It is more difficult to postulate the range of the ungrouped proteins (Dm95F, DmNC, and AcHM), as each of these potential classes is known from only one organism as yet. However, the structure of the trees suggests that proteins related to Dm95F and AcHM will indeed be found in many or most organisms. Nodes *A–C* in Figs. 2 and 3 mark the earliest possible point of the radiation of the eukaryotes on the three major branches. These nodes are where yeast, *Dictyostelium,* and/or metazoans separate on each of the major branches. The actual divergence of the eukaryotes could occur later, if these nodes actually mark gene duplication events, but it cannot happen earlier, as long as the root of the tree is not within one of the strongly supported clusters (placement of the root within these clusters seems unlikely, considering the strength of the bootstrap support for these clusters and the nonmotor structural similarities of proteins within these clusters as seen in Fig. 1). Examination of Figs. 2 and 3 shows that all three of these ungrouped proteins diverge from the rest of the myosins well before nodes A, B, or C. This topology implies that these proteins were also present in a very early eukaryote. If they were present in an early eukaryote and have been retained by at least one modern organism, it seems likely that they have been retained by other organisms as well. These arguments depend on the assumption that inconsistencies in the molecular clock are small enough that branching position for the these sequences as presented in Figs. 2 and 3 is not grossly incorrect. There is no information about the rate of the molecular clock for these branches beside the total branch lengths, and thus, these predictions must be treated with caution. However, the lab of M. Mooseker has recently reported pig homologues of both Dm95F and the partial *Drosophila* sequence mentioned above (Mooseker 1993), demonstrating that these two proteins will be found at least in many metazoan organisms.

While the structure of the trees may suggest that Dm95F and AcHM are ancient proteins, the unusually large length of the DmNC branch prevents the application of these arguments to this protein. The great length of the DmNC branch suggests either that the root of the myosin tree is along this branch or that the molecular clock for DmNC has been running at a much accelerated rate. As mentioned above, widely varying rates of evolution can confound both distance matrix and maximum parsimony methods (Felsenstein, 1988), leading to the possibility that DmNC evolved from a relatively recent gene duplication and was given an artifactually early divergence by both methods. Our earlier analysis had hinted that DmNC might be most closely related to Dm95F. This weak relationship has not held up to the addition of new sequences to the analysis. However, the long DmNC branch length does appear to be most likely caused by altered mutational constraints instead of particularly ancient divergence because of what appears to be the specific degradation of certain generally well-conserved myosin sequences in DmNC (including the replacement of G with Y in the ATP binding site).

Conclusions from Analysis of Relationships within Classes

Myosin I. Examination of Figs. 2 and 3 suggests that most organisms contain several distinct types of myosin I. The existence of classic myosin I proteins (those with membrane binding domains, Src homology 3 [SH3] domains, and proline-rich regions) in yeast, *Dictyostelium,* and *Acanthamoeba* at the time of the last analysis had suggested that these proteins would be found in most or all eukaryotic organisms (Goodson and Spudich, 1993). Consistent with this prediction, a chicken protein that has the classic myosin I tail structure and clusters with classic myosin I in phylogenetic analysis of motor domains, has now been identified (EMBL accession X70400;

Figs. 2 and 3). The gene duplication leading to the two *Acanthamoeba* proteins (AcIB and AcIC) appears to have occurred near the divergence of the eukaryotes as indicated in Fig. 2, point *A*. This topology suggests that AcIB and AcIC may represent two subclasses of classic myosin I that are present in many organisms. However, the poor bootstrap values in this part of the maximum parsimony tree show that the exact order of classic myosin I relationships cannot be determined by this analysis.

Brush border–related myosin I proteins (which lack the SH3 and proline-rich regions) were originally found in intestinal microvilli, and they are an example of myosin proteins that many people expected to be limited to metazoans. The brush border myosin I branch apparently splits off from the classic myosin I branch well before the divergence of the eukarytotes (Fig. 2; see also the high bootstrap values for these nodes in Fig. 3), suggesting that proteins related to brush border myosin I were present in ancient eukaryotic cells and are thus likely to be found in many modern eukaryotic organisms. Despite the predictions from the tree topology, no proteins that clearly group with the brush border proteins have yet been found in nonmetazoan organisms. However, brush border myosin I proteins have now been found in nonmicrovillar metazoan cell types, as shown in Figs. 2 and 3, consistent with the prediction that these proteins are involved in basic cell processes. *Dictyostelium* myosin IA has a similar tail structure, but it groups in most bootstrap trials with the classic myosin I proteins. One might interpret this topology to suggest that the original myosin I also lacked the SH3 domain, and that an early gene duplication led to the existence of a brush border–related myosin I branch and a DdIA branch, with the DdIA branch eventually giving rise to the classic myosin I proteins. However, other possibilities do exist (e.g., DdIA and brush border myosin I could have independently lost an ancestral SH3). In any case, the topology of the myosin I branch, if taken at face value, suggests that the common ancestor of most eukaryotes contained at least one classic myosin I protein, a brush border–like protein, and a DdIA-like protein.

Closer examination of the brush border branch suggests that vertebrates contain at least three subclasses (paralogues) of brush border myosin I: the intestinal brush border proteins (represented by bovine and chicken proteins), a subclass represented by the rat and mouse neuronal proteins RnMyr1 and Mu1α, and a subclass of proteins represented by the bovine BovIB protein. Comparison of branch lengths suggests that more than three paralogous proteins may be represented in these sequences. If one compares the mammal-chicken divergence seen on this branch with those seen on the other branches, one finds that the distances on the brush border myosin I branch are much larger. Either the molecular clock sped up for the brush border–related proteins, or the chicken and bovine brush border proteins are members of subclasses that have been separate longer than cows and chickens. In fact, the maximum parsimony method fails to cluster these proteins in $>75\%$ of bootstrap trials (Fig. 3; inclusion of additional myosin I sequences in the maximum parsimony analysis greatly changed the results, as BovBB and CkBB clustered in 99% of trials in our earlier analysis). In addition, the well-supported divergence of the two mouse proteins Mu1α and MuIL before the divergence of RnMyr1 and Mu1α suggests that an additional gene duplication occurred at least before the divergence of mice and rats. The large number of paralogous myosin I proteins shows that caution should be used when trying to compare two examples of "myosin I protein" purified from different organisms or tissues. Thus, from the sequences in this analysis alone, it appears likely that most eukaryotic organisms will contain at least two types of myosin I (a classic myosin I, a brush border–like myosin I, and possibly a DdIA-like protein), and vertebrates will have more than three types of the brush border myosin I subclass alone.

Myosin II. A great deal of work has been done on the evolutionary relationships of myosin II sequences. Most of this analysis has been based on tail sequence and has focused on relationships between developmentally specific vertebrate isoforms (Stedman et al., 1990; Stewart et al., 1991; Moore et al., 1992, 1993). Our results, based on motor domain sequence, are generally consistent with the results of these tail-based analyses. This agreement is interesting because it suggests that myosin heads and tails have not exchanged across evolutionary time frames. Myosin heads have been attached to new tails, but heads and tails, once connected, have not interchanged. In addition, all of the domain swapping leading to new head–tail combinations seems to have happened very early in evolution (before the metazoan radiation, as indicated by points *A–C*).

There was an expectation within the myosin field that at least some of the unconventional myosins arose recently by addition of new tails onto conventional myosin. This assumption was partly due to a common model of myosin function, which assumes that head domains of myosin proteins are essentially interchangeable units, with the specific functions of the protein being determined by the tail. The strong agreement between head and tail phylogeny (or head phylogeny and tail structure) is instead consistent with the possibility that myosin heads and tails are in some way functionally coupled—a given tail may need a motor with specific biophysical characteristics to function properly. The biophysical requirements for a motor pulling a vesicle along an actin filament may be very different from those of a motor acting in a muscle. One could argue that such swapping events might not have occurred at a high enough rate to have been observed, even if there was no selective advantage against them. However, swapping events do not seem to have occurred even between closely related isoforms, even though there are clear examples of concerted evolution in some regions of these proteins (Moore et al., 1992, 1993; for further discussion see Goodson and Spudich, 1993).

Figs. 2 and 3 can also contribute to the understanding of the relationships between smooth and striated muscle myosins. It is well known that vertebrate smooth muscle myosin is more similar to nonmuscle myosin than to striated muscle myosin, both in sequence and in biochemical characteristics. This fact suggests that smooth muscle is more "primitive" and led many to assume that smooth muscle is the older or "original" form of muscle and that striated muscle evolved from it. However, examination of Figs. 2 and 3 shows that this is not likely to be the case: the smooth muscle and striated muscle myosins branch independently from nonmuscle myosin. Thus, smooth muscle myosin cannot be the progenitor of striated muscle myosin, and, in fact, the two types of muscle myosin do not appear to be related, except through nonmuscle myosin. The independent origin of the muscle myosins suggests that the two types of muscle tissue may also be independently derived from nonmuscle tissue, and that any similarities between these tissues may be the result of convergent evolution.

A second surprising conclusion is that smooth muscle myosin seems to have arisen relatively recently, apparently after the divergence of lines leading to flies and vertebrates (Figs. 2 and 3 show that vertebrate smooth and nonmuscle myosins split after the divergence of fly and vertebrate nonmuscle myosins). This suggests that smooth muscle tissue (at least of the types found in vertebrates) may also have arisen recently, possibly after the protostome-deuterostome divergence. Although this may seem unlikely, morphological studies of *Drosophila* support this idea, showing that all muscles identified in this well-characterized organism are striated muscles, including the gut muscles (Crossley, 1978). It will be very interesting to follow this question as myosins from more organisms and smooth muscle tissue types are sequenced.

Which Myosin Was First?

This question cannot be definitively answered by this type of analysis, but several lines of reasoning do give a few hints. The first two hints are derived from the assumption that the molecular clock has been consistent across the main branches of the tree. Examination of Fig. 2 shows that this assumption is clearly not absolutely valid, but the general level of symmetry suggests that it may be reasonable. If one compares the distance in Fig. 2 from points *A–C* (the earliest possible points of the divergence of the eukaryotes on each of the three major branches, as explained above) to the terminal nodes (modern sequences), one sees that the distances are very similar, as one would expect under conditions of a consistent molecular clock. The point at which the three major classes separate is marked by the letter *U.* If one compares the distances *A-U, B-U,* and *C-U,* one sees that the distance of *A-U* is approximately twice the distance of *B-U* or *C-U.* This discrepancy suggests that myosin II and myosin V (the "dilute" class of unconventional myosin) are actually more closely related to each other than to myosin I. This relationship is especially intriguing in light of the apparent similarity in their tail structures (it should be pointed out that the coiled-coil tail sequences of myosin V proteins are not noticeably more related to those of myosin II proteins than to those of many nonmyosin coiled-coil proteins, although a sophisticated analysis has not been published).

The second molecular clock–dependent hint is derived from determination of the midpoint of the myosin tree. The midpoint is the hypothetical position of the connection of the root (ancestral myosin) to the tree under the assumption that the rate of sequence change has been consistent across all branches. If one includes DmNC in the calculation, the root is along the DmNC branch. If one removes it from consideration on the basis of the hypothesis that it has undergone an acceleration of its molecular clock, the midpoint is in the vicinity of point M in Fig. 2. Both of these hypothetical root points are surrounded by the divergence of several unconventional myosin proteins. Hypothesizing the position of the root cannot determine what the ancestral myosin looked like, but it does emphasize that it is quite possible that conventional myosin is a relative newcomer.

A third hint is provided by the identification of myosin light chains (thus far we have discussed only myosin heavy chains, but all known myosin heavy chains bind at least one light chain). The myosin II light chains are derived from calmodulin (Moncrief, Kretsinger, and Goodman, 1990; Collins, 1991), which suggests that at one time myosin may have had calmodulin light chains. The biochemically characterized members of the myosin V and brush border myosin I branches still use calmodulin as their light chains (reviewed in Mooseker, 1993), suggesting that the binding of conventional myosin light chains is a derived characteristic and that these unconventional myosins may somehow be closer to the ancestral state of myosin. It is interesting to point out that these two unconventional myosins straddle the hypothetical root connection (*M*) estimated from the midpoint of the tree.

A final hint involves consideration of the interactions of the nonmotor regions. The interactions involved in assembling myosin II filaments (initial assembly of a heavy chain dimer, followed by assembly of these dimers into filaments) are quite complicated. One might imagine that the first myosin proteins had more simple interactions, such as those seen (or expected from) the tails of some of the unconventional myosin proteins (for review of tail interactions see Pollard et al., 1991; Mooseker, 1993). It at least seems reasonable to postulate that dimerization (as seen in myosin V) was ancestral to filament formation (as seen in myosin II).

None of these hints suggest anything specific about the nature of the original myosin protein. However, they do emphasize that it is quite possible (even likely?) that conventional myosin, which is generally thought of as being the main actin-based motor in eukaryotic cells, was not a part of the earliest and most fundamental actin bases processes in ancient eukaryotic cells. One might then suggest that it may still not be an integral part of these processes. Genetic experiments suggest that myosin II is indeed dispensable for cell life and for many actin-based cellular functions in both *Dictyostelium* and yeast, including cell motility (De Lozanne and Spudich, 1987; Rodriguez and Patterson, 1990). In contrast, the MYO2 gene (a myosin V) is essential for yeast cell viability (Johnston, Prendergast, and Singer, 1991). However, it is clear that both yeast and *Dictyostelium* cells missing particular myosin I proteins have minor (if any) defects (Jung and Hammer, 1990; Titus, Wessels, Spudich and Soll, 1993; Goodson and Spudich, manuscript in preparation). This may suggest that myosin I proteins are not involved in essential processes, but *Dictyostelium* is known to have many members of the myosin I class (Titus, Warrick and Spudich, 1989; Urrutia, Jung, and Hammer, 1993). Because of possible functional overlap, it may be difficult to truly assess the functional importance of myosin I proteins.

Conclusions

There are several specific conclusions about the myosin family to be drawn from this work. First, phylogenetic analysis of myosin motor domains shows that the old classification of myosin into conventional and unconventional classes is insufficient. Instead, myosin proteins cluster into at least three groups: myosin I, myosin II, and myosin V (by the new nomenclature), with the two main types of unconventional myosin being as divergent from each other as from conventional myosin. Three myosin sequences do not cluster with each other or with the three main groups, suggesting that they may each represent additional classes of myosin. It seems likely that more classes of myosin are waiting to be identified. Secondly, it is likely that most eukaryotic organisms (including protists, plants, and fungi) contain at least members of these three main classes. Examination of tree structure also suggests that several of the subclasses of the main classes and the potential classes that are as yet represented by only one sequence will also be present in most eukaryotes, though these predictions are contingent on relative consistency of the molecular clock across the different branches. A third conclusion is that myosin heads and tails have not swapped in evolutionary time frames. This result is consistent with the possibility that myosin heads and tails are functionally coupled and questions the idea that heads are interchangeable units, with the unique function being determined by the tail. Finally, we see that myosin II, instead of being the progenitor of all myosins, may instead have appeared relatively recently. It is impossible to tell from this analysis what were the characteristics of the ancestral myosin protein. However, hints from the structure of the distance matrix tree, the identity of myosin light chains, and the interactions of the myosin tail domains may suggest that the unconventional myosins are likely to have more ancestral characteristics.

A number of more general conclusions about the broad utility of phylogenetic analysis of protein families can also be obtained from this work. First, one sees that such analyses can be very useful for cell biology. With phylogenetic analysis, the cell biologist can get an idea of the number of related proteins acting in a given cell, as well as how ubiquitous they are likely to be. The hypothesized range of a protein is useful both because it gives information about function (ubiquitous proteins are likely to be involved in more universal functions) and because it alerts people to look for proteins that are expected to be in a given organism but have not been found yet. Another useful application of phylogenetic analysis to cell biology is that it can allow one to determine whether two proteins from two organisms are orthologues or paralogues. A good example can be seen with the bovine and chicken brush border myosins I: these proteins have been thought to be orthologues, but the branch lengths suggest otherwise. More definitive predictions can be made when robust patterns of protein divergence do not match the expected pattern of organism divergences. This very important question is difficult to address by any other means.

This type of analysis is useful for evolutionary studies, but not in a way limited to the history of the protein family at hand: as shown with the smooth and striated muscle myosins, study of protein families can have implications for the origin of organs and tissues. Protein families can also be useful for the study of domain swapping and the mechanisms for the generation of diversity: here we see that the main diversity of the myosin family was apparently generated before the radiation of the organisms on the crown of the eukaryotic tree, although gene duplication and specialization has been occurring through recent times. Protein families should also provide a unique opportunity for studies of the factors affecting the rates of the molecular clock. We have not addressed this very interesting issue in any depth, but a comparative study of different branches of a protein family could be considered to be better controlled than a study of several unrelated proteins. Under what conditions do paralogues mutate or not mutate at similar rates? Do specific organisms appear to be consistently associated with altered clocks? Can abnormal branch lengths be used to identify proteins with unusual constraints or altered functions? DmNC would appear to be a good candidate for a protein with altered constraints. Yeast myosin II might be a less extreme example.

Yet another important use of phylogenetic analysis in a nonevolutionary context could be in the study of relationships between structure and function. First, the structure and branch lengths of the trees can help to suggest when it is and is not reasonable to infer biochemical activities of uncharacterized proteins from the activities of related, characterized ones. Secondly, identification of proteins and parts of proteins that have altered molecular clocks could be very useful for correlating specific structures with specific biochemical functions. One might expect that proteins with altered clocks would have many of their changes concentrated in regions that correspond to domains involved in lost or gained functions. In families of proteins where the general function of regions is already known, identification of regions that have undergone changes in constraint in particular family members could be used to suggest that these family members have undergone changes in the functions associated with these regions. Such an analysis has recently been done for the HSP70 members (Hughes, 1993). In the opposite way, information on biochemical differences between members of a protein family with unknown structure/function relationships could be coupled to the identification of domains that have undergone a change in constraints to help to assign functions to particular parts of these proteins.

The analysis presented here should be useful for people studying the cell biology and biochemistry of myosin proteins, but will also hopefully be a useful starting point for people more interested in studying evolutionary processes. As the database of protein sequences grows larger and the analysis of biological problems grows more sophisticated, many fields beyond evolution will gain by the application of molecular phylogenetic techniques to protein families. The results of these cross-disciplinary endeavors can only be enriched by greater communication between people studying the biological properties of proteins and those studying their evolution.

Acknowledgements

This work would not have been possible without the generous advice, support, and encouragement of James A. Spudich. Hans Warrick is gratefully acknowledged for helpful discussions and careful reading of the manuscript.

This work was supported by National Institutes of Health grant GM46551 to James A. Spudich.

References

Cheney, R. E., M. A. Riley, and M. S. Mooseker. 1993. Phylogenetic analysis of the myosin superfamily. *Cell Motility and the Cytoskeleton.* 24:215–223.

Collins, J. H. 1991. Myosin light chains and troponin C: structural and evolutionary relationships revealed by amino acid sequence comparisons. *Journal of Muscle Research and Cell Motility.* 12:3–25.

Crossley, C. 1978. The morphology and development of the *Drosophila* muscular system. *In* The Genetics and Biology of *Drosophila.* Academic Press Ltd., London. 499–550.

De Lozanne, A., and J. A. Spudich. 1987. Disruption of the *Dictyostelium* myosin heavy chain gene by homologous recombination. *Science.* 236:1086–1091.

Emerson, P., and S. I. Bernstein. 1987. Molecular genetics of myosin function. *Annual Review of Biochemistry.* 56:695–726.

Felsenstein, J. 1988. Phylogenies from molecular sequences: inference and reliability. *Annual Review of Genetics.* 22:521–565.

Felsenstein, J. 1989. PHYLIP—Phylogeny Inference Package (Version 3.2). *Cladistics.* 5:164–166.

Feng, D. F., and R. F. Doolittle. 1990. Progressive alignment and phylogenetic tree construction of protein sequences. *In* Molecular Evolution: Computer Analysis of Protein and Nucleic Acid Sequences. Academic Press, Inc., San Diego, CA. 375–387.

George, E. L., M. B. Ober, and C. P. Emerson. 1989. Functional domains of the *Drosophila melanogaster* muscle myosin heavy-chain gene are encoded by alternatively spliced exons. *Molecular and Cellular Biology.* 9:2957–2974.

Goodson, H. V., and J. A. Spudich. 1993. Molecular evolution of the myosin family: relationships derived from comparisons of amino acid sequences. *Proceedings of the National Acadamy of Sciences USA.* 90:659–663.

Higgins, D. G., A. J. Bleasby, and P. M. Sharp. 1992. CLUSTAL V: improved software for multiple sequence alignment. *Computer Applications in the Biosciences.* 8:189–191.

Hughes, A. L. 1993. Nonlinear relationships among evolutionar rates indentify regions of functional divergence in heat-shock 70 genes. *Molecular Biology and Evolution.* 10:243–255.

Johnston, G. C., J. A. Prendergast, and R. A. Singer. 1991. The *Saccharomyces cerevisiae* MYO2 gene encodes an essential myosin for vectorial transport of vesicles. *Journal of Cell Biology.* 113:539–551.

Jung, G., and J. Hammer III. 1990. Generation and characterization of *Dictyostelium* cells deficient in a myosin I heavy chain isoform. *Journal of Cell Biology.* 110:1955–1964.

Kimura, M. 1983. The Neutral Theory of Molecular Evolution. Cambridge University Press, Cambridge, United Kingdom.

Mabuchi, I., and M. Okuno. 1977. The effect of myosin antibody on the division of starfish blastomeres. *Journal of Cell Biology.* 74:251–263.

Matsuoka, R., K. W. Beisel, M. Furutani, S. Arai, and A. Takao. 1991. Complete sequence of human cardiac α-myosin heavy chain gene and amino acid comparison to other myosins based on structural and functional differences. *American Journal of Medical Genetics.* 41:537–547.

Moncrief, N. D., R. H. Kretsinger, and M. Goodman. 1990. Evolution of EF-hand calcium modulated proteins. I. Relationships based on amino acid sequences. *Journal of Molecular Evolution.* 30:522–562.

Moore, L. A., W. E. Tidyman, M. J. Arrizubieta, and E. Bandman. 1992. Gene conversions within the skeletal myosin multigene family. *Journal of Molecular Biology.* 223:383–387.

Moore, L. A., W. E. Tidyman, M. J. Arrizubieta, and E. Bandman. 1993. The evolutionary relationship of avian and mammalian myosin heavy chain genes. *Journal of Molecular Evolution.* 36:21–30.

Mooseker, M. 1993. A multitude of myosins. *Current Biology.* 3:245–248.

Pollard, T. P., S. K. Doberstein, and H. G. Zot. 1991. Myosin-I. *Annual Review of Physiology.* 53:653–681.

Rodriguez, J. R., and B. M. Patterson. 1990. Yeast myosin heavy chain mutant: maintenance of the cell type specific budding pattern and the normal deposition of chitin and cell wall components requires an intact myosin heavy chain gene. *Cell Motility and the Cytoskeleton.* 17:301–308.

Stedman, H. H., M. Eller, E. H. Jullian, S. H. Fertels, S. Sarkar, J. E. Sylvester, A. M. Kelly, and N. A. Rubinstein. 1990. The human embryonic myosin heavy chain. Complete primary structure reveals evolutionary relationships with other developmental isoforms. *Journal of Biological Chemistry.* 265:3568–3576.

Stewart, A. F., M. B. Camoretti, D. Perlman, M. Gupta, S. Jakovcic, and R. Zak. 1991. Structural and phylogenetic analysis of the chicken ventricular myosin heavy chain rod. *Journal of Molecular Evolution.* 33:357–366.

Titus, M. A., H. M. Warrick, and J. A. Spudich. 1989. Multiple actin-based motor genes in *Dictyostelium. Cell Regulation.* 1:55–63.

Titus, M. A., D. Wessels, J. A. Spudich, and D. Soll. 1993. The unconventional myosin encoded by the myoA gene plays a role in *Dictyostelium* motility. *Molecular Biology of the Cell.* 4:233–246.

Urrutia, R. A., G. Jung, and J. A. Hammer. 1993. The *Dictyostelium* myosin I-E heavy chain gene encodes a truncated isoform that lacks sequences corresponding to the actin binding site in the tail. *Biochimica et Biophysica Acta.* 1173:225–229.

Warrick, H. M., and J. A. Spudich. 1987. Myosin structure and function in cell motility. *Annual Review of Cell Biology* 3:379–421.

Diversity of Myosin-based Motility: Multiple Genes and Functions

Allison Weiss, D. C. Ghislaine Mayer, and Leslie A. Leinwand

Albert Einstein College of Medicine, Department of Microbiology and Immunology, Bronx, New York 10461

Introduction

Myosin is a ubiquitous motor protein found in all eukaryotes where it performs functions of movement as diverse as cytokinesis, muscle contraction, phagocytosis, and vesicular transport. The recent expansion in the number of identified myosin genes (many of which are of unknown function) has resulted in the organization of a myosin superfamily with at least seven distinct classes (Cheney, Riley, and Mooseker, 1993). The extensively studied "conventional" myosin molecule, classified as myosin II, will hereafter be referred to as myosin. This molecule consists of a pair of heavy chains and two pairs of nonidentical light chains. Myosin heavy chains (MYH) are encoded by multigene families in all vertebrates examined to date (Emerson and Bernstein, 1987), and are subject of the current review.

The first part of this review will focus on the functional classification and distribution patterns of MYH isoforms in mammalian muscle fibers. These studies have established a foundation for molecular and genetic analysis of MYH gene families for several mammalian species, and support an ultimate goal of understanding how complex forms of regulation of the MYH gene family can be translated into functional changes in motility. Ongoing genetic studies to determine the full extent of the human skeletal MYH gene family will be discussed below, as will the molecular basis for functional distinctions among of MYH isoforms. The final section of this manuscript will discuss the presence of conventional sarcomeric myosin in what appear to be unconventional, nonmuscle locations. The potential functional implications of expression of these "muscle-specific" genes in somewhat surprising cellular locations will be discussed.

Muscle Fiber Type Diversity

It was apparent to microscopists early on that skeletal muscle is comprised of a mosaic of morphologically distinct single fibers (Ranvier, 1874). These fibers can be classified as slow (type 1) or fast (type 2) with regard to both their speed of contraction and their ATPase histochemical staining patterns (see Pette and Staron, 1990; Gauthier, 1991). In myofibrillar ATPase staining, muscle sections are histochemically stained after a preincubation at acidic or alkaline pH, and variations in the pH range of the staining procedure (Brooke and Kaiser, 1970) enable resolution of subclasses within types 1 and 2. For example, in the rat, type 2 muscle fibers have been further subdivided into subtypes 2A and 2B by ATPase staining. The pH sensitivity of ATPase staining is largely due to the ATPase activity of myosin. Thus, this type of assay can be used to assess the myosin content of the muscle fiber (see Table I, and Pette and Staron, 1990).

The actual combination of components required to confer the unique properties of a given fiber type are unknown. However, physiological variation among fibers that are morphologically similar undoubtedly reflects, in part, the underlying patterns of

TABLE I
Mammalian Skeletal MYH Isoform Characteristics

MYH isoform	ATPase* staining	Specific antibody	Clones	
			cDNA	Genomic
Embryonic	pH 4.3	+	H(1, 2) M(3)	H(2) R(4)
Perinatal	pH 4.3	+	H(5, 22) M(3)	H(6) R(7)
Adult IIa	pH 9.6	+	R(8) M(9)	
Adult IIb	pH 4.5	+	R(8) M(9)	
Adult IId/2X	pH 4.5	+	H(10) R(11, 12)	H(6)
Extraocular	pH 9.6	–		R(13)
Beta/Slow	pH 4.3 pH 4.5	+	H(10) R(14)	H(16, 19) R(21) M(15)
Alpha (cardiac-specific)	pH 4.3 pH 10.5	+	R(18)	H(16, 20, 17) R(21) M(15)

*The pH of the preincubation after which the fiber stains darkly (i.e., the activity is not inhibited) is indicated. (*H*) Human; (*R*) rat; (*M*) mouse. References: 1. Eller et al., 1989; 2. Karsch-Mizrachi et al., 1989; 3. Weydert et al., 1985; 4. Strehler et al., 1986; 5. Feghali and Leinwand, 1989; 6. Yoon et al., 1992; 7. Periasamy et al., 1984; 8. Nadal-Ginard et al., 1982; 9. Weydert et al., 1983; 10. Saez and Leinwand, 1986; 11. Ogata et al., 1993; 12. De Nardi et al, 1993; 13. Wieczorek et al., 1985; 14. Kraft et al., 1989; 15. Gulick et al., 1991; 16. Saez et al., 1987; 17. Liew et al., 1990; 18. McNally et al., 1989*a*; 19. Jaenicke et al., 1990; 20. Yamauchi-Takihara, 1989; 21. Lompre et al., 1984; 22. Karsch-Mizrachi et al., 1990.

expression of genes encoding various contractile proteins. Early biochemical studies revealed up to eight different MYH species as indicated by their different electrophoretic mobilities on high resolution protein gels (see Termin, Staron, and Pette, 1989; LaFramboise, Dood, Guthrie, Moretti, Schiaffini, and Ontell, 1990). The generation of antibodies against MYH proteins enabled the simultaneous application of ATPase staining and immunohistochemistry on serial sections of groups of muscle fibers. This facilitated the localization of MYH isoforms to individual muscle fibers in mixed populations, and began a reconciliation of histochemical profiles of fibers with specific MYH isoforms (see Gauthier, 1991). This kind of analysis allowed the identification of a third type of fast MYH, 2X.

MYH Gene Expression and Isoform Distribution Patterns

Molecular cloning has further refined studies of MYH localization as well as patterns of gene expression (see Table I). In situ hybridization analysis on serial muscle sections using gene-specific antisense probes from the 3′ untranslated regions (UTRs) of various MYH isoforms has been carried out in parallel with immunohistochemical techniques. In the rat, in situ hybridization data support immunohistochemical definitions of fiber subpopulations 2A, 2B, and 2X in that only certain combinations of MYH's are found within individual fibers. For example, combinations of transcripts representing genes 2X and 2A, or 2X and 2B, and less frequently 2A and beta/slow have been localized to single fibers. However, 2A and 2B appear never to be colocalized within single fibers in the muscle groups studied (DeNardi et al., 1993). Studies of different MYH transcripts in rat and mouse indicate that they display specific regional distributions throughout development, and that different transcripts accumulate in different subpopulations of muscle fibers depending on the muscle being studied (DeNardi et al., 1993; Ontell, Ontell, Sopper, Mallonga, Lyons, and Buckingham, 1993).

Human Skeletal MYH Gene Family

At present, the number of human skeletal MYH genes is predicted to be at least eight (see Table I and text below). Human MYH loci have been assigned to three different chromosomes. The two known cardiac genes (alpha and beta) are tightly linked and located on chromosome 14 (Saez et al., 1987), and a nonmuscle MYH gene is located on chromosome 22 (Saez, Myers, Shows, and Leinwand, 1990). The human skeletal MYH gene family is clustered on chromosome 17, a region syntenic to a location on mouse chromosome 11 (Leinwand, Fournier, Nadal-Ginard, and Shows, 1983). Despite their tight linkage in the genome, the skeletal MYH genes appear to be independently regulated.

A physical map represented by a contiguous set (contig) of nine overlapping yeast artificial chromosomes (YACs) containing six MYH genes has been generated for the human skeletal MYH gene family (Yoon et al., 1992). Sequence data from four of these genes reveal that they encode embryonic, perinatal and two adult fast MYH isoforms (Yoon et al., 1992). The remaining two genes in the contig are as of yet uncharacterized. Additional sequence data from an independently isolated human genomic clone suggests that two additional skeletal MYH genes of unknown location exist (J. Gundling, unpublished data). Thus, the full extent of the human MYH gene family remains to be defined.

Evolutionary Comparisons of MYH Sequences

The complexity of the MYH gene family raises the question of whether individual genes provide unique functions in the muscle in which they are expressed, or whether these genes are functionally redundant. While quantitative differences in the ATPase activities of some myosins have been demonstrated, the majority of the sarcomeric myosins have not been individually characterized. Sequence comparisons are a useful approach for understanding the relationships among members of this gene family.

The overall primary structure of MYH genes is remarkably conserved across species. The locations of intron/exon borders are frequently maintained interspecifically (Strehler, Strehler-Page, Perriard, Periasamy, and Nadal-Ginard, 1986; Jaenicke et al., 1990). One exception includes a fusion of the last two exons of a rat gene into a single final exon in the chicken (see Moore, Arrizubieta, Tidyman, Herman, and Bandman, 1993). The 5′ UTRs of MYH genes are usually comprised of two exons with the coding region beginning in exon three. The 3′ UTRs tend to contain divergent sequences and are extremely useful for generating gene-specific probes for nucleic acid hybridization studies (see Saez and Leinwand, 1986).

Evolutionary studies that compare the primary structure of the known mammalian MYH genes are often based on the subdivision of the myosin molecule into two functionally and structurally distinct domains: the amino-terminal "head" or motor domain in which ATPase and actin binding activities reside, and a less conserved carboxy-terminal "tail" or rod domain required for assembly of myosin into filaments and higher order structures such as sarcomeres in skeletal muscle.

Comparisons of MYH genes from evolutionarily distinct species indicate that the head domain is more highly conserved then the rod domain (see Warrick and Spudich, 1987). Despite the absence of direct sequence homology, certain structural features of the rod domain are preserved including the presence of a heptad repeat in which a hydrophobic residue is present in the first and fourth positions of the repeat (McLachlan and Karn, 1982). Conserved higher order repeats involving distribution of positively and negatively charged residues are thought to be required for the lateral interactions of myosin molecules during thick filament formation. (MacLachlan and Karn, 1982).

MYH gene comparisons among mammalian species suggest that the mammalian MYH gene family arose from multiple duplications of an ancestral MYH gene (see Moore et al. 1993; Stedman, Eller, Julian, Fentels, Sarkar, Sylvester, Kelly, and Rubenstein, 1990). Interspecific comparisons of mammalian MYH genes that have similar regulation patterns (such as a comparison between the human embryonic and rat embryonic genes) reveal orthologous relationships between these genes. In contrast, inter- and intraspecific comparisons of the fast MYH chicken genes reveal that they are more similar to each other than to any mammalian gene of similar regulation (Moore et al., 1992). These studies suggest that the chicken MYH gene family is the result of concerted evolution of genes that arose from duplications that occurred subsequent to the avian/mammalian split (Moore et al., 1993). Likewise, the differences in the genetic structure between the avian and mammalian MYH families suggest that they may have evolved different means of regulating muscle-specific genes and myogenesis as well.

At present, no functional distinctions among fast MYH isoforms are known. Amino acid sequence comparisons can be used to explore the molecular basis for functional differences among MYH isoforms, and to create a model framework for site-directed mutagenesis studies (McNally, Kraft, Bravo-Zehnder, Taylor, and Leinwand, 1989*b*). When comparisons of human and rat MYHs from both the skeletal and cardiac gene clusters are made, regions of absolute sequence identity are relatively evenly distributed across the molecule (Fig. 1*A*). However, when the comparison is restricted to the skeletal MYH genes from the skeletal gene clusters only, there are more regions of absolute sequence identity, particularly in the head region of the molecule (Fig. 1 *B, left of arrow*). This indicates a closer evolutionary relationship between the genes within a chromosomal cluster, which is consistent with a gene duplication mechanism for generating diversity. Despite information that can be gained by sequence comparisons, genetic and biochemical analysis of pure populations of MYH will be required for ultimately understanding the functional impact of the different muscle MYH isoforms.

Coexpression of Muscle and Nonmuscle Cytoskeletal Proteins

Skeletal muscle myosin self-associates to form filaments, which are further organized into sarcomeres, the basic structural unit of skeletal muscle. Under most circumstances the myosin found in skeletal muscle is both structurally and enzymatically

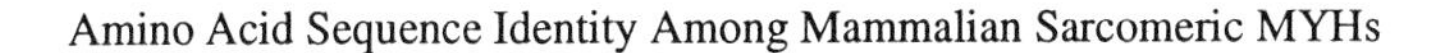

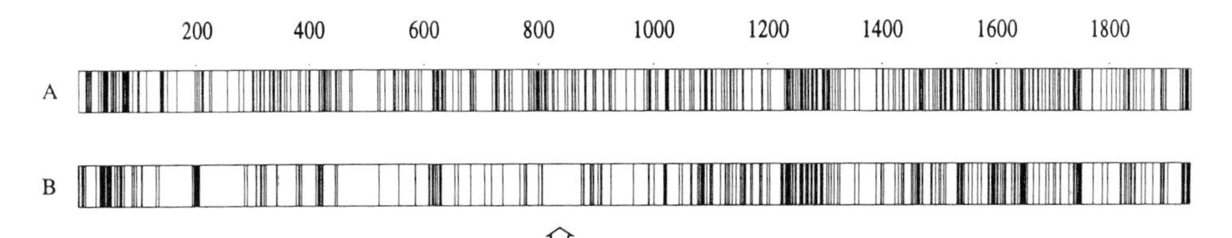

Figure 1. Amino acid sequence identity among mammalian skeletal MYH genes. Sequences of the MYH genes indicated below were compared by the GCG Pileup program and used to generate a schematic model of the MYH protein. Open areas indicate that all isoforms have identical amino acids at that position. Vertical lines indicate amino acid positions where at least one isoform has a variant residue. Arrow indicates mysosin head/rod junction. (*A*) Amino acid sequence comparisons were derived from human embryonic (Eller et al., 1989), human beta (Saez and Leinwand, 1986), human MYHa (Saez and Leinwand, 1986), human MYHas8 (I. M. Karsch-Mizrachi, unpublished data), human perinatal (Karsch-Mizrachi et al., 1990), rat embryonic (Weydert et al., 1985), rat beta (Kraft et al., 1989), and rat 2d (DeNardi et al., 1993) genes. (*B*) Includes above sequences excluding human and rat beta/slow isoforms. Note: sequence comparisons in head region are limited by availability of sequence data from the 5′ ends of MYH genes. The head region in *B* represents comparisons of human perinatal and embryonic, and rat embryonic genes only.

distinct from nonmuscle myosin. The myosin in nonmuscle cells generally does not form recognizable higher order structures, and its ATPase activity is significantly lower than the myosin found in muscle.

Until recently, nonmuscle and muscle cells were thought to be largely nonoverlapping in terms of their contractile protein composition. These differences may reflect the functionally distinct motile processes in which they participate. Nonmuscle myosins are required for diverse functions ranging from cell surface receptor capping and cytokinesis to cellular morphogenesis. In contrast, muscle myosin performs a single specialized function, that of muscle contraction. However, cell types have been described that appear to have characteristics of both muscle and nonmuscle cells (Table II). For example, testicular stromal cells, fat-storing cells of the liver and a category of reticular cell in lymph nodes and spleen all have phenotypic markers characteristic of muscle (Sappino, Schurch, and Gabbiani, 1990). Further, the expression of these markers can be modulated physiologically, suggesting that phenotypic differences might correspond to functional properties of the cells. This modulation is perhaps best exemplified by wound healing, where fibroblasts during experimental wound healing gradually develop α-smooth muscle actin-containing microfilaments, which disappear when the wound is healed (Sappino et al., 1990).

Other examples of coexpression of multiple types of contractile proteins exist. A recent report demonstrated transient expression of a brain myosin isoform in muscle cells of neonatal mice (Murakami, Trenkner, and Elzinga, 1993). Sarcomeric myosin has been demonstrated in the conduction system of the adult heart of various organisms, including humans (Kuro-o, Tsuchimochi, Ueda, Takaku, and Yazaki, 1986; Komuro, Nomoto, Sugiyama, Kurabayashi, Takaku, and Yazaki, 1987). It is not known whether these cells are of mesodermal or neural crest origin, but they appear to represent a cell type that is intermediate between muscle and nonmuscle cells. Several lines of evidence indicate that these cells are clearly distinct from the

TABLE II
Expression of Muscle Cytoskeletal Proteins in Nonmuscle Cells

	Desmin	α-smooth muscle actin	Sarcomeric MYH	Smooth muscle MYH	Refs.
Testicular stromal cells	+	+	ND	+	*A, B*
Ovarian thecal cells	ND	+	ND	ND	*C, D*
Stromal cells of the rat intestinal villi	+	+	ND	+	*E, B, F*
Liver Ito cells	+	+	+	–	*G, H, I*

(*A*) Benzonana et al., 1988; (*B*) Skalli et al., 1986; (*C*) Czernobilisky et al., 1989; (*D*) O'Shea, 1970; (*E*) Kaye et al., 1968; (*F*) Skalli and Gabbiani, 1988; (*G*) Yokoi et al., 1984; (*H*) Bellardini et al., 1988; (*I*) Ogata et al., 1993. (ND) Not determined.

surrounding cardiac musculature despite the striated appearance of the contractile proteins. First, they express a type of gap junction protein, connexin 40 (Cx40), which is not normally expressed in myocytes, as well as Cx43 and Cx45, the latter two being characteristic of ventricular myocytes (Kanter, Laing, Beau, Beyer, and Saffitz, 1993). Second, these cells express a form of desmin, an intermediate filament protein found in all muscle cell types, that is distinct from the desmin found in cardiac muscle (Virtanen, 1990). These cells also express the brain creatine kinase isoform. Moreover, cardiac Purkinje fibers express a splice variant of dystrophin mRNA that is predominantly found in the brain (Bies, Friedman, Roberts, Perryman, and Caskey, 1992). Given these examples of coexpression of muscle and nonmuscle proteins, our notion of the cytoskeleton and nonmuscle cell motility may need to be expanded.

Myofibroblasts in the Liver

Some of the cells described above which appear to have characteristics of both muscle and nonmuscle cells have been called myofibroblasts. One class of myofibroblasts can be found in the liver, and is associated with disease and fibrosis. These cells are thought to arise from Ito cells (also known as fat-storing cells, pericytes or perisinusoidal cells), which were first described by Ito and Nemoto (1952). They are localized in the space of Disse, between the hepatocytes and the endothelial cells, where their extensions are in contact with endothelial cells. There are 3 to 6 Ito cells per 100 hepatocytes. Two of their primary characteristics are the storage of vitamin A droplets and the production of liver extracellular matrix proteins, such as types I and IV collagen and laminin (De Leuw, McCarthy, Geerts, and Knook, 1984; Wake, 1971). These cells also express desmin, an intermediate filament protein that is normally found in muscle cells (Yokoi et al., 1984). Desmin has since been used as a marker for the identification of Ito cells.

Ramadori, Veit, Schwogler, Dienes, Knittel, Rieder, and Meyer zum Buschenfelde, (1990) showed that diseased human livers contain transitional cells with a phenotype intermediate between smooth muscle cells and Ito cells. They contain indented nuclei, large rough endoplasmic reticulum, microfilament bundles, fat droplets and stain positive for desmin. Several lines of evidence suggest that Ito cells convert to myofibroblasts during liver fibrogenesis. First, in normal liver, Ito cells are found in larger numbers than myofibroblasts, whereas in cirrhotic liver the situation is reversed (Mak, Leo, and Lieber, 1984; Mak and Lieber, 1988; Geerts, Lazou, De Bleser, and Wisse, 1991). Second, Ito cells express several proteins normally found in muscle such as desmin, α-smooth muscle actin, sarcomeric myosin, and a smooth muscle-specific cyclic GMP-dependent protein kinase (Bellardini et al., 1988; Schmitt-Graff, Kruger, Bochard, Gabbiani, and Denk, 1991; Ogata et al., 1993; Joyce, DeCamilli, and Boyles, 1984). Third, treatment of myofibroblasts with vitamin A or insulin and indomethacin reverts their phenotype to that of fat-storing cells (Margis and Brojevic, 1989). As Ito cells were further characterized, it became apparent that they share additional characteristics with muscle cells.

The localization of Ito cells in the perivascular region, and their close contact with the endothelial cells led to the suggestion that these cells may be involved in the regulation of perisinusoidal blood flow (Pinzani, Failli, Ruocco, Casini, Milani, Baldi, Giotti, and Gentilini, 1992). Consistent with this hypothesis is their response to vasoconstrictors such as endothelin-1, thrombin, and angiotensin II by an intracellular calcium wave. This hypothesized function would predict specialized contractile behavior, such as the observation that Ito cells can contract on collagen plates while hepatocytes of the same density are unable to do so (Rockey, Housset, and Friedman, 1993).

Sarcomeric MYH Expression in Liver Ito Cells

Recently, in collaboration with Dr. Marcos Rojkind (Albert Einstein College of Medicine), we have begun a detailed investigation into the cytoskeletal protein composition of rat Ito cell lines. Surprisingly, in addition to the expected nonmuscle myosin heavy chain, skeletal MYH was also expressed in Ito cells lines (Ogata et al., 1993). One of the MYH genes was identified as adult skeletal fast IId, a gene which is normally expressed only in adult skeletal muscle fibers. In addition, the perinatal MYH isoform was also present in these cells (C. Saez, unpublished observation). Immunohistochemical studies suggest that the sarcomeric MYH does not interact with the actin cytoskeleton or with the nonmuscle myosin heavy chain in these cells. Rather, the sarcomeric MYH forms dense patches around the nucleus and the periphery of the cells (Fig. 2). It is tempting to speculate that these myosins might

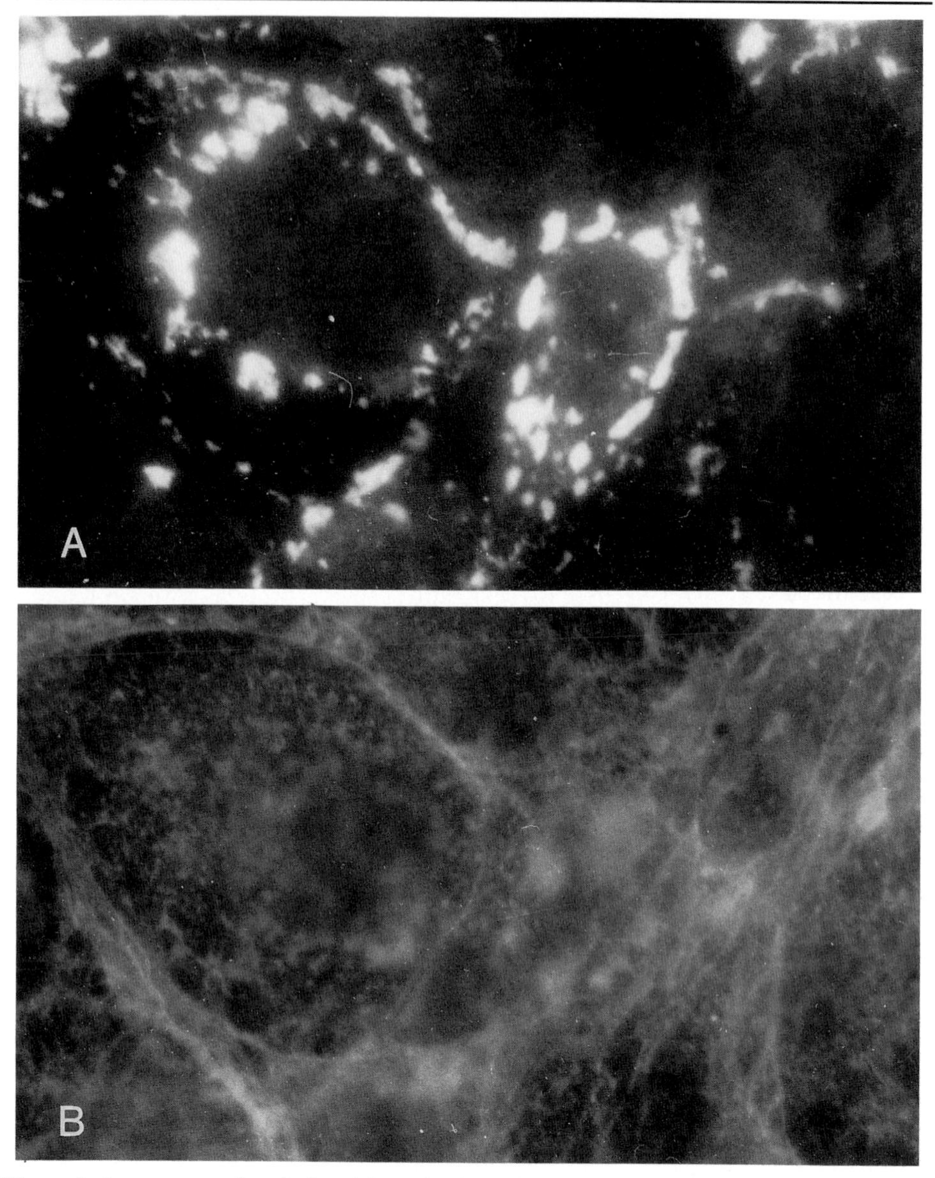

Figure 2. Immunocytochemical staining of cultured rat Ito cells (Ogata et al., 1993) with an anti-sarcomeric MYH-specific monoclonal antibody (F59 antibody; Miller et al., 1989; *A*), and with rhodamine phalloidin to detect actin filaments (*B*).

provide the contractile force for regulating sinusoidal blood flow. Further experimentation is required to answer this question and is currently under investigation.

The study of these cells has provided a framework for reconsidering our concepts of muscle and nonmuscle cytoskeletons. It may be that expression of muscle contractile proteins in nonmuscle settings represents an evolutionary adaptation that permits specialized motile behavior. In addition to expanding our understanding of motility, these observations raise a number of very interesting questions. For example, the study of muscle-specific gene expression has identified myogenic regulators and muscle-specific transcription factors that determine the activity of many muscle-specific genes. It will be interesting to determine whether the mechanisms for expressing a skeletal muscle MYH gene in a nonmuscle cell are the same as in a muscle cell. In addition, it seems that only a small subset of the full program of sarcomeric proteins are expressed in these intermediate cell types. What mechanism determines the selection of genes to activate? It will also be interesting to determine the full range of cytoskeletal molecules expressed in these cells. For example, if a sarcomeric MYH is expressed, which myosin light chain is utilized? These and other questions can be addressed experimentally and should advance our understanding of contractile protein gene expression and myosin-based motiliy.

Acknowledgements

We would like to acknowledge Brian Lu and Ken Krauter for their assistance with sequence comparisons, and Karen Vikstrom and Alyson Kass-Eisler for critical reading of the manuscript. We especially appreciate the contribution of unpublished data by Claudia Saez.

A. Weiss is supported by training grant T32CA09173. Research in this manuscript is supported by an NIH grant (GM29090) to L. A. Leinwand.

References

Bellardini, G., M. Falllani, G. Biagini, F. B. Bianchi, and E. Pisi. 1988. Desmin and actin in the identification of Ito cells and in monitoring their evolution to myofibroblasts in experimental liver fibrosis. *Virchows Archives B. Cell Pathology.* 56:45–49.

Benzonana, G., O. Skalli, and G. Gabbiani. 1988. Correlation between the distribution of smooth muscle or nonmuscle myosins and α-smooth muscle actin in normal and pathological soft tissues. *Cell Motility and the Cytoskeleton.* 11:260–274.

Bies, R. D., D. Friedman, R. Roberts, M. B. Perryman, and C. T. Caskey. 1992. Expression and localization of dystrophin in human Purkinje fibers. *Circulation.* 86:147–153.

Brooke, M. H., and K. K. Kaiser. 1970. Muscle fiber types: how many and what kind? *Archives of Neurology.* 23:369–379.

Cheney, R. E., M. A. Riley, and M. S. Mooseker. 1993. Phylogenetic analysis of the myosin superfamily. *Cell Motility and the Cytoskeleton.* 24:215–223.

Czernobilisky, B., E. Shezen, B. Lifschitz-Mercer, M. Fogel, A. Luzon, N. Jacob, O. Skalli, and G. Gabbiani. 1989. Alpha smooth muscle actin (α-SM actin) in normal human ovaries, in ovarian stromal hyperplasia and in ovarian neoplasms. *Virchows Archives B Cell Pathology.* 57:55–61.

De Leuw, A. M., S. B. McCarthy, A. Geerts, and D. L. Knook. 1984. Purified rat liver fat-storing cells in culture divide and contain collagen. *Hepatology.* 4:392–403.

DeNardi, C. A. Ausoni, P. Moretti, L. Gorza, M. Velleca, M. Buckingham, and S. Schiaffino. 1993. Type 2X myosin heavy chain is coded by a muscle fiber type-specific and developmentally regulated gene. *Journal of Cell Biology.* 123:823–835.

Eller, M., H. H. Stedman, J. E. Sylvester, S. H. Fertels, N. A. Rubenstein, A. M. Kelly, and S. Sarkar. 1989. Nucleic acid sequence of a human embryonic myosin heavy chain cDNA. *Nucleic Acids Research.* 17:3591–3592.

Emerson, C. P., and S. I. Bernstein. 1987. Molecular genetics of myosin. *Annual Review of Biochemistry.* 56:695–726.

Feghali, R., and L. A. Leinwand. 1989. Molecular and genetic characterization of a developmentally regulated human perinatal myosin heavy chain. *Journal of Cell Biology.* 108:1791–1797.

Gauthier, G. F. 1991. Skeletal muscle fiber types. *In* Myology: Basic and Clinical. A. G. Engel, and B. Q. Banker, editors. McGraw-Hill Book Company, NY. 255–282.

Geerts, A., J. M. Lazou, P. De Bleser E., and W. Wisse. 1991. Tissue distribution, quantitation and proliferation kinetics of fat-storing cells in carbon tetrachloride-injured rat liver. *Hepatology.* 13:1193–1202.

Gulick, J., A. Subramanian, J. Neuman, and J. Robbins. 1991. Isolation and characterization of the mouse cardiac myosin heavy chain genes. *Journal of Biological Chemistry.* 266:9180–9185.

Ito, I., and M. Nemoto. 1952. Ueber die Fufferschen Stenenzellen und die enwand der menschilchen Leber. *Okajima Folia Anatomica Japonira.* 24:243–258.

Jaenicke, T., K. W. Diederich, W. Haas, J. Schleich, P. Lichter, M. Fordt, A. Bach, and H-P. Vosberg. 1990. The complete sequence of the human beta-myosin heavy chain gene and a comparative analysis of its product. *Genomics.* 8:194–206.

Joyce, N. C., P. DeCamilli, and J. Boyles. 1984. Pericytes, like vascular smooth muscle cells are immunocytochemically positive for cyclic GMP dependent protein kinase. *Microvasculature Research.* 28:206–219.

Kanter, H. L., J. G. Laing, S. L. Beau, E. C. Beyer, and J. E. Saffitz. 1993. Distinct pattern of connexin expression in canine Purkinje fibers and ventricular muscle. *Circulation Research.* 72:1124–1131.

Karsch-Mizrachi, I., M. Travis, H. Blau, and L. A. Leinwand. 1989. Expression and DNA sequence analysis of a human embryonic skeletal muscle myosin heavy chain gene. *Nucleic Acids Research.* 17:6167–6179.

Karsch-Mizrachi, I. M., R. Feghali, T. B. Shows, and L. A. Leinwand. 1990. Generation of a full-length human perinatal myosin heavy chain encoding cDNA. *Gene.* 89:289–294.

Kaye, G. I., N. Lane, and P. R. Pascal. 1968. Colonic pericryptal fibroblast sheat: replication, migration and cytodifferentiation of a mesenchymal cell-system in adult tissue. II. Fine structural aspects of rabbit and human colon. *Gastroenterology.* 60:515–536.

Komuro, I., K. Nomoto, T. Sugiyama, M. Kurabayashi, F. Takaku, and Y. Yazaki. 1987. Isolation and characterization of myosin heavy chain isozymes of the bovine conduction system. *Circulation Research.* 61:859–865.

Kraft, R., M. Bravo-Zehnder, D. A. Taylor, and L. A. Leinwand. 1989. Complete nucleotide sequence of full length cDNA for rat beta myosin heavy chain. *Nucleic Acids Research.* 17:7529–7530.

Kuro-o, M., H. Tsuchimochi, S. Ueda, F. Takaku, and Y. Yazaki. 1986. Distribution of cardiac myosin isozymes in human conduction system. *Journal of Clinical Investigation.* 77:340–347.

LaFramboise, W. A., M. J. Dood, R. D. Guthrie, P. Moretti, S. Schiaffini, and M. Ontell. 1990. Electrophoretic separation of type 2X myosin heavy chain in rat skeletal muscle. *Biochemica Biophysica Acta.* 1035:109–112.

Leinwand, L. A., R. E. K. Fournier, B. Nadal-Ginard, and T. B. Shows. 1983. Multigene family for sarcomeric myosin heavy chain in mouse and human DNA: localization on a single chromosome. *Science.* 221:766–769.

Liew, C. C., M. J. Sole, K. Yamouchi-Takihara, B. Kellam, D. H. Anderson, L. Lin, and J. C. Liew. 1990. Complete sequence and organization of the human cardiac B-myosin heavy chain gene. *Nucleic Acids Research.* 18:3647–3651.

Lompre, A. M., B. Nadal-Ginard, and V. Mahdavi. 1984. Expression of the cardiac ventricular alpha and beta myosin heavy chain genes is developmentally and hormonally regulated. *Journal of Biological Chemistry.* 259:6437–6446.

Mak, K. M., and C. S. Lieber. 1988. Lipocytes and transitional cells in alcoholic liver disease: A morphogenetic study. *Hepatology.* 8:1027–1033.

Mak, K. M., M. A. Leo, and C. S. Lieber. 1984. Alcoholic liver injury in baboons: transformation of lipocytes and transitional cells. *Gastroenterology.* 87:188–200.

Margis, R., and R. Brojevic. 1989. Retinoid mediated induction of the fat-storing cells phenotype in a liver connective tissue cell line (GRX). *Biochimica Biophysica Acta.* 1011:1–5.

McLachlan, A. D., and J. Karn. 1982. Periodic charge distribution in the myosin rod amino acid sequence match cross-bridge spacings in muscle. *Nature.* 299:226–231.

McNally, E. M., K. M. Gianola, and L. A. Leinwand. 1989*a*. Complete nucleotide sequence of full length cDNA for rat alpha cardiac myosin heavy chain. *Nucleic Acids Research.* 17:7527–7528.

McNally, E. M., R. Kraft, M. Bravo-Zehnder, D. Taylor, and L. A. Leinwand. 1989*b*. Full length alpha and beta cardiac myosin heavy chain sequences: comparison suggest a molecular basis for functional differences. *Journal of Molecular Biology.* 210:665–671.

Miller, J. B., S. B. Teal, and F. E. Stockdale. 1989. Evolutionarlily conserved sequences of striated muscle myosin heavy chain isoforms. Epitope mapping by cDNA expression. *Journal of Biological Chemistry.* 264:13122–13130.

Moore, L. A., M. J. Arrizubieta, W. E. Tidyman, L. A. Herman, and E. Bandman. 1992. Analysis of the chicken fast myosin heavy chain family: localization of isoform-specific antibody epitopes and regions of divergence. *Journal of Molecular Biology.* 225:1143–1151.

Moore, L. A., W. E. Tidyman, M. J. Arrizubieta, and E. Bandman. 1993. The evolutionary relationship of avian and mammalian myosin heavy chain genes. *Journal of Molecular Evolution.* 36:21–30.

Murakami, N., E. Trenkner, and M. Elzinga. 1993. Changes in expression of nonmuscle myosin heavy chain isoforms during muscle and nonmuscle tissue development. *Developmental Biology.* 157:19–27.

Nadal-Ginard, B., R. M. Medford, H. T. Nguyen, M. Periasamy, M. Wydro, D. Horning, R. Gubits, L. I. Garfinkel, D. F. Wieczorek, E. Bekesi, and V. Madhavi. 1982. Structure and regulation of a mammalian sarcomeric myosin heavy chain gene. *In* Muscle Development: Molecular and Cellular Control. M. L. Pearson and H. F. Epstein, editors. Cold Spring Harbor Laboratory Press, Cold Spring Harbor, NY. 143–168.

Ogata, I., C. G. Saez, P. Greenwel, M. L. Ponce, A. Geerts, L. Leinwand, and M. Rojkind. 1993. Rat liver fat-storing cell lines express sarcomeric myosin heavy chain mRNA and protein. *Cell Motility and the Cytoskeleton.* 26:125–132.

Ontell, M., M. P. Ontell, M. M. Sopper, R. Mallonga, G. Lyons, and M. Buckingham. 1993. Contractile protein gene expression in primary myotubes of embryonic mouse hindlimb muscles. *Development.* 117:1435–1444.

O'Shea, J. D. 1970. An ultrastructural study of smooth-muscle like cells in the theca externa of the ovarian follicle of the rat. *Anatomical Record.* 167:127–140.

Periasamy, M., D. F. Wieczorek, and B. Nadal-Ginard. 1984. Characterization of a developmentally regulated perinatal myosin heavy chain gene expressed in skeletal muscle. *Journal of Biological Chemistry.* 259:13573–13578.

Pette, D., and R. S. Staron. 1990. Cellular and molecular diversities of mammalian skeletal muscle fibers. *Reviews of Physiology, Biochemistry, and Pharmacology.* 116:2–47.

Pinzani, M., P. Failli, C. Ruocco, A. Casini, S. Milani, E. Baldi, A. Giotti, and P. Gentilini. 1992. Fat-storing cells as liver specific pericytes. *Journal of Clinical Investigation.* 90:642–646.

Ramadori, G., T. Veit, S. Schwogler, H. P. Dienes, T. Knittel, H. Rieder, and K. H. Meyer zum Buschenfelde. 1990. Expression of the gene α-smooth muscle-actin isoform in rat liver and in rat fat-storing (ITO) cells. *Virchows Archives B Cell Pathology.* 59:349–357.

Ranvier, L. 1874. De quelques faits relatifs a l'histologie et a la physiologie des muscles stries. *Archives de Physiologie Normale et Pathologique.* 1:5.

Rockey, D. C., C. N. Housset, and S. L. Friedman. 1993. Activation-dependent contractility of rat hepatic lipocytes in culture and in vivo. *Journal of Clinical Investigation.* 92:1795–1864.

Saez, C. G., J. C. Myers, T. B. Shows, and L. A. Leinwand. 1990. Human nonmuscle myosin mRNA: generation of transcript diversity through alternate polyadenylation. *Proceedings of the National Acadamy of Sciences, USA.* 87:1164–1168.

Saez, L. J., K. M. Gianola, E. M. McNally, R. Feghali, R. Eddy, T. B. Shows, and L. A. Leinwand. 1987. Human cardiac myosin heavy chain genes and their linkage in the genome. *Nucleic Acids Research.* 15:5443–5459.

Saez, L. J., and L. A. Leinwand. 1986. Characterization of diverse forms of myosin heavy chain expressed in adult human skeletal muscle. *Nucleic Acids Research.* 14:2951–2969.

Sappino, A. P., W. Schurch, and G. Gabbiani. 1990. Biology of disease: Differentiation repertoire of fibroblastic cells: expression of cytoskeletal proteins as marker of phenotypic regulation. *Laboratory Investigation.* 63:144–161.

Schmitt-Graff, A., S. Kruger, F. Bochard, G. Gabbiani, and H. Denk. 1991. Modulation of alpha-smooth muscle actin and desmin expression in perisinusoidal cells of normal and diseased livers. *American Journal of Pathology.* 138:1233–1242.

Skalli, O., P. Ropraz, A. Trzeciak, G. Benzonana, D. Gilessen, and G. Gabbiani. 1986. A monoclonal antibody for α-smooth muscle actin, a new probe for smooth muscle differentiation. *Journal of Cell Biology.* 103:2787–2796.

Skalli, O., and G. Gabbiani. 1988. The biology of the myofibroblast: relationship to wound contraction and fibrocontractive diseases. *In* The Molecular and Cellular Biology of Wound Repair. R. A. F. Clark, and P. M. Henson, editors. Plenum Publishing Corp., NY. 373–402.

Stedman, H. H., M. Eller, E. H. Julian, S. H. Fentels, S. Sarkar, J. E. Sylvester, A. M. Kelly, and N. A. Rubenstein. 1990. The human embryonic myosin heavy chain: complete primary structure reveals evolutionary relationships with other developmental isoforms. *Journal of Biological Chemistry.* 265:3568–3576.

Strehler, E. E., M. A. Strehler-Page, J. C. Perriard, M. Periasamy, and B. Nadal-Ginard. 1986. Complete nucleotide and encoded amino acid sequence of a mammalian myosin heavy chain gene: evidence against intron-dependent evolution of rod. *Journal of Molecular Biology.* 190:291–317.

Termin, A., R. S. Staron, and D. Pette. 1989. Myosin heavy chain isoforms in histochemically defined fiber types of rat muscle. *Histochemistry.* 92:453–457.

Virtanen, I., O. Narvanen, and L. E. Thornell. 1990. Monoclonal antibody to desmin purified from cow Purkinje fibers reveals a cell-type specific determinant. *Federation of European Biochemical Societies.* 267:176–178.

Wake, K. 1971. "Sternzellen": Perisinusoidal cells with reference to storage of vitamin A. *American Journal of Anatomy.* 132:429–462.

Warrick, H. M., and J. A. Spudich. 1987. Myosin structure and function in cell motility. *Annual Review of Cell Biology.* 3:379–421.

Weydert, A., P. Daubas, I. Lazaridis, P. Barton, I. Garner, D. P. Leader, F. Bonhomme, J. Catalan, D. Simon, J. Guenet, F. Gross, and M. Buckingham. 1985. Genes for skeletal muscle myosin heavy chain are clustered and are not located on the same mouse chromosome as cardiac myosin heavy chain. *Proceedings of the National Acadamy of Sciences, USA.* 82:7183–7187.

Weydert, A., P. Daubas, M. Caravatti, A. Minty, G. Bugaisky, A. Cohen, B. Robert, and M. Buckingham. 1983. Sequential accumulation of mRNAs encoding different myosin heavy chain isoforms during skeletal muscle development in vivo detected with a recombinant plasmid as identified as coding for an adult fast myosin heavy chain from mouse skeletal muscle. *Journal of Biological Chemistry.* 258:13867–13874.

Wieczorek, D. F., M. Periasamy, G. Butler-Brown, R. G. Whalen, and B. Nadal-Ginard. 1985. Co-expression of multiple myosin heavy chain genes, in addition to a tissue-specific one, in extraocular musculature. *Journal of Cell Biology.* 101:618–629.

Yamauchi-Takihara, K., M. J. Sole, J. Liew, D. Ing, and C. C. Liew. 1989. Characterization of human cardiac myosin heavy chain genes. *Proceedings of the National Acadamy of Sciences, USA.* 86:3504–3508.

Yokoi, Y., T. Namihisa, H. Kuroda, I. Komatsu, A. Miyazaki, S. Wanatabe, and K. Usui. 1984. Immunocytochemical detection of desmin in fat-storing cells (Ito cells). *Hepatology.* 4:709–714.

Yoon, S., S. H. Seiler, R. Kucherlapati, and L. A. Leinwand. 1992. Organization of the human skeletal myosin heavy chain gene cluster. *Proceedings of the National Academy of Science, USA.* 89:12078–12082.

The Actin Protein Superfamily

Christine Fyrberg, Liza Ryan, Lisa McNally, Maura Kenton, and Eric Fyrberg

Department of Biology, The Johns Hopkins University, Baltimore, Maryland 21218

The last two years have witnessed the discovery of several novel actin-related proteins having primary sequences that are only 35–55% identical to those of conventional nonmuscle actins. This low level of similarity, and other characteristics to be described below, strongly suggest that they have roles distinct from those of conventional actin isoforms. Several of these actin-related proteins are now known to occur in diverse organisms, strongly indicating that, whatever their precise functions, they are essential for most eukaryotic cells.

Actin is a highly conserved protein that has been investigated intensively for decades. It was once hypothesized, most notably by Bray (1972) that actins from all sources had identical primary sequences. That theory was modified, however, when Gruenstein and Rich (1975) demonstrated that actin isolated from chicken brain had a slightly different tryptic peptide map from that isolated from chicken muscle, and when Garrels and Gibson (1976) demonstrated that chickens harbored at least three actin isoforms that could be separated using two-dimensional gel electrophoresis. Finally, Vandekerckhove and Weber (1978) directly sequenced mammalian muscle and nonmuscle actin isoforms, and thereby demonstrated that they have primary sequences that differed in ~25/375 positions.

In ensuing years, actin genes from diverse organisms were isolated and sequenced. It thus became apparent that animal actin isoforms were highly similar (Bray [1972] wasn't so far off the mark), seldom differing by more than 3–7% when analyzed by pairwise comparisons (Hightower and Meagher, 1986). Actins from plants and protozoans proved to be more divergent, having 11–30% divergence when analyzed by pairs, but apparently never tolerated peptide insertions, as all of the isoforms have lengths of 374–376 amino acids (Hightower and Meagher, 1986; Greslin, Loukin, Oka, and Prescott, 1988; Cupples and Pearlman, 1986; Ben Amar, Pays, Tebabi, Dero, Seebeck, Steinart, and Pays, 1988).

The discoveries of more distant actin relatives thus were unexpected, and have generated considerable excitement. Remarkably, the majority of findings made to date were serendipitous, in that cDNAs and genes that encode these proteins were recovered during random sequencing projects, or in the course of isolating unrelated DNA sequences. For example, during a polymerase chain reaction-based screen for the *Schizosaccharomyces pombe* tropomyosin gene, Lees-Miller, Henry, and Helfman (1992) isolated a gene, called *act2,* encoding a protein of 427 residues that bore only 35–50% similarity to conventional actin isoforms and, as might be surmised from the

Address correspondence to Eric Fyrberg, Department of Biology, The Johns Hopkins University, Charles and 34th Streets, Baltimore, MD 21218.

fact that it is over 50 residues longer than conventional actins, contains several peptide insertions. During the course of isolating a gene referred to as *PRP9* from *Saccharomyces cerevisiae* Schwob and Martin (1992) recovered a gene, also called *act2,* that encoded a protein of 391 amino acids that was 47% identical to conventional nonmuscle actin isoforms. Several additional cDNAs that specify actin-related proteins were discovered during random sequencing of expressed messenger RNAs of *Caenorhabditis elegans* and humans (Adams et al., 1992; Waterston et al., 1992; McCombie et al., 1992).

Because they have been discovered very recently, relatively little is known about the functions of these novel actin-related proteins, but several important clues have been discovered. Lees-Miller et al. (1992) and Schwob and Martin (1992) demonstrated by gene knockout experiments that both the *S. pombe* and *S. cerevisiae* actin-related genes were essential for viability. Schwob and Martin (1992) have furthermore inferred from the phenotype of haploid yeast strains having only one functioning *act2* gene that the gene product is involved in bud formation. However, it is my opinion that this conclusion is rather equivocal, as mutations in any of a number of genes that regulate the yeast cell cycle could confer a similar phenotype. Neither of these yeast actin-related proteins are likely to copolymerize with bona fide actins, because both have peptide insertions that will almost certainly interfere with the relevant interactions, based upon the structural models of monomeric and polymeric actin proposed by Kenneth Holmes and his collaborators (Kabsch, Mannherz, Suck, Pai, and Holmes, 1990; Holmes, Popp, Gebhard, and Kabsch, 1990).

Thus, the two actin-related proteins discovered in yeasts have peptide insertions that almost certainly prevent their polymerization, yet both retain overall similarities (35–50% amino acid identity) to conventional actins. As such, either protein could in theory interact with, but not polymerize with, monomeric or polymeric actin, and thus regulate some aspect of actin assembly dynamics or filament morphology. That hypothesis is consistent with two recent observations. The first is that homologs of these two proteins now have been found in several organisms, including *C. elegans* (D. Helfman, personal communication), *Drosophila* (Fyrberg and Fyrberg, 1994, our unpublished results), and mammals (Tanaka et al., 1992) indicating that they participate in fundamental processes. Second, steady state levels for both of these proteins are strongly substoichiometric, relative to those of conventional nonmuscle actins (Tanaka et al., 1992, Fyrberg and Fyrberg, 1994, our unpublished observations), suggesting that they are unlikely to be major structural components, and perhaps indicating that they instead regulate actin filament dynamics.

The biochemical function of a third member of the actin related family of proteins is known in more detail. This protein, referred to as actin-RPV (Lees-Miller, Helfman, and Schroer, 1992), or centractin (Clark and Meyer, 1992) has 376 residues and is roughly 54% identical to mammalian nonmuscle actin isoforms. Actin-RPV has been found to be part of the dynactin complex, a protein assembly that facilitates transport of vesicles along microtubules by cytoplasmic dynein (Lees-Miller et al., 1992; Paschal, Holzbauer, Pfister, Clark, Meyer, and Vallee, 1993). This is a remarkable finding for those interested in either cellular physiology or evolution, as it may be viewed as a cooperative venture of microtubule- and microfilament-based motility systems (see Goldstein and Vale, 1992). Precisely what role is played by actin-RPV in the dynactin complex is not known, however electron microscopy of the partially purified components indicates that it forms short stalks of uniform length (T. Schroer, personal communication). Perhaps these stalks help to link vesicles to the motor complex. As in the case of proteins specified by the *S. pombe* and *S. cerevisiae act2* genes, the role of actin-RPV is probably fundamental, because it is known to be present in nematodes, *Drosophila,* and vertebrates (D. Helfman, personal communication, our unpublished observations).

How many more actin-related proteins remain to be characterized, and how widely distributed are they? These basic questions must be answered before the roles of these proteins can be conceptualized accurately. We have addressed this issue by attempting to isolate all of the actin-related protein encoding genes from a single organism, the genetically tractable eukaryote *Drosophila melanogaster.* We synthesized degenerate oligonucleotide primers for two peptides, GDGVTH and VLSGGTT, that are highly conserved in all conventional actins as well as in all actin-related proteins characterized to date. Both peptides are located within the ATP-binding pocket, where the need for stereochemical constancy presumably is greater than elsewhere within actin. Intervening *Drosophila* genomic DNA was amplified using the polymerase chain reaction, cloned in vector pUC19, and sequenced using universal primers and the method of Sanger, Nicklen, and Coulson, (1977). From nearly 300 clones we found, in addition to the six genes encoding conventional *Drosophila* actin isoforms (Fyrberg, Bond, Hershey, Mixter, and Davidson, 1981), only four genes that specify actin-related proteins. Two of these encode, based upon their degrees of amino acid similarities, the homologs of the *S. pombe* and *S. cerevisiae* genes described above. A third gene is clearly homologous to actin-RPV. The fourth is, we believe, a gene without precedent. It is like actin-RPV in that it encodes a protein of 376 residues having no peptide insertions. However, the encoded protein sequence is not, based upon its sequence, an RPV isoform. Remarkably, the mRNA and protein appear to accumulate only within males, and preliminary observations suggest that it is expressed in testes. Perhaps this novel protein facilitates the activity of dynein in flagellated spermatozoa, hence serves a role somewhat analogous to that of actin-RPV with respect to cytoplasmic dynein.

In sum, our research and that of several other laboratories strongly suggest that the number of actin-related proteins is small, probably only three or four in most organisms, and also demonstrates that homologs of these proteins occur in a wide variety of phyla. The emerging picture is one of only a few proteins, each of which is involved in fundamental functions and conserved over vast evolutionary distances.

Is there any evidence for even more distant relatives of actin? The answer to the question is yes, and the story just as surprising as that described earlier. The case for divergent evolution is not quite as clear cut as for the actin-related proteins discussed above, but strong arguments for that mechanism can be marshalled nevertheless.

This part of the actin protein superfamily story is based entirely upon x-ray crystallography. As was described earlier in this manuscript, Kenneth Holmes and his collaborators (Kabsch et al., 1990; Holmes et al., 1990) solved the structures of monomeric and polymeric actin using improved area detectors and computer fitting programs. The atomic structures provided a number of surprises, two of which are highly germane to actin evolution. The first is that monomeric actin is comprised of four subdomains, two of which (one and three) are structurally similar despite their lacking any obvious primary sequence conservation (Kabsch et al., 1990). This finding strongly suggests that actin evolution entailed the duplication of a prototypical subunit. The second, more surprising finding is that this same basic structure is characteristic of sugar kinases, most notably hexokinase, the model enzyme for the "induced fit" mode of substrate binding (Steitz, Fletterick, Anderson, and Anderson, 1976; reviewed by Branden, 1990). Hexokinase, like actin, is an ATPase. However, actin and hexokinase functions are rather distinct. Actin splits ATP so as to implement its assembly into a filament (Pollard, 1990), whereas hexokinase phosphorylates glucose at the start of the glycolytic pathway (Middleton, 1990). Because the structural relatedness of the two proteins is not reflected in their primary sequences, it is not absolutely certain whether they are products of divergent or convergent evolution. However, it certainly is plausible that actin evolution entailed the duplication and divergence of part of its structure. The same divergent mechanism is the most likely explanation for the similarity of actin and hexokinase because, as Bork, Sander, and Valencia (1992) have pointed out, while the similarity of the ATP-binding pockets could be dictated by stereochemical limitations, hence occur by convergence, it is very difficult for that same mechanism to explain the structural similarity of the complete molecules.

An even more remarkable finding followed the solution of the 44-kD (ATP-binding) subunit of hsc70. This protein uses the energy liberated by ATP breakdown to chaperone proteins across membranes or to uncoat clathrin-coated vesicles (Gething and Sambrook, 1992). Hence, the non-ATPase portion of the molecule binds any number of ligands. When David McKay and collaborators (Flaherty, DeLuca-Flaherty, and McKay 1990; Flaherty, McKay, Kabsch, and Holmes, 1991) analyzed the structure of the 44-kD fragment they found that it is very similar to that of actin. 241 of 376 amino acids proved to be structurally equivalent when the two structures were superimposed. Of these, 39 were identical in the two proteins, and 56 additional amino acids were of very similar character, though not identical. However, the two proteins are effectively unrelated by their primary sequences. As in the case of actin comparisons to sugar kinases, one cannot be absolutely sure whether the actin similarity to molecular chaperonins is due to convergent or divergent evolution without further analysis, but the more compelling case can be constructed for divergent evolution.

These just described structural similarities have been used to construct an algorithm for recognizing additional members of the actin superfamily (Bork et al., 1992). In the course of detailed structural comparisons of actin, hsc70, and hexokinase these researchers found five small peptide motifs whose "characters" are conserved in all three proteins. By searching the Swiss-Prot database for proteins having these motifs, they discovered that several prokaryotic proteins involved in driving the cell cycle and stabilizing plasmids are likely to be members of the actin protein superfamily. If this finding is ultimately confirmed it demonstrates that the actin protein superfamily predates the divergence of eukaryotes and prokaryotes despite the fact that actin does not occur in prokaryotes.

In conclusion, both gene cloning and x-ray crystallography have contributed to our understanding of the actin protein superfamily. The recently discovered actin-related proteins described in the first portion of this paper are clearly products of divergent evolution, and it will be very interesting to detail their cellular roles. Although actin is clearly related to the sugar kinases and certain molecular chaperonins, the case for divergent evolution is not quite as certain, although Bork et al. (1992) have put forth compelling arguments. In any event, actin and its relatives are now interesting subjects for a variety of evolutionary studies, some of which may well inform us further as to physiological mechanisms.

References

Adams, M. D., J. M. Kelley, J. D. Gocayne, M. Dubnick, M. H. Polymeropoulos, H. Xiao, C. R. Merril, A. Wu, B. Olde, R. F. Moreno, A. R. Kerlavage, W. R. McCombie, and J. C. Venter. 1991. Complementary DNA sequencing: expressed sequence tags and human genome project. *Science.* 252:1651–1656.

Ben Amar, M. F., A. Pays, P. Tebabi, B. Dero, T. Seebeck, M. Steinert, and E. Pays. 1988. Structure and transcription of the actin gene of *Trypanosoma brucei. Molecular and Cellular Biology.* 8:2166–2176.

Bork, P., C. Sander, and A. Valencia. 1992. An ATPase domain common to prokaryotic cell cycle proteins, sugar kinases, actin, and hsp70 heat shock proteins. *Proceedings of the National Academy of Sciences, USA.* 89:7290–7294.

Branden, C.-I. 1990. Founding fathers and families. *Nature.* 346:607–608.

Bray, D. 1972. Cytoplasmic actin: a comparative study. *Cold Spring Harbor Symposium on Quantitative Biology.* 37:567–571.

Clark, S. W., and D. I. Meyer. 1992. Centractin is an actin homologue associated with the centrosome. *Nature.* 359:246–250.

Cupples, C. G., and R. E. Pearlman. 1986. Isolation and characterization of the actin gene from *Tetrahymena thermophila Proceedings of the National Academy of Sciences, USA.* 83:5160–5164.

Flaherty, K. M., C. DeLuca-Flaherty, D. B. McKay. 1990. Three-dimensional structure of the ATPase fragment of a 70 K heat-shock cognate protein. *Nature.* 346:623–628.

Flaherty, K. M., D. B. McKay, W. Kabsch, and K. C. Holmes. 1991. Similarity of the three-dimensional structures of actin and the ATPase fragment of a 70-kDa heat shock cognate protein. *Proceedings of the National Academy of Sciences, USA.* 88:5041–5045.

Fyrberg, E. A., B. J. Bond, N. D. Hershey, K. S. Mixter, and N. Davidson. 1981. The actin genes of *Drosophila:* protein coding regions are highly conserved but intron positions are not. *Cell.* 24:107–116.

Fyrberg, C., and E. Fyrberg. 1994. A Drosophila homolog of the *Schizosaccharomyces pombe* act2 gene. *Biochemical Genetics.* In press.

Garrels, J. I., and W. Gibson. 1976. Identification of multiple forms of actin. *Cell.* 9:793–805.

Gething, M. J., and J. Sambrook. 1992. Protein folding in the cell. *Nature.* 355:33–42.

Goldstein, L. S. B., and R. D. Vale. 1992. New cytoskeletal liasons. *Nature.* 359:193–194.

Greslin, A. F., S. H. Loukin, Y. Oka, and D. M. Prescott. 1988. An analysis of the macronuclear actin genes of oxytricha. *DNA.* 7:529–536.

Gruenstein, E., and A. Rich. 1975. Non-identity of muscle and non-muscle actins. *Biochemical Biophysical Research Communication.* 64:472–477.

Hightower, R. C., and R. B. Meagher. 1986. The molecular evolution of actin. *Genetics.* 114:315–332.

Holmes, K. C., D. Popp, W. Gebhard, and W. Kabsch. 1990. Atomic model of the actin filament. *Nature.* 347:44–49.

Kabsch, W., H. G. Mannherz, D. Suck, E. F. Pai, and K. C. Holmes. 1990. Atomic structure of the actin:DNase I complex. *Nature.* 347:37–43.

Lees-Miller, J. P., G. Henry, and D. M. Helfman. 1992. Identification of *act2,* an essential gene in the fission yeast *Schizosaccharomyces pombe* that encodes a protein related to actin. *Proceedings of the National Academy of Sciences, USA.* 89:80–83.

Lees-Miller, J. P., D. M. Helfman, and T. A. Schroer. 1992. A vertebrate actin-related protein is a component of a multisubunit complex involved in microtubule-based vesicle motility. *Nature.* 359:244–246.

McCombie, W. R., M. D. Adams, J. M. Kelley, M. G. Fitzgerald, T. R. Utterback, M. Khan, M. Dubnick, A. R. Kerlavage, J. C. Venter, and C. Fields. 1992. *Caenorhabditis elegans* expressed sequence tags identify gene families and potential disease gene homologues. *Nature and Genetics.* 1:124–131.

Middleton, R. J. 1990. Hexokinases and glucokinases. *Biochemical Society Transactions.* 18:180–183.

Paschal, B. M., E. L. F. Holzbaur, K. K. Pfister, S. Clark, D. I. Meyer, and R. B. Vallee. 1993. Characterization of a 50-kDa polypeptide in cytoplasmic dynein preparations reveals a complex with p150^{Glued} and a novel actin. *Journal of Biological Chemistry.* 20:15318–23.

Pollard, T. D. 1990. Actin. *Current Opinions in Cell Biology.* 2:33–37.

Sanger, F., S. Nicklen, and A. R. Coulson. 1977. DNA sequencing with chain-terminating inhibitors. *Proceedings of the National Academy of Sciences, USA.* 74:5463–5467.

Schwob, E., and R. P. Martin. 1992. New yeast actin-like gene required late in the cell cycle. *Nature.* 355:179–182.

Steitz, T. A., R. J. Fletterick, W. F. Anderson, and C. M. Anderson. 1976. High resolution x-ray structure of yeast hexokinase, an allosteric protein exhibiting a non-symmetric arrangement of subunits. *Journal of Molecular Biology.* 104:197–222.

Tanaka, T., F. Shibasaki, M. Ishikawa, N. Hirano, R. Sakai, J. Nishida, T. Takenawa, and H. Hirai. 1992. Molecular cloning of bovine actin-like protein, actin2. *Biochemical Biophysical Research Communication.* 187:1022–1028.

Vandekerckhove, J., and K. Weber. 1978. Mammalian cytoplasmic actins are the product of at least two genes and differ in primary structure in at least 25 identified positions from skeletal muscle actins. *Proceedings of the National Academy of Sciences, USA.* 75:1106–1110.

Waterston, R., C. Martin, M. Craxton, C. Huynh, A. Coulson, L. Hillier, R. Durbin, P. Green, R. Shownkeen, N. Halloran, M. Metzstein, T. Hawkins, R. Wilson, M. Berks, Z. Du, K. Thomas, J. Thiery-Mieg, and J. Sulston. 1992. A survery of expressed genes in *Caenorhabditis elegans. Nature and Genetics.* 1:114–123.

Chapter 4

Structure, Function, Evolution of Plasma Membrane Proteins

Ion Channels of Microbes

Yoshiro Saimi, Boris Martinac*, Robin R. Preston‡, Xin-Liang Zhou, Sergei Sukharev, Paul Blount, and Ching Kung

*Laboratory of Molecular Biology and Department of Genetics, University of Wisconsin, Madison, Wisconsin 53706; *Department of Pharmacology, The University of Western Australia, Nedlands, Western Australia 6009; ‡Department of Physiology, Medical College of Pennsylvania, Philadelphia, Pennsylvania 19129*

The presence of pathways for the entry and exit of solutes distinguish biological membrane from artificial lipid bilayer. Among these pathways are membrane channels that allow passive flow of solutes along their individual electrochemical gradients. Channel activities and channel-gene sequences in microbes, reviewed here, indicate an early emergence in evolution. As is evident from other chapters of this volume, microbes, which encompass many eucaryotes and all procaryotes, are vastly more diverse than plants and animals combined. Besides our interest in microbial biology and channel evolution, we would like to study the molecular details of ion-channel structures and functions. Thus, most of the microbial species examined with patch clamp are those that are genetically tractable and that have been used extensively in biochemical or molecular-biological research. We will review work on *Paramecium, Dictyostelium,* budding and fission yeasts, *Escherichia coli,* and *Bacillus subtilis.*

Ion-channel activities discovered in these organisms are summarized in Table I. Before the advent of patch clamp, electrical measurements could only be made reliably on giant cells in vivo, or by incorporating channel proteins into planar lipid bilayers in vitro. The patch-clamp method not only allows us to examine the activities of individual ion channels (Hamill, Marty, Neher, Sakmann, and Sigworth, 1981) but also allows the study of small cells. Few modifications of standard solutions, electrodes, and amplifiers are needed to record from microbes. Rather, the challenge is to generate appropriate preparations of microbial membranes that can form gigaohm seals with patch-clamp pipets. For example, the *Paramecium* surface is smothered with cilia. Yeast membrane is overlaid with a cell wall, and different yeasts differ in their wall components. *E. coli* has two membranes, with a cell wall sandwiched between the outer and the inner membrane. No single method can be applied to all these preparations. Specific methods used to study channels from these microbes can be found in Saimi, Martinac, Delcour, Minorsky, Gustin, Culbertson, Adler, and Kung (1992) and references cited under different organisms below.

Paramecium

A paramecium swims by means of 5,000 incessantly beating cilia. Like those of metazoa, each cilium consists of a dynein-ATPase powered 9 + 2 assembly of sliding

TABLE I
Microbial Ion-Channel Activities

Species	Current	Activation	Selectivity	Observed as mac.*	Observed as mic.* (pS)	Ref.[§§]
Paramecium tetraurelia	$I_{Ca(d)}$	depolarization	Ca^{2+}	+	+‡(1.6)	1
	$I_{K(d)}$	depolarization	K^+	+		2
	$I_{K(Ca,d)}$	Ca^{2+}, depolarization	K^+	+	+(150)	3,9
	$I_{Na(Ca)}$	Ca^{2+}§	Na^+	+	+(19)	4
	$I_{Mg(Ca)}$	Ca^{2+}	Mg^{2+}	+		5
	$I_{Ca(h)}$	hyperpolarization	Ca^{2+}	+		6
	$I_{K(Ca,h)}$	Ca^{2+}, hyperpolarization	K^+	+	+(72)	7
	$I_{M(mech)}$	mechanical‖	M^{2+}	+		8
	$I_{K(mech)}$	mechanical¶	K^+	+		9
	I_M	depolarization	M^+		+(40)	10
	$I_{M(ATP)}$	ATP	M^+		+(160)	10
	I_{Cl}	Ca^{2+}	Cl^-		+(18)	10
Dictyostelium discoideum	$I_{K(dI)}$	depolarization	K^+		+(11)	11
	$I_{K(dII)}$	depolarization	K^+		+(6)	11
	$I_{Ca(h)}$	hyperpolarization	Ca^{2+}		+(3)	11
Saccharomyces cerevisiae	$I_{K(d)}$	depolarization	K^+	+	+(20)	12
	$I_{M(mech)}$	mechanical	$M^+ > A^-$	+	+(36)	13
Schizosaccharomyces pombe	$I_{M(mech)}$	mechanical	$M^+ > A^-$		+(180)	14
	$I_{?(v)}$**	voltage	?		+(<153)	15
Escherichia coli	porins‡‡	voltage	$M^+ \geq A^-$		+(200)	16
	MscS	mechanical	$M^+ \leq A^-$		+(1,000)	17
	MscL	mechanical	$M^+ = A^-$		+(2,500)	18
Bacillus subtilis	$I_{(mech)}$	mechanical	$M^+ = A^-$		+(100–3,000)	19

*Macroscopic currents (mac.) recorded with a two-electrode voltage clamp on *Paramecium* and with a patch clamp in the whole-cell mode for yeast; microscopic currents (mic.) by excised patch mode.
‡Recorded from ciliary membrane material in planar lipid bilayers. (Ehrlich et al., 1984).
§Ca^{2+} activation apprently through direct channel-calmodulin interaction.
‖Located in the soma (nonciliary) membrane at the anterior of the *Paramecium* cell.
¶Located in the soma (nonciliary) membrane at the posterior of the *Paramecium* cell.
**Poorly selective, exhibiting several conductances.
‡‡Porins are the only channel proteins, the x-ray structure of which is known.
§§References: 1. Oertel et al. (1977); Brehm and Eckert (1978); Ehrlich et al. (1984). 2. Machemer and Ogura (1979); Satow and Kung (1980*a*). 3. Satow and Kung (1980*b*); Preston et al. (1990*b*). 4. Saimi and Kung (1980); Saimi (1986). 5. Preston (1990). 6. Preston et al. (1992*a,b*). 7. Preston et al. (1990*a*). 8. Ogura and Machemer (1980); Satow et al. (1983). 9. Ogura and Machemer (1980). 10. Martinac et al. (1988); Saimi, Martinac, Preston, unpublished observations. 11. Mueller and Hartung (1990). 12. Gustin et al. (1986); Bertl et al. (1992, 1993). 13. Gustin et al. (1988). 14. Zhou and Kung (1992). 15. Vacata et al. (1992). 16. Schindler and Rosenbusch (1978); Berrier et al. (1991); Delcour et al. (1989*b*, 1991). 17. Martinac et al. (1987). 18. Sukharev et al. (1993, 1994). 19. Zoratti et al. (1990); Martinac et al. (1992).

microtubules enclosed in a ciliary membrane that is continuous with the soma membrane. The *Paramecium* membrane is excitable and ciliary motility is governed by membrane electrogeneses. The avoiding reaction (Jennings, 1906), for example, correlates with a Ca^{2+}-based action potential, and the resultant intraciliary increase of Ca^{2+} causes the cilia to beat in a reversed direction to propel the cell backward (Eckert and Naitoh, 1972). Hyperpolarization triggers other ion currents and a more rapid ciliary beat in a near-normal direction, causing the cell to spurt forward (Naitoh and Eckert, 1973). Cyclic AMP is thought to play an important role in this spurting response (Bonini, Gustin, and Nelson, 1986; Nakaoka and Machemer, 1990). Recently, Schultz and co-workers (Schultz, Klumpp, Benz, Schurhoff-Goeters, and Schmid, 1992) suggested that, the K^+ efflux underlying membrane hyperpolarization and cAMP increase in *Paramecium* might be effected by the same entity—the ciliary adenylate cyclase. In support of this hypotheses, they found the purified cyclase, upon incorporation into planar lipid bilayer, to be associated with a 320-pS, nonselective conductance.

Because the cell is large, some 300 μm in length, *Paramecium* has been investigated with the classical two-electrode voltage clamp as well as patch clamp. Experiments with *P. tetraurelia* using these two methods revealed at least 13 types of ion currents: $I_{Ca(d)}$, $I_{Ca(h)}$, $I_{M(mech)}$, $I_{K(d)}$, $I_{K(h)}$, $I_{K(Ca,d)}$, $I_{K(Ca,h)}$, $I_{K(mech)}$, $I_{Na(Ca)}$, $I_{Mg(Ca)}$, I_{Cl}, I_M, $I_{M(ATP)}$ (Table I). This diversity echoes the anatomical and functional complexity of this highly differentiated one-cell animal (Jurand and Selman, 1969) that has to move, recoil, adapt, endocytose, excrete, secrete, divide, mate, autogamize, et cetera (Goertz, 1988).

When one examines *Paramecium* bathed in various external solutions under a two-electrode voltage clamp, membrane depolarizations can trigger at least five currents. (*a*) $I_{Ca(d)}$, a depolarization-activated Ca^{2+} current: this current activates rapidly in response to voltage and then inactivates in a Ca^{2+}-dependent manner (Oertel et al., 1977; Brehm and Eckert, 1978). Deciliation abolishes this current, indicating that the channels are located in the Ca^{2+} ciliary membrane and not the body membrane (Dunlap, 1977; Machemer and Ogura, 1979). Pawn mutations abolish or reduce this current (Kung and Eckert, 1972; Oertel et al., 1977). (*b*) $I_{K(d)}$, a rapid K^+ current activated by the depolarization that inactivates slowly (Machemer and Ogura, 1979; Satow and Kung, 1980*a*). (*c*) $I_{Na(Ca)}$, an Na^+ current activated by internal Ca^{2+}, presumably from the above Ca^{2+} current (Saimi and Kung, 1980; Saimi, 1986). A subclass of calmodulin mutants called "fast-2" has little or none of this Na^+ current (Saimi, 1986; Kink, Maley, Preston, Ling, Wallen-Friedman, Saimi, and Kung, 1990; Kung, Preston, Maley, Ling, Kanabrocki, Seavey, and Saimi, 1992). (*d*) $I_{K(Ca,d)}$, a K^+ current, also activated by internal Ca^{2+} (Satow and Kung, 1980*b*, Preston, Wallen-Friedman, Saimi, and Kung, 1990*b*). A different subclass of calmodulin mutants called "pantophobiacs" shows large reductions in this current (Saimi, Hinrichsen, Forte, and Kung, 1983, Kink et al., 1990; Preston, Saimi, Amberger, and Kung, 1990*b*). This current is greatly enhanced in the *teaA* mutant, (Hennessey and Kung, 1987; Preston et al., 1990*c*). (*e*) $I_{Mg(Ca)}$, a Mg^{2+}-specific current activated by Ca^{2+} (Preston, 1990). This novel current, first ever reported in any organism, has a selectivity of $Mg^{2+} > Mn^{2+} = Co^{2+} > Sr^{2+} = Ba^{2+} \gg Ca^{2+}$. It is affected in some of the calmodulin mutants (Preston, unpublished results).

Hyperpolarizations under voltage clamp also reveal at least five currents. (*a*) $I_{Ca(h)}$, an unusual Ca^{2+} current. Unlike other Ca^{2+} channels, this permeability is not only impermeant to, but actually blocked by Ba^{2+} (Preston et al., 1992*a*). It is also blocked by amiloride and shows a Ca^{2+}-dependent inactivation (Preston *et al.*, 1992*b*). Nakaoka and Iwatsuki (1992) found a sustained component of a current that correlates with ciliary beat frequency increase upon hyperpolarizations in *P. caudatum*. This current seems to have some of the characteristics of $I_{Ca(h)}$ of *P. tetraurelia*. (*b*) $I_{K(h)}$, a hyperpolarization-activated K^+ current blockable by quinidine (Oertel, Schein, and Kung, 1978; Preston et al., 1990*a*). (*c*) $I_{K(Ca,h)}$, a K^+ current activated after $I_{Ca(h)}$. This current is kinetically and pharmacologically distinct from $I_{K(Ca,d)}$ (Preston et al., 1990*a,c*). Calmodulin mutations of the pantophobiac type reduce or eliminate this current, while some of the fast-2 type calmodulin mutations and *rst* cause an early activation of it (Preston et al., 1990*b,c*). (*d*) $I_{Na(Ca)}$. This current, induced by $I_{Ca(h)}$, is believed to flow through the same Na^+ channel activated by Ca^{2+} from $I_{Ca(d)}$, because the Na^+ currents at either polarizations are affected by the fast-2 type of calmodulin mutations in identical manners (Saimi, 1986; Preston, Kink, Hinrichsen, Saimi, and Kung, 1991). (*e*) $I_{Mg(Ca)}$. This current also most likely flows through the same Mg^{2+} channel activated by $I_{Ca(d)}$, since the Mg^{2+} currents observed at opposite polarizations have the same ion selectivity and blockage pattern (Preston, 1990).

When a *Paramecium* strikes an obstacle in its swim path, it backs up. Mechanical impacts on the anterior end of electrode-pinned cells elicit a depolarization prior to the action potential. This receptor potential was found to be based on Ca^{2+} (Eckert, Naitoh, and Friedman, 1972). Although nearly all divalent cations were able to carry $I_{Ca(mech)}$, the corresponding receptor current under voltage clamp, Ca^{2+} presumably is the natural carrier, given the normal extra- and intracellular distributions of ions (Ogura and Machemer, 1980, Satow et al., 1983). Touching the posterior of a *Paramecium* causes it to spurt forward. This mechanical stimulation induces a hyperpolarization by means of a K^+ efflux, $I_{K(mech)}$ (Naitoh and Eckert, 1973; Ogura and Machemer, 1980).

The activities of two types of divalent-cation passing channels were reported, when a fraction of *Paramecium* ciliary membrane was incorporated into planar lipid bilayers, (Ehrlich et al., 1984): (*a*) a voltage-independent 29 pS channel that permeates Mg^{2+} slightly better than Ba^{2+} and Ca^{2+}. This channel might be equivalent to the anterior mechanoreceptor channel. (*b*) A voltage-dependent 1.6 pS channel that favors Ba^{2+} and Ca^{2+} greatly over Mg^{2+}. Its activities are affected by calmodulin antagonists (Ehrlich, Jacobson, Hinrichson, Sayrer, and Forte, 1988). This channel current might be the microscopic equivalent of $I_{Ca(d)}$.

The application of patch clamp to surface blisters or detached cilia from *Paramecium* allowed recording of activities from many different types of ion channel. These include at least four types of K^+ channels, a Ca^{2+} channel, a Na^+ channel, a Cl^- channel (I_{Cl}), 2 cation-nonspecific channels (I_M), one of which can be activated by ATP ($I_{M(ATP)}$) (Martinac et al., 1988; Saimi, Martinac, Preston, unpublished observations). The activities of two of these channels have been documented extensively. (*a*) A Ca^{2+}-dependent K^+ channel of 72 pS conductance, which is more active upon hyper- than depolarization (Saimi and Martinac, 1989). This appears to be the microscopic equivalent of $I_{K(Ca,h)}$. Limited proteolytic digestion activates this channel and makes it independent of Ca^{2+} and voltage (Kubalski, Martinac, and Saimi, 1989). (*b*) A 19-pS Ca^{2+}-dependent Na^+ channel, which depends on Ca^{2+}-calmodulin for its activation (Saimi and Ling, 1990). This is apparently the channel responsible for the macroscopic $I_{Na(Ca)}$ above. For details on these two types of channels, see Preston et al., 1991.

The widely studied voltage-gated K^+-channel genes in the *Shaker* family, including the closely related members: *Shaker, Shal, Shab* and *Shaw,* have been found in various species, ranging from insects to mammals (Salkoff, Baker, Butler, Covarrubias, Pak, and Wei, 1992). Such genes have recently been discovered in jellyfish by polymerase chain reactions and library searches (Jegla, Laraque, Gallin, Grigoriev, Spencer, Salkoff, 1993). Similar techniques, however, failed to discover such homologs in *Paramecium* and budding yeast. Recent studies also revealed a loosely connected group of channel genes, only remotely related to the *Shaker* in terms of sequences. However, these channels have a structural motif recognizably *Shaker*-like: with six putative membrane-crossing helices (S1–S6) and a putative pore forming region (H5). Members of this broader group include the cyclic nucleotide-gated cation channels in vertebrate rod outer segments (Kaupp et al., 1989) and olfactory cilia, the *eag* channel of *Drosophila* (Warmke, Drysdale, and Ganetsky, 1991; Bruggemann, Pardo, Stuhmer, and Pongs, 1993), and the inward rectifier K^+ channels of the mustard *Arabidopsis* (Schachtman, Schroeder, Lucas, Anderson, and Gaber, 1992). These channels differ in their conductances, ion selectivity and gating mechanisms. Recently, Jegla et al. (1993) discovered several genes in *Paramecium* that apparently encode channels of this group. Although the activities of the corresponding channel proteins have yet to be determined, their work shows that ancestral forms of this extended family have apparently originated before the divergence of protozoa and metazoa.

Dictyostelium

Dictyostelium discoideum is a slime mold that grows as individual amoebae. Upon starvation, the amoebae aggregate to form a multicellular fruiting body using cAMP as an intercellular signal. Aggregating amoebae, alternating in time and in location with respect to neighbors in a swarm, release and respond to cAMP. Among other effects, binding of cAMP appears to stimulate a Ca^{2+} uptake (Bumann, Malchow, and Wurster, 1984) and a K^+ release (Aeckerle, Wurster, and Malchow, 1985).

In their first reported patch-clamp study of *Dictyostelium* cells, Mueller, Malchow, and Hartung (1986) found a 9-pS channel that probably passes K^+. More recently, Mueller and Hartung (1990) reexamined the membranes of amoebic cells after induction of differentiation by nutrient removal. In the cell-attached configuration, they consistently observed the activities of three types of unit conductance: DI, an 11-pS K^+ conductance activated upon depolarization; DII, a 6-pS K^+ conductance activated upon depolarization; and HI, a 3-pS Ca^{2+}-passing conductance activated upon hyperpolarization. Based on probability of encounter, channel densities appear low: $\sim 0.1/\mu m^2$ for DI and HI and $\sim 0.02/\mu m^2$ for DII.

Saccharomyces Cerevisiae

This budding yeast has been used extensively in modern biological research. As expected, Ca^{2+}, K^+, Na^+, Mg^{2+}, and several trace metals are required for its growth. Ca^{2+}, in particular, is needed for cell-cycle progression (Iida, Sakaguchi, Yagawa, and Anraku, 1990). At least two CDC proteins, products of **c**ell-**d**ivision **c**ycle genes, have the EF-hand structures indicative of their being Ca^{2+} binding proteins (Baum, Furlong, and Byers, 1986; Miyamoto, Okya, Ohsumi, and Anraku, 1987).

The plasma membrane of *S. cerevisiae* can form a gigaseal with patch-clamp pipettes after enzymatic removal of the cell wall. Patch-clamp investigation showed activities of at least two types of ion channels. (*a*) A 20-pS channel is activated upon depolarization above a threshold of ~ -10 mV but that is not activated by hyperpolarization. This channel is highly specific for K^+ over Na^+ and is blocked by the usual K^+-channel blockers, tetraethylammonium and Ba^{2+} (Gustin et al., 1986). Bertl et al. (1992, 1993) found that the activities of this channel at different voltages are affected in different manners by Ca^{2+} in the cytoplasmic side. Surprisingly, Ramirez, Vacata, McCusker, Haber, Mortimer, Owen, and Lecar (1989) assigned an electric activity to a K^+ channel, which appears at $> +100$ mV or < -100 mV in a wild type. In a mutant defective in the structural gene for the plasma-membrane H^+-ATPase, however, they found this activity to appear at lower voltage, in the presence of > 50 mM ATP. The mutation appears to make this K^+ channel sensitive to cytoplasmic ATP. It is difficult to see how an altered ATPase can affect the channel in this manner. (*b*) Stretch forces exerted on the membrane activate a 36-pS channel, which is blockable by 10 μM gadolinium applied to the cytoplasmic side. This channel passes a variety of ions including Ca^{2+}, but prefers cations over anions. This channel also shows an adaptation to a sustained stretch force. Interestingly, this adaptation depends on voltage. It takes place when force is applied when the membrane is polarized (cytoplasmic negative); a force applied during depolarization (positive) activates these channels but does not inactivate them. In whole-cell recording, previously inactivated channels can apparently recover during depolarization and force application. Experiments with whole-spheroplasts of different sizes showed that the channel-activating stimulus is a tension along the plane of the membrane and not a pressure perpendicular to it (Gustin et al., 1988). This stretch force is most conveniently applied by a suction in the on-cell patch mode or a pressure in the whole-cell mode. Changing bath osmolarity, can also activate these channels recorded in the whole-cell mode (Zhou, Gustin, and Martinac, unpublished observation). A failure to elicit the whole-cell macroscopic mechanosensitive current anticipated by microscopic currents seen in patches excised from snail neurons, in particular, cautioned us against the possible artifactual origin of mechanically induced activities in general (Morris and Horn, 1991). However, the expected macroscopic currents have been recorded from *S. cerevisiae* spheroplasts and were found to saturate at higher stretch forces, as expected of a finite number of openable channels in a yeast cell and not expected of membrane breakdown and leakage (Gustin et al., 1991). Although the function(s) of the mechanosensitive channels is not yet clear, it seems reasonable to suppose that they play a role in osmoregulation. Yeasts show clear physiological responses to osmotic stress (Blomberg and Adler, 1989). An increase in external osmolarity is followed by an internal osmolarity adjustment. The downstream events of this osmosensing signal transduction pathway include a protein phosphorylation cascade similar to the MAP kinase and MAP-kinase kinase pathway (Brewster, de Valoir, Dwyer, Winter, and Gustin, 1993). Whether ion channels are involved, especially in the upstream events such as sensing the osmotic changes, remains to be determined.

Schizosaccharomyces Pombe

All unicellular fungi are called yeasts. However, unicellularity is not a profound trait; fission yeasts and budding yeasts apparently separated very early in evolution.

Among their many differences, they differ in cell shape and wall compositions. Treatment of the fission yeast, *Schizosaccharomyces, pombe,* with zymolyase causes protoplast protuberances to appear at the pointed end of the cell. These blebs can readily seal onto patch-clamp pipettes. The best documented ion channel from these blebs is a mechanosensitive channel (Zhou and Kung, 1992). Currents through one or two such channels and ensemble currents of tens of units have been recorded from excised patches and whole-blebs, respectively. Under suction, the open probability of these channels is biased by membrane voltage, both in activation and in adaptation. Like those in *S. cerevisiae,* these channels can be blocked by submillimolar gadolinium. However, the *S. pombe* channels are more conductive (180 pS) and more selective ($P_{K+}/P_{Cl-} = 3.6$) than their *S. cerevisiae* cousins. In addition, depolarization inactivates the *S. pombe* channel while it activates the *S. cerevisiae* channel, during force application. There are also reports of activities of other ion channels in the *S. pombe* plasma membrane. Vacata, Hofer, Larson, and Lecar (1992) noted the activity of a voltage-gated poorly selective channel of four conductance levels, and Zhou and Kung (1992) intimated three types of voltage-activated channels besides the mechano-sensitive one in this membrane. More detailed reports are anticipated.

Escherichia Coli

Escherichia coli, like other Gram-negative bacteria, has two cell membranes separated by a peptidylglycan cell wall forming a complex collectively referred to as the cell envelope. The outer membrane is specialized, having an outer monolayer composed of lipopolysaccharides instead of the usual phospholipids. The inner membrane is the true cytoplasmic membrane that houses the machinery for electron transport, nutrient uptake, chemoreception, et cetera.

Some 10^5 porin molecules constitute a major portion of the outer membrane of an *E. coli* cell. The major porins: OmpC, OmpF, and PhoE, are more than 70% homologous in their amino-acid sequences (Jeanteur, Lakey, and Pattus, 1991). Although trimeric, each ~30-kD porin monomer is an individual aqueous channel that can pass hydrophilic solutes up to 600 m.w. X-ray crystallography showed that each monomeric subunit consists of a 16-stranded antiparallel β-barrel enclosing a pore (Cowan, Schermer, Rummel, Steiert, Ghosh, Pauptit, Jansonius, and Rosenbusch, 1992). A loop between β strand 6 and 7 folds into the pore, constricting the pore at about half the height of the barrel to 7 × 11 Å in an elliptical cross section. This constriction apparently defines the conductance and selectivity of porin. Biochemical studies led to the view that the outer membrane acts as an inert molecular sieve with porins being the static pores (Benz, 1988; Nikaido, 1992). Such a sieve should have little electric resistance. Yet, the surfaces of giant spheroplasts or giant cells of *E. coli* (Buechner, Delcour, Martinac, Adler, and Kung, 1990; Saimi et al., 1992) can form gigaohm seals with patch-clamp pipettes (Martinac et al., 1987; Kubalski, Martinac, Adler, and Kung, 1993; Sukharev et al., 1993) indicating that the porins in the patch are closed. Either porins are dynamic and are often closed but openable in vivo, or the patch-clamp manipulation induces porin closure. Studies in planar bilayers or with reconstituted liposomes under patch clamp showed that porins can be closed by membrane polarization (Schindler and Rosenbusch, 1978; Morgan, Lansdale, and Adler, 1990; Berrier et al., 1991). Delcour et al. (1989*b,* 1991) found the activities of wild-type or a point-mutant OmpC to be closable by pipette voltage, when reconstituted in liposomes and examined under a patch clamp.

However, the activities of OmpC examined in this manner differ from those in planar bilayer for unknown reasons. Ascribing a physiological function to voltage gating of porins is difficult based on current knowledge. However, there is a Donnan potential across the outer membrane. If periplasmic charges are organized and are intimately associated with the inner monolayer of the outer membrane, the potential across this membrane, as experienced by porins, may be larger than can be estimated (Stock, Rauch, and Roseman, 1977). Major contributors to these charges are the membrane-derived oligosaccharides (MDO's) in the periplasm. Interestingly, Delcour, Martinac, Adler, and Kung (1992) showed recently that MDO not only reduces the OmpC currents but also promotes their cooperative closures.

A suction delivered to a patch activates at least two types of channels in giant spheroplasts: a **m**echano-**s**ensitive **c**hannel of a **s**maller conductance (MscS, ~1 nS, Martinac et al., 1987) and a mechano-sensitive channel of **l**arge conductance (MscL, ~2.5 nS, Sukharev et al., 1993, 1994). By the standard of metazoan channels these are both channels of extraordinarily large conductances. Szabo, Petronilli, Guerra, and Zoratti (1990) reported *E. coli* mechanosensitive currents of many different size classes besides the two encountered by Sukharev et al. (1993, 1994). Differences in strains, procedures, and interpretations of channel substate behavior may account for some of the differences.

Mild suction or pressure greatly increases the open probability of MscS in a manner predictable by a Boltzmann distributions, in which the applied mechanical energy partitions the channel between its open and closed conformations. An osmotic change in the solution bathing an inside-out patch can also activate this channel (Martinac et al., 1992). Membrane depolarization favors the open state. This channel has a conductance of ~1 nS, but visits substates of lower conductance. The dwell time in these substates is much longer in a mutant lacking the major lipoprotein (Kubalski et al., 1993). It favors the passage of anions slightly over cations and its kinetic behavior is affected by the ionic species (Martinac et al., 1987). MscL is activated upon suctions higher than that needed to activate MscS. It also shows a much faster open/close kinetics than does MscS (Sukharev et al., 1993).

E. coli mechanosensitive channels in membrane fragments have been reconstituted into asolectin liposomes and found to retain the same functions observed in spheroplasts, including their ability to be activated by mild suction (Delcour, Martinac, Adler, and Kung, 1989*a,* Sukharev et al., 1993). This functional reconstitution opens the way to a biochemical purification of the mechanosensitive channel proteins. Buechner et al. (1990) reported that mutants lacking individual porin species retain a full complement of mechanosensitive conductances, arguing against MscS and MscL being porins. Unlike channels gated by ligands or voltage, no mechanosensitive channel proteins or their corresponding genes had been identified until the recent report by Sukharev et al. (1994). These authors identified the MscL protein and sequenced the corresponding gene, *mscL,* that has no homology to any known channel genes. Because this is the first mechanosensitive channel gene to be cloned, much more work is needed to see how mechanosensitive channels are related in molecular terms.

A biophysical approach has been taken to determine how force applied to the membrane is transduced to the channel protein. Cell membranes are asymmetric; the inner monolayers have more phospholipids with negatively charged head groups than do the outer monolayers (Rothman and Lenard, 1977). The outer membrane of

E. coli is even more asymmetric, because its outer monolayer is of an unusual composition. An amphipathic molecule is expected to partition differentially into the two monolayers because of this asymmetry. The uneven insertion of amphipaths leads to local membrane buckling, which is readily visible in blood cells, as predicted by the bilayer couple hypothesis (Sheetz and Singer, 1974). Such a perturbation is expected to create a mechanical stress force in the membrane. Martinac, Adler, and Kung (1990) showed that a variety of amphipaths indeed activate mechanosensitive channels of *E. coli.* Furthermore, the time courses and effectiveness in channel activation by different amphipaths correspond to their known solubilities into lipid. In the study of animal cells, it is believed that the gating force is transmitted to the mechanosensitive channels through the cytoskeleton (Sachs, 1988). The amphipath experiment and the reconstitution experiment in *E. coli* suggest that the channel responds to stress forces in the lipid bilayer instead. Indeed, Sukharev et al. (1993, 1994) have shown that MscS and MscL can be solubilized in a detergent and reconstituted into liposomes of foreign lipids, yet retain their mechanosensitive-channel characteristics. Thus, bacterial channels appeared to be gated by tension transduced exclusively via the lipid bilayer. In *E. coli,* the outer membrane is anchored to the peptidylglycan (cell wall) which provides rigidity. Digestion of this peptidylglycan with lysozyme activates rather than inactivates the channel (Buechner et al., 1990; Martinac et al., 1992). It seems that peptidylglycan, the equivalent of cytoskeleton here, normally restrains the outer membrane, attenuating rather than transmitting the mechanical forces in the membrane (Martinac, 1992). Membrane deformation can also be induced by mutations. The most abundant protein in the outer membrane is the 7.2-kD lipoprotein which is embedded in the inner monolayer and does not protrude into the outer monolayer. This lipoprotein is also a major covalent link between the outer membrane and the peptidylglycan. Kubalski et al. (1993) showed that the outer membrane of a mutant lacking this major lipoprotein is much less able to transmit stretch forces to the mechanosensitive channel. A bulky amphipath (lysolecithin) but not smaller amphipaths can restore this ability. The forces in the bilayer that gate the channel resulting from the relative but simultaneous expansion and compression of the two monolayers have been treated in a theoretical model by Markin and Martinac (1991).

The location of the *E. coli* mechanosensitive channels remains controversial, however. Martinac and coworkers (1987, 1992; Buechner et al., 1990) argued that MscS is located in the outer membrane for several reasons. Giant spheroplasts have an outer membrane, visible by electron microscopy or immunofluorescent microscopy. It seems unlikely that the patch-clamp pipette passes through the outer membrane and is then still able to seal onto the inner membrane. Digestion of excised inside-out patches with lysozyme added to the bath activates the channel. Should the channel be in the inner membrane, the peptidylglycan would have been facing the pipette side, inaccessible to lysozyme. Also, mutational deletion of lipoproteins that are clearly located in the outer membrane, strongly affects the behavior of these mechanosensitive channels, as described above (Kubalski et al., 1993). Finally, recent patch-clamp experiments on protoplasts bounded only by the cytoplasmic membrane (Birdsell and Cota-Robles, 1967) revealed the activities of a new set of channels but not the activities of porins, MscS or MscL (Kubalski, Martinac, Adler, and Kung, 1992). On the other hand, Berrier, Coulombe, Houssin, and Ghazi (1989) observed mechanosensitive-channel activities in 50% of the patches from liposomes incorporated with inner-membrane material but only 7% of the patches from those incorporated with outer-membrane material and concluded that the channel is localized in the inner membrane. However a perfect separation of inner and outer membranes by gradient centrifugation cannot be achieved (Osborn, Gander, Parisi, and Carson, 1972) and a direct comparison assumes the same efficiency of incorporation of different material into liposomes and the same degree of channel inactivation in the procedures. Berrier et al. (1989) found the mechanosensitive channel(s) to be of different unit conductances ranging from 140 to 950 pS. Events of even higher conductance were observed, though rarely. The differences between the observations by Berrier et al. (1989) and Delcour et al. (1989*a*), both made on reconstituted liposomes, have yet to be resolved.

As early as the 1950's, it has been observed that hypotonic shock causes *E. coli* to jettison its smaller solutes and ions, while retaining its macromolecules (Britten and McClure, 1962). The pathway has not been identified, but it clearly has an exclusion limit. Berrier, Coulombe, Szabo, Zoratti, and Ghazi, (1992) showed that the loss of ATP, glutamate, and lactose and K^+ from osmotically down-shocked cells could be reduced by gadolinium, an ion that also blocks the mechanosensitive channels that they maintain is located in the inner membrane. The authors concluded that this channel is the pathway for the controlled efflux of solutes upon hypo-osmotic shocks.

Using gradient-fractionated inverted vesicles of *E. coli* plasma membrane fused to planar lipid bilayers Simon, Blobel, and Zimmerberg, (1989) detected a channel that has a conductance of 115 pS when bathed in 45 mM potassium glutamate. This channel prefers anions over cations and was more active upon negative voltages. The authors believe this to be a channel that conducts protein to the periplasm.

Recently, Milkman (1994) discovered a gene in *E. coli* that corresponds to a protein similar to *Shaker*-type K^+ channel in size and in hydropathy plot. The 13-residue peptide that appears to line the pore (the H5 region) shows 60–75% identity with corresponding regions in various eucaryotic *Shaker*-type K^+ channels. The presence of this gene in 15 different wild-type *E. coli* strains suggest that it is functional, although its function has not yet been directly tested yet. If one can rule out importation of this gene from eucaryotes, this finding appears to establish the ancestry of channels with S1–S6 and H5 motif to procaryotes.

Bacillus Subtilis

Gram-positive bacteria have only one membrane, the cytoplasmic membrane. Mechanosensitive channels of large conductances have also been recorded from the membrane of *Streptococcus faecalis* (Zoratti and Petronilli, 1988) and *Bacillus subtilis* (Zoratti et al., 1990; Martinac et al., 1992).

Conclusion

The application of a variety of techniques has clearly shown that microbes, like other organisms, have ion channels. Although microbial ion channels have mostly been described in terms of the currents that they support, channel proteins and their corresponding genes have been identified for bacterial porins and mscL (the mechanosensitive channel of large conductance). A gene predicting protein structures like the *Shaker* K^+ channels has also been found in *E. coli* and several genes apparently corresponding to eag-like cation channels have been discovered in *Paramecium.* Because microbes are vastly more diverse than animals and plants, it is not surprising that different types of channels are encountered in different microbial membranes.

Interestingly, voltage-gated K^+ specific channels seem to have evolved in all branches of eucaryotic organisms, including plants, animals, as well as a protist, a slime mold, and several fungi reviewed here. Voltage-gated, Ca^{2+}-specific channels are found in *Paramecium* and *Dictyostelium,* but not in yeasts. Voltage-gated Na^+ channels have not been encountered in microbes to date. Such Na^+ channels might represent an evolutionary late modification of Ca^{2+} channels, to exploit only the electric and not the ionic effects. To date, there have been no reports of microbial channels directly gated by external ligands. This may not represent a lack of such channels in microbes but the difficulty in not knowing which, among the hundreds of possible ligands, should one present to the microbial surfaces. Gating by internal second messengers is exemplified by four types of Ca^{2+} gated channels in *Paramecium.* Analyses show clearly that these channels are Ca^{2+}-calmodulin gated. This method of gating is apparently not limited to protists, as there are strong hints that it is also employed by higher animals. It would be interesting to see whether other types of channel activities first described in microbes, such as a Mg^{2+}-specific current and hyperpolarization-activated Ca^{2+} current, also have equivalents in higher forms.

Channel gating by stretch force in the lipid bilayer appears to be an ancient invention. Mechanosensitive currents and/or channels are found in *Paramecium,* yeasts, *E. coli* and other bacteria. This fact may also reflect the need of all microbes to deal with osmotic stress. The identification of the protein and the cloning of the gene corresponding to a mechanosensitive channel in *E. coli* heralds a deeper understanding of the molecular basis of mechanosensation.

Acknowledgements

Work in the authors' laboratory was supported by grants from NIH and a grant from the Lucille P. Markey Trust.

References

Aeckerle, S., B. Wurster, and D. Malchow. 1985. Oscillations and cyclic AMP-induced changes of the K^+ concentration in *Dictyostelium discoideum. EMBO Journal.* 4:39–43.

Baum, P., C. Furlong, and B. Byers. 1986. Yeast gene required for spindle pole body duplication: homology of its product with Ca^{2+}-binding proteins. *Proceedings of the National Academy of Sciences, USA.* 83:5512–5516.

Benz, R. 1988. Structure and function of porins from gram-negative bacteria. *Annual Review of Microbiology.* 42:359–393.

Berrier, C., A. Coulombe, C. Houssin, and A. Ghazi. 1989. A patch-clamp study of inner and outer membranes and of contact zones of *E. coli,* fused into giant liposomes. Pressure-activated channels are localized in the inner membrane. *FEBS Letters.* 259:29–32.

Berrier, C., A. Coulombe, C. Houssin, and A. Ghazi. 1991. Fast and slow kinetics of porin channels from *Escherichia coli* reconstituted into giant liposomes and studied by patch-clamp. *FEBS Letters.* 306:251–256.

Berrier, C., A. Coulombe, I. Szabo, M. Zoratti, and A. Ghazi. 1992. Gadolinium ion inhibits loss of metabolites induced by osmotic shock and large stretch-activated channels in bacteria. *European Journal of Biochemistry.* 206:559–565.

Bertl, A., D. Gradmann, and C. L. Slayman. 1992. Calcium- and voltage-dependent ion channels in *Saccharomyces cerevisiae. Philosophical Transactions of the Royal Society of London B.* 338:63–72.

Bertl, A., C. L. Slayman, and D. Gradmann. 1993. Gating and conductance in an outward-rectifying K^+ channel from the plasma membrane of *Saccharomyces cerevisiae. Journal of Membrane Biology.* 132:183–199.

Birdsell, D. C., and E. H. Cota-Robles. 1967. Production and ultrastructure of lysozyme and ethylenediaminetetracetate-lysozyme spheroplasts of *Escherichia coli. Journal of Bacteriology.* 93:427–437.

Blomberg, A., and L. Adler. 1989. Roles of glycerol and glycerol-3-phosphate dehydrogenase (NAD^+) in acquired osmotolerance of *Saccharomyces cerevisiae. Journal of Bacteriology.* 171:1087–1092.

Bonini, N., M. Gustin, and D. L. Nelson. 1986. Regulation of ciliary motility by membrane potential in *Paramecium:* a role for cyclic AMP. *Cell Motility of the Cytoskeleton.* 6:256–272.

Brehm, P., and R. Eckert. 1978. Calcium entry leads to inactivation of calcium channel in *Paramecium. Science.* 202:1203–1206.

Brewster, J. L., T. de Valoir, N. D. Dwyer, E. Winter, and M. C. Gustin. 1993. An osmosensing signal transduction pathway in yeast. *Science.* 259:1760–1763.

Britten, R. J., and F. T. McClure. 1962. The amino acid pool of *Escherichia coli. Bacteriological Review.* 26:292–335.

Bruggemann, A., L. A. Pardo, W. Stuhmer, and O. Pongs. 1993. Ether-a-go-go encodes a voltage-gated channel permeable to K^+ and Ca^{2+} and modulated by cAMP. *Nature.* 365:445–448.

Buechner, M., A. H. Delcour, B. Martinac, J. Adler, and C. Kung. 1990. Ion channel activities in the *Escherichia coli* outer membrane. *Biochimica et Biophysica Acta.* 1024:111–121.

Bumann, J., D. Malchow, and B. Wurster. 1984. Attractant-induced changes and oscillations of the extracellular Ca^{2+} concentration in suspensions of differentiating *Dictyostelium* cells. *Journal of Cell Biology.* 98:173–178.

Cowan, S. W., T. Schermer, G. Rummel, M. Steiert, R. Ghosh, R. A. Pauptit, J. N. Jansonius, and J. P. Rosenbusch. 1992. Crystal Structures explain functional properties of two *E. coli* porins. *Nature.* 358:727–733.

Delcour, A. H., J. Adler, and C. Kung. 1991. A single amino acid substitution alters conductance and gating of OmpC porin of *Escherichia coli. Journal of Membrane Biology.* 119:267–275.

Delcour, A. H., J. Adler, C. Kung, and B. Martinac. 1992. Membrane-derived oligosaccharides (MDO's) promote closing of a *E. coli* porin channel. *FEBS Letters.* 304:216–220.

Delcour, A. H., B. Martinac, J. Adler, and C. Kung. 1989*a.* Modified reconstitution method used in patch-clamp studies of *Escherichia coli* ion channels. *Biophysical Journal.* 56:631–636.

Delcour, A. H., B. Martinac, J. Adler, and C. Kung. 1989*b.* Voltage-sensitive ion channel of *Escherichia coli. Journal Membrane Biology.* 112:267–275.

Dunlap, K. 1977. Localization of calcium channels in *Paramecium caudatum. Journal of Physiology.* 271:119–134.

Eckert, R., and Y. Naitoh. 1972. Bioelectric control of locomotion in the ciliates. *Journal of Protozoology.* 18:237–243.

Eckert, R., Y. Naitoh, and K. Friedman. 1972. Sensory mechanisms in *Paramecium*. I. Two components of the electric response to mechanical stimulation of the anterior surface. *Journal of Experimental Biology*. 56:683–694.

Ehrlich, B., A. Finkelstein, M. Forte, and C. Kung. 1984. Voltage-dependent calcium channels from *Paramecium* cilia incorporated into planar lipid bilayers. *Science*. 225:427–428.

Ehrlich, B., A. Jacobsen, R. Hinrichsen, L. Sayre, and M. Forte. 1988. *Paramecium* calcium channels are blocked by a family of calmodulin antagonists. *Proceedings of the National Academy of Sciences, USA*. 85:5718–5722.

Goertz, H.-D. 1988. Paramecium. Spring-Verlag, Berlin. 444 pp.

Gustin, M. C., B. Martinac, Y. Saimi, M. R. Culbertson, and C. Kung. 1986. Ion channels in yeast. *Science*. 233:1195–1197.

Gustin, M. C., F. Sachs, W. Sigurdson, A. Ruknudin, C. Bowman, C. E. Morris, and R. Horn. 1991. Single-channel mechanosensitive currents. *Science*. 253:800–802.

Gustin, M. C., X.-L. Zhou, B. Martinac, and C. Kung. 1988. A mechanosensitive ion channel in the yeast plasma membrane. *Science*. 242:762–765.

Hamill, O. P., A. Marty, E. Neher, B. Sakmann, and F. J. Sigworth. 1981. Improved patch-clamp techniques for high-resolution current recordings from cells and cell-free membrane patches. *Pfluegers Archiv*. 391:85–100.

Hennessey, T. M., and C. Kung. 1987. A calcium-dependent potassium current is increased by a single-gene mutation in *Paramecium*. *Journal of Membrane Biology*. 98:145–155.

Iida, H., S. Sakaguchi, Y. Yagawa, and Y. Anraku. 1990. Cell cycle controls by Ca^{2+} in *Saccharomyces cerevisiae*. *Journal of Biological Chemistry*. 265:21216–21222.

Jeanteur, D., J. H. Lakey, and F. Pattus. 1991. The bacterial porin superfamily: sequence alignment and structure prediction. *Molecular Microbiology*. 5:2153–2164.

Jegla, T. J., K. Larocque, W. Gallin, N. Grigoriev, A. N. Spencer, and L. Salkoff. 1993. Cloning of potassium channels from early metazoan and ciliate protozoans. *Journal of General Physiology*. 102:11*a*. (Abstr.)

Jennings, H. S. 1906. Behavior of Lower Organisms. Indiana University Press, Bloomington, Indiana. 366 pp.

Jurand, A., and G. G. Selman. 1969. The Anatomy of *Paramecium aurelia*. Macmillan/St. Martin's Press, London. 218 pp.

Kaupp, U. B., T. Niidome, T. Tanabe, S. Terada, W. Bönigk, W. Stühmer, N. J. Cook, K. Kangawa, H. Matsuo, T. Hirose, T. Miyata, and S. Numa. 1989. Primary structure and functional expression from complementary DNA of the rod photoreceptor cyclic GMP-gated channel. *Nature*. 342:762–766.

Kink, J. R., M. E. Maley, R. R. Preston, K.-Y. Ling, M. A. Wallen-Friedman, Y. Saimi, and C. Kung. 1990. Mutations in *Paramecium* calmodulin indicate functional differences between the COOH-terminal and the NH_2-terminal lobes *in vivo*. *Cell*. 62:165–174.

Kubalski, A., B. Martinac, and Y. Saimi. 1989. Proteolytic activation of a hyperpolarization- and calcium-dependent potassium channel in *Paramecium*. *Journal of Membrane Biology*. 112:91–96.

Kubalski, A., B. Martinac, K.-Y. Ling, J. Adler, and C. Kung. 1993. Activities of a mechanosensitive ion channel in an *E. coli* mutant lacking the major lipoprotein. *Journal of Membrane Biology*. 131:151–160.

Kubalski, A., B. Martinac, J. Adler, and C. Kung. 1992. Patch clamp studies on *Escherichia coli* inner membrane in vivo. *Biophysical Journal.* 61:513*a.* (Abstr.)

Kung, C., and R. Eckert. 1972. Genetic modification of electric properties in an excitable membrane. *Proceedings of the National Academy of Sciences, USA.* 69:93–97.

Kung, C., R. R. Preston, M. E. Maley, K.-Y. Ling, J. A. Kanabrocki, B. R. Seavey, and Y. Saimi. 1992. *In vivo Paramecium* mutants show that calmodulin orchestrates membrane responses to stimuli. *Cell Calcium.* 13:413–425.

Machemer, H., and A. Ogura. 1979. Ionic conductances of membranes in ciliated and deciliated *Paramecium. Journal of Physiology.* 296:49–60.

Markin, V. S., and B. Martinac. 1991. Mechanosensitive ion channels as reporters of bilayer expansion: a theoretical model. *Biophysics Journal.* 60:1120–1127.

Martinac, B. 1992. Mechanosensitive ion channels: biophysics and physiology. *In* CRC Thermodynamics of Membrane Receptors and Channels. M. B. Jackson, editor. CRC Press, Boca Raton, FL. 327–352.

Martinac, B., J. Adler, and C. Kung. 1990. Mechanosensitive ion channels of *E. coli* activated by amphipaths. *Nature.* 348:261–263.

Martinac, B., M. Buechner, A. H. Delcour, J. Adler, and C. Kung. 1987. A pressure-sensitive ion channel in *Escherichia coli. Proceedings of the National Academy of Sciences, USA.* 84:2297–2301.

Martinac, B., A. H. Delcour, M. Buechner, J. Adler, and C. Kung. 1992. Mechanosensitive ion channels in bacteria. *In* Advances in Comparative and Environmental Physiology. Vol. 10. F. Ito, editor. Springer-Verlag, Berlin. 3–18.

Martinac, B., Y. Saimi, M. C. Gustin, and C. Kung. 1988. Ion channels of three microbes: *Paramecium,* yeast and *Escherichia coli. In* Calcium and Ion Channel Modulation A. D. Grinnell, D. Armstrong, and M. B. Jackson, editor. Plenum Publishing Corp., New York. 415–430.

Milkman, R. 1994. An *E. coli* homologue of eukaryotic potassium channel proteins. *Proceedings of the National Academy of Sciences, USA.* In press.

Miyamaoto, S., Y. Okya, Y. Ohsumi, and Y. Anraku. 1987. Nucleotide sequence of the CLS4 (CDC24) gene of *Saccharomyces cerevisiae. Gene.* 54:125–132.

Morgan, H., J. T. Lansdale, and G. Adler. 1990. Polarity-dependent voltage-gated porin channels from *Escherichia coli* in lipid bilayer membrane. *Biochimica et Biophysica Acta.* 1021:175–181.

Morris, C. E., and R. Horn. 1991. Failure to elicit neuronal macroscopic mechanosensitive currents anticipated by single-channel studies. *Science.* 251:1246–1249.

Mueller, U., and K. Hartung. 1990. Properties of three different ion channels in the plasma membrane of the slime mold *Dictyostelium discoideum. Biochimica et Biophysica Acta.* 1026:204–212.

Mueller, U., D. Malchow, and K. Hartung. 1986. Single ion channels in the slime mold *Dictyostelium discoideum. Biochimica et Biophysica Acta.* 857:287–290.

Naitoh, Y., and R. Eckert. 1973. Sensory mechanisms in *Paramecium.* II. Ionic basis of the hyperpolarizing mechanoreceptor potential. *Journal of Experimental Biology.* 59:53–65.

Nakaoka, Y., and H. Machemer. 1990. Effects of cyclic nucleotides and intracellular Ca on voltage-activated ciliary beating in *Paramecium. Journal of Comparative Physiology.* 166:401–406.

Nakaoka, Y., and K. Iwatsuki. 1992. Hyperpolarization-activated inward current associated with the frequency increase in ciliary beating of *Paramecium. Journal of Comparative Physiology.* 170:723–727.

Nikaido, H. 1992. Porins and specific channels of bacterial outer membranes. *Molecular Microbiology.* 6:435–442.

Oertel, D., S. J. Schein, and C. Kung. 1977. Separation of membrane currents using a *Paramecium* mutant. *Nature.* 268:120–124.

Oertel, D., S. J. Schein, and C. Kung. 1978. A potassium conductance activated by hyperpolarization in *Paramecium. Journal of Membrane Biology.* 43:169–185.

Ogura, A., and H. Machemer. 1980. Distribution of mechano-receptor channels in *Paramecium* surface membrane. *Journal of Comparative Physiology.* 135:233–242.

Osborn, M. J., J. E. Gander, E. Parisi, and J. Carson. 1972. Mechanism of assembly of the outer membrane of *Salmonella typhimurium. Journal of Biological Chemistry.* 247:3962–3972.

Preston, R. R. 1990. A magnesium current in *Paramecium. Science.* 250:285–288.

Preston, R. R., J. A. Kink, R. D. Hinrichsen, Y. Saimi, and C. Kung. 1991. Calmodulin mutants and Ca^{2+}-dependent channels in *Paramecium. Annual Review of Physiology.* 53:309–319.

Preston, R. R., Y. Saimi, and C. Kung. 1990*a*. Evidence for two K^+ currents activated upon hyperpolarization of *Paramecium tetraurelia. Journal of Membrane Biology.* 115:41–50.

Preston, R. R., M. A. Wallen-Friedman, Y. Saimi, and C. Kung. 1990*b*. Calmodulin defects cause the loss of Ca^{2+}-dependent K^+ currents in two pantophobiac mutants of *Paramecium tetraurelia. Journal of Membrane Biology.* 115:51–60.

Preston, R. R., Y. Saimi, E. Amberger, and C. Kung. 1990*c*. Interactions between mutants with defects in two Ca^{2+}-dependent K^+ currents of *Paramecium tetraurelia. Journal of Membrane Biology.* 115:61–69.

Preston, R. R., Y. Saimi, and C. Kung. 1992*a*. Calcium current activated upon hyperpolarization of *Paramecium tetraurelia. Journal of General Physiology.* 100:233–251.

Preston, R. R., Y. Saimi, and C. Kung. 1992*b*. Ca-dependent inactivation of the calcium current activated upon hyperpolarization of *Paramecium tetraurelia. Journal of General Physiology.* 100:253–268.

Ramirez, J. A., V. Vacata, J. H. McCusker, J. E. Haber, R. K. Mortimer, W. G. Owen, and H. Lecar. 1989. ATP-sensitive K^+ channels in a plasma membrane hyperpolarization-ATPase mutant of the yeast *Saccharomyces cerevisiae. Proceedings of the National Academy of Sciences, USA.* 86:7866–7870.

Rothman, J. E., and J. Lenard. 1977. Membrane asymmetry. The nature of membrane asymmetry provides clues to the puzzle of how membranes are assembled. *Science.* 195:743–753.

Sachs, F. 1988. Mechanical transduction in biological systems. *CRC Critical Review of Biomedical Engineering.* 16:141–169.

Saimi, Y. 1986. Calcium-dependent sodium currents in *Paramecium:* mutational manipulations and effects of hyper- and depolarization. *Journal Membrane Biology.* 92:227–236.

Saimi, Y., R. D. Hinrichsen, M. Forte, and C. Kung. 1983. Mutant analysis shows that the

Ca^{2+}-induced K^+ current shuts off one type of excitation in *Paramecium. Proceedings of the National Academy of Sciences, USA.* 80:5112–5116.

Saimi, Y., and C. Kung. 1980. A Ca-induced Na^+ current in *Paramecium. Journal of Experimental Biology.* 88:305–325.

Saimi, Y., and K.-Y. Ling. 1990. Calmodulin activation of calcium-dependent sodium channels in excised membrane patches of *Paramecium. Science.* 249:1441–1444.

Saimi, Y., and B. Martinac. 1989. Calcium-dependent potassium channel in *Paramecium* studied under patch clamp. *Journal of Membrane Biology.* 112:79–89.

Saimi, Y., B. Martinac, A. H. Delcour, P. V. Minorsky, M. C. Gustin, M. R. Culbertson, J. Adler, and C. Kung. 1992. Patch clamp studies of microbial ion channels. *In* Methods in Enzymology. Vol. 207. B. Rudy and L. E. Iverson, editors. Academic Press, New York. 681–691.

Salkoff, L., K. Baker, A. Butler, M. Covarrubias, M. D. Pak, and A. Wei. 1992. An essential "set" of K^+ channel conserved in flies, mice, and humans. *Trends in Neuroscience.* 15:161–166.

Satow, Y., and C. Kung. 1980*a*. Membrane currents of pawn mutants of the *pwA* group in *Paramecium tetraurelia. Journal of Experimental Biology.* 84:57–71.

Satow, Y., and C. Kung. 1980*b*. Ca-induced K^+-outward current in *Paramecium tetraurelia. Journal of Experimental Biology.* 88:293–303.

Satow, Y., A. D. Murphy, and C. Kung. 1983. The ionic basis of the depolarizing mechanoreceptor potential of *Paramecium tetraurelia. Journal of Experimental Biology.* 103:253–264.

Schachtman, D. P., J. I. Schroeder, W. J. Lucas, J. A. Anderson, and R. F. Gaber. 1992. Expression of an inward-rectifying potassium channel by the *Arabidopsis KAT1* cDNA. *Science.* 258:1654–1658.

Schindler, H., and J. P. Rosenbusch. 1978. Matrix protein from *Escherichia coli* outer membranes forms voltage-controlled channels in lipid bilayers. *Proceedings of the National Academy of Sciences, USA.* 75:3751–3755.

Schultz, J. E., S. Klumpp, R. Benz, W. J. Ch. Schurhoff-Goeters, and A. Schmid. 1992. Regulation of adenylyl cyclase from *Paramecium* by an intrinsic potassium conductance. *Science.* 255:600–603.

Simon, S. M., G. Blobel, and J. Zimmerberg. 1989. Large aqueous channels in membrane vesicles derived from the rough endoplasmic reticulum of canine pancreas or the plasma membrane of *Escherichia coli. Proceedings National Academy of Sciences, USA.* 86:6176–6180.

Sheetz, M. P., and S. J. Singer. 1974. Biological membranes as bilayer couples. A molecular mechanism of drug-erythrocyte interaction. *Proceedings National Academy of Sciences, USA.* 71:4457–4461.

Stock, J. B., B. Rauch, and S. Roseman. 1977. Periplasmic space in *Salmonella typhimurium* and *Escherichia coli. Journal Biological Chemistry.* 252:7850–7861.

Sukharev, S. I., B. Martinac, V. Y. Arshavsky, and C. Kung. 1993. Two types of mechanosensitive channels in the *Escherichia coli* cell envelope: solubilization and functional reconstitution. *Biophysical Journal.* 65:177–183.

Sukharev, S. I., P. Blount, B. Martinac, F. R. Blattner, and C. Kung. 1994. *mscL* alone encodes a functional large-conductance mechanosensitive channel in *E. coli. Nature.* In press.

Szabo, I., V. Petronilli, L. Guerra, and M. Zoratti. 1990. Cooperative mechanosensitive ion channels in *Escherichia coli. Biochemical and Biophysical Research Communication.* 171:280–186.

Vacata, V., M. Hofer, H. P. Larsson, and H. Lecar. 1992. Ionic channels in the plasma membrane of *Schizosaccharomyces pombe:* evidence from patch-clamp measurements. *Journal of Bioenergetics and Biomembranes.* 24:43–53.

Warmke, J., R. Drysdale, and B. Ganetzky. 1991. A distinct potassium channel polypeptide encoded by the *Drosophila eag* locus. *Science.* 252:1560–1562.

Zhou, X.-L., and C. Kung. 1992. A mechanosensitive ion channel in *Schizosaccharomyces pombe. EMBO Journal.* 11:2869–2875.

Zoratti, M., and V. Petronilli. 1988. Ion-conducting channels in a gram-positive bacterium. *FEBS Letters.* 240:105–109.

Zoratti, M., V. Petronilli, and I. Szabo. 1990. Stretch-activated composite ion channels in *Bacillus subtilis. Biochemical and Biophysical Research Communication.* 168:443–450.

Using Sequence Homology to Analyze the Structure and Function of Voltage-gated Ion Channel Proteins

H. Robert Guy and Stewart R. Durell

Laboratory of Mathematical Biology, National Cancer Institute, National Institutes of Health, Bethesda, Maryland 20892

Introduction

Much effort in membrane biophysics and electrophysiology during the latter half of this century has been directed toward understanding the structure and function of membrane channel proteins. Exquisitely sensitive methods have been developed to measure currents that flow through a single channel and kinetics of conformational changes that open and close these channels. Unfortunately, little is known about the three-dimensional structure of membrane proteins because they are difficult to isolate and crystalize; porins from the outer membrane of gram negative bacteria are the only integral membrane channel structures that have been determined experimentally (Cowan, Schirmer, Rummel, Steiert, Ghost, Pauptit, Jansonius, and Rosenbusch, 1992; Weiss, Abele, Weckesser, Welte, Schiltz, and Schulz, 1991). Efforts to relate function to structure were stimulated in the 1980's by cloning and sequencing nucleic acids that code for channel proteins, by expressing these sequences in cells such as *Xenopus* oocytes, and by using sophisticated biophysical methods such as patch clamps to measure functional properties of both normal and mutated channels. These experimental efforts for voltage-gated channels have been complemented by theoretical efforts to model the structure of ion channels. The experimental and theoretical processes are iterative: initial experimental data are used to develop preliminary models, which are then tested experimentally and often modified or refined on the basis of the new experimental results. Knowledge obtained by comparing homologous sequences has been pivotal in developing structural models of voltage-gated channels and in designing mutagenesis experiments to analyze their structure-function relationships. The primary goal of this manuscript is to review how this information is being used experimentally and theoretically.

Models of Transmembrane Topology and Functional Mechanisms

Our first steps in modeling a membrane protein are to predict which segments span the membrane, to predict the secondary structure of those segments, and to predict which segments of the protein form functionally important mechanisms or sites such as voltage sensors, activation gates, inactivation gates, ion selectivity filters, and drug or toxin binding sites (Durell and Guy, 1992). K^+ channels are comprised of four identical or homologous subunits (MacKinnon, 1991); whereas, the pore-forming portions of Na^+ and Ca^{2+} channels are formed by a single polypeptide chain that contains four homologous repeats (Noda, Shimizu, Tanabe, Takai, Kayano, Ikeda, Takahashi, Nakayama, Kanaoka, Minamino, Kangawa, Matsuo, Raftery, Hirose, Notake, Inayama, Hayashida, Miyata, and Numa, 1984; Noda, Ikeda, Kayano, Suzuki, Takeshima, Kurasaki, Takahashi, and Numa, 1986; Tanabe, Takeshima, Mikami, Flockerzi, Takatiashi, Kangawa, Kojima, Matsuo, Hirose, and Numa, 1987), each of which is postulated to have a transmembrane topology similar to that illustrated in Fig. 1. Although the schematic in Fig. 1 is based on our more recent models of K^+ channels (Durell and Guy, 1992), the general topology is almost the same that we proposed for each repeat of sodium channels (Guy and Seetharamulu, 1986). Many features of our models were proposed originally by Numa's group who first determined the sequences. They postulated that each repeat has six transmembrane segments, which they labeled S1–S6, that the positively charged S4 segment acts as the voltage sensor for activation gating (Noda et al., 1984; Noda et al., 1986), and that the segment linking homologous repeat III to IV forms the Na^+ channel's inactivation gate (Noda et al., 1986). Our primary contribution was an unorthodox hypothesis that a sequentially short segment between S5 and S6, which we now call the P segment (also called SS1-SS2 and H5), forms a hairpin structure that transverses the outer part of the transmembrane region and that forms the ion

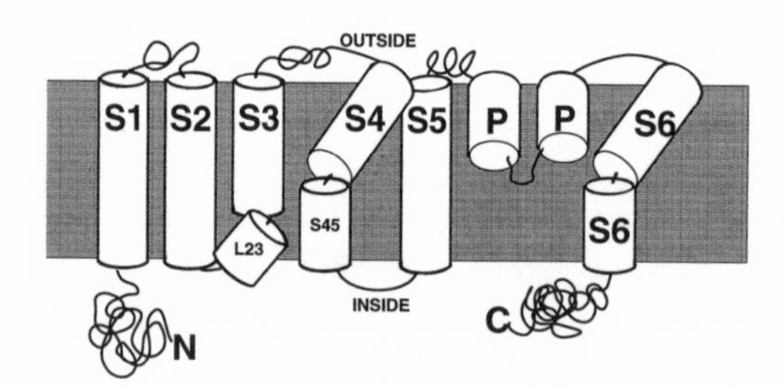

Figure 1. Postulate transmembrane topology for one voltage-gated channel subunit (K^+ channel) or homologous repeat (Na^+ and Ca^{2+} channels). Cylinders represent α helices.

selective portion of the pore. Although this model was entirely hypothetical when proposed, many of its features have been confirmed experimentally. There is now evidence that each of the putative extracellular loops of K^+ channels are exposed on the extracellular surface (Chua, Tytgat, Liman, and Hess, 1992; MacKinnon and Yellen, 1990; Shen, Chen, Boyer, and Pfaffinger, 1993), that the NH_2- and COOH-terminal ends are on the cytoplasmic side of the membrane (Liman, Tytgat, and Hess, 1992; Zagotta, Hosti, and Aldrich, 1990), and that the central part of the P segment is accessible to tetraethylammonium (TEA) which enters and blocks the pore from the inside (Choi et al., 1993; Hartmann et al., 1991; Yellen et al., 1991).

Fig. 2 is a schematic representation of our predictions (Guy, 1988; Guy and Seetharamulu, 1986) about which sequential segments of Na^+ channel proteins form functional sites or mechanisms previously postulated by membrane biophysicists. Activation gating had been postulated to be triggered by movement of several charged "voltage sensors" through the membrane's electric field (Armstrong, 1981). The S4's were postulated to be the voltage sensors because they have an unusual pattern of positively charged residues at every third position separated by hydrophobic residues. S4's were placed in the transmembrane region so they could respond to voltage changes. The activation gate had been postulated to occlude the cytoplasmic entrance of the pore when the channel closed and then to move out of the way when it opens (Armstrong, 1981; Hille, 1977). The segments linking S4's to S5's were postulated to be the activation gate because they are linked to S4 and because they were located near the cytoplasmic channel entrance in the model. The channel's ion selective region or "selectivity filter" had been postulated to be a narrow negatively charged region near the pore's extracellular entrance (Armstrong, 1981; Hille, 1975). The P segments were postulated to form the selectivity filter because their negatively charged residues could balance some of the positively charged residues of S4 and because they could form a narrow region near the pore's extracellular entrance.

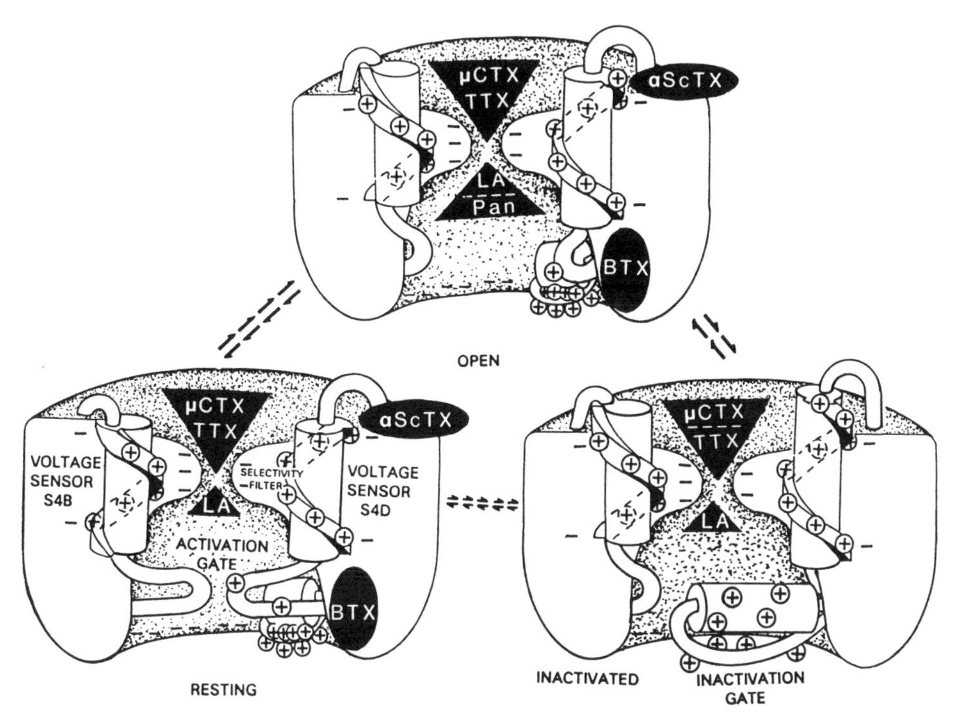

Figure 2. Early model postulating which segments of Na^+ channel sequences form previously postulated mechanisms or binding sites. Postulated binding sites for tetrodotoxin (*TTX*), μ conotoxin (μ*CTX*), local anesthetics (*LA*), pancuronium (*Pan*), batrachotoxin (*BTX*), and α scorpion toxin (α*ScTX*) are shown in black. Predictions that have been confirmed or supported experimentally are that the voltage sensors are formed by S4's, the inactivation gate by the segment linking repeat III to repeat IV, the selectivity filter by P's, and the TTX binding site by charged residues near the ends of the P's. Reproduced from Guy (1988).

Tetrodotoxin (TTX) and saxitoxin (STX) had been postulated to block the extracellular entrance of Na^+ channels and bind to negatively charged carboxyl groups (Hille, 1975). The COOH-terminal ends of the P segments were postulated to form these toxin binding sites because they have an appropriate charge and location in the model. The inactivation gate had been postulated to be a positively charged protein moiety that enters and blocks the intracellular entrance to the pore when the channel is in an open or activated conformation (Armstrong, 1981). The cytoplasmic segment linking homologous repeats III to IV was postulated to form the inactivation gate because this positively charged segment is well conserved between eel and rat brain Na^+ channels (Noda et al., 1986) and could be located near the cytoplasmic channel entrance in the models.

The predictions that the P segments form the ion selective portion of the pore and binding sites for toxins have been verified for both Na^+ and K^+ channels. Mutations of the charged residues near the COOH-terminal ends of the Na^+ channel's P segments alter channel conductance, selectivity, and blockade by TTX and STX as anticipated by the model (Noda et al., 1986; Pusch, Noda, and Conti, 1991; Terlau, Heinemann, Stühmer, Pusch, Conti, Imoto, and Numa, 1991). Mutations near the beginning and end of the P segment of K^+ channels affect blockade of the pore from the outside by TEA and charybdotoxin (MacKinnon and Yellen, 1990), mutations near the middle affect blockade of the pore from the inside by TEA (Choi et al., 1993, Hartmann et al., 1991; Yellen et al., 1991), and mutations of glycines in the latter part of P eliminate the selectivity of the pore (Heginbotham, Lu, Abramson, and Mackinnon, 1994).

A strategy that has proven effective in analyzing P segments is to identify homologous proteins with known sequences that have different pore properties and then demonstrate that those properties can be transferred by mutating residues of the P segment of one protein to aligned residues of the other protein. One position of the alignment of P segments is occupied by glutamic acid in all four homologous repeats of the Ca^{2+} channel α subunit; whereas, in the Na^+ channel it is occupied by glutamic acid in repeat I, by aspartic acid in repeat II, by lysine in repeat III and by alanine in repeat IV. Heinemann, Terlau, Stühmer, Imoto, and Numa (1992) showed that mutating the lysine and alanine to glutamic acid in Na^+ channels created a channel with pore properties very similar to those of Ca^{2+} channels. Cardiac Na^+ channels are much less sensitive to TTX than other Na^+ channels and are blocked by Zn^{2+} at much lower concentrations. Cardiac channels have a cysteine at a position just preceding the COOH terminal of the P segment of repeat I; the analogous position is occupied by phenylalanine or tyrosine in most other Na^+ channels. Mutating the phenylalanine or tyrosine to cysteine dramatically decreases blockade by TTX or STX and increases blockade by Zn^{2+}; whereas, mutating the cardiac cysteine to phenylalanine or tyrosine has the opposite effect (Backx, Yue, Lawrence, Marban, and Tomaselli, 1992; Heinemann, Terlau, and Imoto, 1992; Satin, Kyler, Chen, Bell, Cribbs, Fozzard, and Rogart, 1992). The first part of the P segment of cation-selective cyclic nucleotide-gated channels is quite similar to the analogous sequence of K^+ channels. However, the latter part of the P segment is very different in the two families; in fact, a deletion occurs in the latter portion of the cyclic nucleotide-gated channel P segment (Guy, Durell, Warmke, Drysdale, and Ganetzky, 1991). Heginbotham, Abramson, and MacKinnon (1992) deleted a glycine and tyrosine in the *Shaker* K^+ channel at the position of the apparent natural deletion and found that the properties of the mutated pore resembled those of the cyclic nucleotide channel; i.e., it was permeant to both Na^+ and K^+ and was blocked by relatively low concentrations of Ca^{2+}. Stühmer, Rupperberg, Schroter, Sakmann, Stocker, Giese, Perschke, Baumann, and Pongs (1989*b*) identified a series of rat K^+ channels with very different affinities for extracellular blockade of the pore by TEA (TEA blocks K^+ channels at two distinct sites, one accessible from the outside and one accessible from the inside). MacKinnon and Yellen (1990) showed that these binding affinities could be duplicated quite well by mutating a residue near the end of the P segment in *Shaker* K^+ channels from *Drosophila* to the residue at the corresponding position in the rat channels. Hartmann et al. (1991) identified two K^+ channels, Drk1 and Ngk2, with different single channel conductances and different affinities for blockade by both intracellular and extracellular TEA. They then performed a chimera experiment in which the P segment and some adjacent residues of Drk1 were mutated into the Ngk2 sequence and the analogous segment of Ngk2 was mutated into the Drk1 sequence. They found that all three properties of the pore of the Ngk2-Drk1(P) mutant were identical to those of the Drk1 channel, and likewise that all properties of the Drk1-Ngk2(P) chimera were identical to those of the Ngk2 channel.

Our predictions concerning the activation and inactivation gating mechanisms have also been supported by mutagenesis experiments. Almost all mutations in S4's of Na^+ and K^+ channels alter activation gating and those that reduce the magnitude of the positive charge tend to reduce the voltage dependency of the gating (Papazian, Timpe, Jan, and Jan, 1991; Stühmer, Conti, Suzuki, Wang, Noda, Yahagi, Kubo, and Numa, 1989*a*). Mutations and antibody binding to the segment linking repeats III and IV of Na^+ channels alter inactivation (Papazian et al., 1991; Vassilev, Scheuer,

```
SHAK    SIPDAFWWAVVTMTTVGYGDMTP
SLO     SYWICVYFLIVTMSTVGYGDVYC
ECO     SLMTAFYFSIETMSTVGYGDIVP
ROMK1   GMTSAFLFSLETQVTIGYGFRFV
EAG     MYVTALYFTMTCMTSVGFGNVAA
KAT1    RYVTALYWSITTLTTTGYGDFHA
cGMP    KYVYSLYWSTLTLTTIG--ETPP
         :   :    :    :    :
         P1  P5   P10  P15  P20
```

Figure 3. Sequences of P segments from distantly related K^+ channels and a cyclic nucleotide-gated channel. Sequences are from Tempel et al. (1987); Kaupp et al. (1989); Atkinson et al. (1991); Warmke et al. (1991); Anderson et al. (1992); Ho et al. (1993); Kubo et al. (1993); and Milkman and McKane Bridges (1993). Segments are numbered beginning with the first proline and ending with the last proline in the *Shaker* P segment. Residues in bold occur in three or more sequences.

and Catterall, 1988). Mutagenesis experiments have also identified an inactivation segment at the NH_2-terminal of some K^+ channels and have shown that peptides with the sequence of, or similar to, this segment can mimic inactivation by plugging the intracellular entrance of the pore (Zagotta et al., 1990). Remarkably, a peptide with the sequence of the putative Na^+ inactivation gate can also mimic the effect of the K^+ inactivation particle; i.e., it can restore inactivation to K^+ channels that do not inactivate (Patton, West, Catterall, and Goldin, 1993), even though the K^+ and Na^+ inactivation segments exhibit no apparent homology.

The hypothesis that the P segment forms the ion selective portion of the pore and spans only the outer portion of the membrane was contrary to the prevalently held view that, because the membrane lipid alkyl region is extremely hydrophobic, most transmembrane segments of integral membrane proteins are hydrophobic alpha helices that span the entire transmembrane region and that they are oriented approximately orthogonally to the plane of the membrane. In my view, this postulate is likely to be valid for membrane proteins, such as the photosynthetic reaction center and bacteriorhodopsin, in which all transmembrane segments are exposed to lipid and which do not have water-filled pores. It is not true for porins, for which the transmembrane segments are beta strands, many of which have polar side chains that extend into the large water-filled pore (Cowan et al., 1992; Weiss et al., 1991). For voltage-gated and homologous channels, the hydrophobic helix hypothesis is likely to be valid for the faces of those helices that are in contact with lipid; however, these large proteins probably have additional transmembrane or partially transmembrane segments that have no contact with lipid and that are partially exposed to water. The physio-chemical constraints for these segments may be similar to those of segments in soluble proteins because their residues are involved only in protein–protein and protein-water interactions. If this supposition is correct, segments that form the narrowest part of the pore need not span the entire membrane, they need not be orthogonal to the plane of the membrane, they need not be comprised almost exclusively of hydrophobic residues, and they need not have regular secondary structures such as alpha helices or beta strands. The difficulty with this view in modeling membrane protein structures is that it is very difficult to distinguish on the basis of a single sequence alone whether a particular segment forms part of the central core of the transmembrane portion of the protein or is part of an extracellular or intracellular domain. Our initial model, thus, was not based on an objective method to distinguish between these possibilities, but rather was based on a subjective effort to construct a structure that contained most of the functional features that had been proposed previously for Na^+ channels. The initial model was based upon a single sequence. Had we had the sequence information that is now available, our task would have been much easier and the model would have been much more compelling. To identify the ion selective region one simply needs to align all the homologous sequences and then identify the segment that is conserved best among the channels that have the same ion selectivity but that is poorly conserved when sequences of channels with different ion selectivities are compared. Thus, information obtained by comparing homologous membrane proteins may prove valuable in identifying functionally important portions of proteins that have "unorthodox" transmembrane topologies.

Using Homology in Developing Three-dimensional Molecular Models

When the models described above were first proposed, many scientists thought that subjective unorthodox structural models were unlikely to be of value and that results of mutagenesis experiments in the absence of detailed structure would not be interpretable. These fears have proved groundless. Mutagenesis experiments have been very effective in verifying many features of voltage-gated channel models, and the models have been valuable in designing the mutagenesis experiments. The unresolved issues now deal more with precision than with validity. It is one thing to accurately predict the transmembrane topology of a protein or to predict which segments will be involved in a particular process and to use mutagenesis experiments to test these predictions; it is quite another to accurately predict the precise positions of all the atoms of a large protein in its different conformations, to predict precisely how these atoms interact with those of drugs and toxins that affect the channels, to predict exactly how the atoms determine which ions pass through the pore and how fast they can do so, and to use mutagenesis experiments to test these predictions. It seems unlikely that it will be possible soon to combine results of mutagenesis experiments with molecular modeling to determine three-dimensional structures of voltage-gated channels with the precision and certainty that would be obtained from a high resolutions crystal structure. Nonetheless, these less precise methods should be continued because: (*a*) it is not clear if and when crystal structures of these channels will be solved; (*b*) it should be possible to use mutagenesis experiments and molecular modeling to improve our knowledge of these proteins, perhaps to the point that rational drug design can be used for some specific sites on the protein; (*c*) the knowledge gained through these experiments may help in isolating and crystallizing the proteins in a stable functional conformation; and (*d*) once a static crystal structure is determined, the mutagenesis results and molecular modeling techniques developed in this effort may help in understanding the protein's dynamic functional mechanisms and may help in modeling the structure of many homologous proteins from one or a few experimentally determined structures.

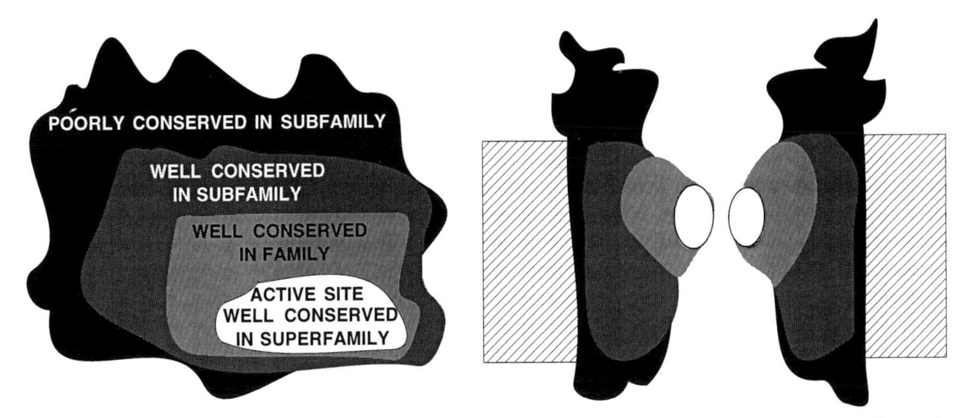

Figure 4. Schematic representation for how the best conserved residues among distantly related proteins tend to cluster together around functionally important sites and tend to be surrounded by successive layers of less well conserved residues in (*A*) water soluble proteins and (*B*) in ion channels.

In developing three-dimensional models of large integral membrane proteins such as the voltage-gated channels, we attempt to satisfy a series of criteria, e.g., we attempt to optimize residue–residue, residue–water, and residue–lipid interactions based on the physiochemical properties of the residues, we use helix packing theory if the transmembrane segments appear to be helical and theories about beta sheets and/or beta barrels if they appear to be beta strands, and we attempt to make our models consistent with available structural and/or functional data. Unfortunately, these criteria, which have been review elsewhere (Durell and Guy, 1992; Guy, 1988, 1990; Guy and Conti, 1990), are typically insufficient to develop a unique model of the protein's general folding pattern.

Additional information can be obtained by comparing homologous sequences that are different evolutionary distances apart and for which different functional features have either been conserved or altered. Fig. 4 illustrates some of the basic concepts that we use in developing models based on the extent of conservation for each position in an alignment of a protein superfamily that share a common functional property. In this analysis, we assume that the most highly conserved residues will cluster together at a functional site or sites shared by all members of the superfamily. Furthermore, the extent to which residues are conserved will decrease as a function of distance from the active sites and will increase as a function of distance into the protein's core form its surface. These concepts appear valid for both soluble and membrane proteins. In the globin superfamily and proteins homologous to bacteriorhodopsin, the best conserved residues cluster around the functionally essential heme or retinal group, the least well conserved residues are on the proteins surface that is furtherest from the active site, and most insertions and deletions are in the loop regions between alpha helices or at the beginnings and ends of the helices (personal observation, see Fig. 5). Although three-dimensional structures are not known for voltage-gated channels or protein's homologous to them, analyses of their sequences suggests that the same types of evolutionary phenomena occur. K^+ channels represent an ancient and highly diversified superfamily of membrane channels from which nonselective cyclic nucleotide-gated and voltage-gated Ca^{2+} channels have evolved (voltage-gated Na^+ channels evolved from Ca^{2+} channels). Many sequences are now known for K^+ channels; these include voltage-gated, calcium-activated, plant, bacterial, and Mg^{2+}-blocked inward rectifying channels. Comparisons of the distantly related families of K^+ channels indicate that the P segments are the best conserved. In fact, they are the only segments that can be aligned reliably; the inward rectifying channels that are blocked by Mg^{2+} have no segments homologous to S1–S4, and comparisons of alignments of S1–S6 (excluding P) between the other families reveal no two consecutive residues that are identical. In addition to assisting in identifying the ion selective portion of the pore, sequence comparisons may aid in modeling its three dimensional structure. When sequences of voltage-gated and calcium-activated K^+ channels are compared, no residues are identical in the first and last part of P, whereas, 9 out of 10 residues are identical in segment P9–P18 (see numbering nomenclature in Fig. 3). This pattern suggests that P9–P18 is the most important part of P for ion selectivity, however, it provides little information about how this segment may fold. Comparison of more distantly related sequences is useful in this respect. In this segment the Mg^{2+}-blocked inward rectifying channels (e.g., ROMK1) have only five residues (two threonines and then a Gly-Tyr-Gly segment) that are identical to those of voltage- and calcium-gated K^+ channels. Some of the differences are very nonconservative substitutions; hydrophobic V_{P9}, M_{P11}, and M_{P19} or V_{P19} are replaced by hydrophilic Glu, Gln, and Arg residues and the negatively charged D_{P18} is replaced by a hydrophobic Phe residue in ROMK1. The two threonines are conserved in nonselective cyclic nucleotide-gated channels as well, which suggests that they are conserved for reasons unrelated to ion selectivity. Also the two threonines and tyrosine are replaced by serine, cystine, and phenylalanine in *Eag* K^+ channel, leaving the two glycines as the only residues that are identical in all sequences. These results suggests that the two glycines may be the most important residues in determining the channels selectivity for K^+. This supposition has been supported by mutagenesis experiments. Heginbotham et al. (1994) mutated every position of the highly conserved portion of the *Shaker* P segment (one substitution at a time) and found that, with the exception of the two glycines, every position can be mutated without substantially altering the channel's selectivity for K^+

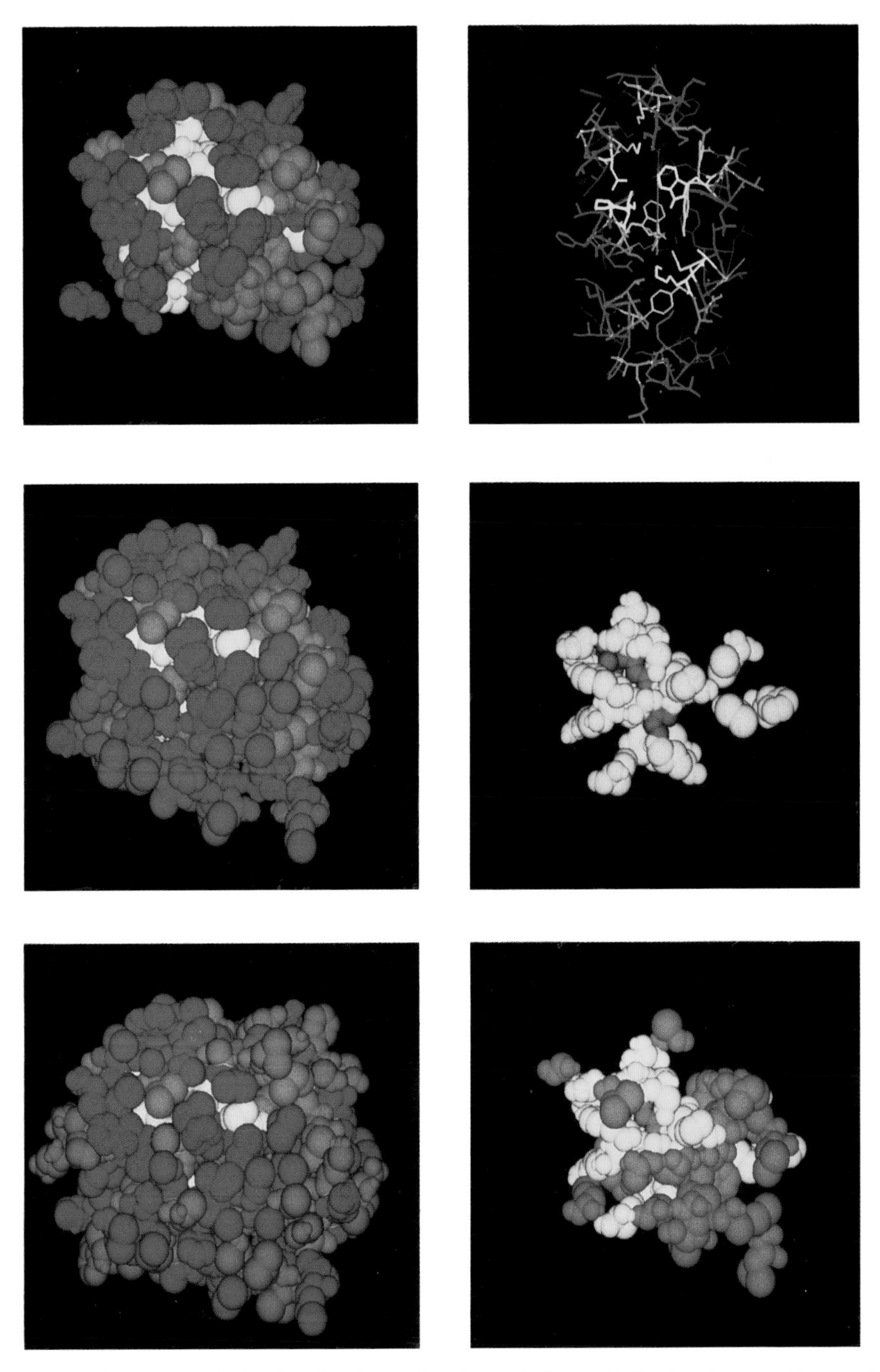

Figure 5. Illustration of the distributions of residues in bacteriorhodopsin according to how well they are conserved among bacteriorhodopsin and eleven of its homologues. (*A*) Cross-section parallel to the membrane. (*B–F*) Side views illustrating the "layering" of

over Na^+; however, almost any substitution at either Gly position makes the pore relatively nonselective among cations.

These findings have important implications about the structure of the ion selective portion of the pore. The mutagenesis results suggest that no side chain is essential for K^+ selectivity and that the highly conserved glycines are involved. If so, then K^+ selectivity may be due to interactions with backbone carbonyl groups in the highly conserved portion of the P segments. Circular dichroism experiments on peptides with the sequences of the *Shaker* P segment indicate that in apolar solvents the first part of P has an alpha helical conformation, whereas the latter part, that includes the highly conserved glycines, has a disordered conformation (Peled and Shai, 1993). In soluble proteins, the most highly conserved glycines tend to be buried in the protein and have backbone conformations that are energetically unfavorable for other residues (Overington, Donnelly, Johnson, Sali, and Blundell, 1992). If this is true for K^+ channels, then the portion of the pore that determines ion selectivity does not have a regular alpha helical or beta strand conformation. These findings and suppositions have led us to develop models in which the first and last parts of P form short alpha helices that are connected by the best conserved part of P which has a more random conformation. The major difficulty with these models is that glycines which are not part of an alpha helix or beta strand introduce a great deal of conformational freedom into the structure, and it has proven difficult to identify a unique structure that best satisfies the experimental results and our modeling criteria. One of these models is illustrated in Fig. 6. In these models, backbone amide oxygens of the conserved glycines are postulated to interact with K^+ ions to form the ion selective portion of the pore. K^+ channels possess several, perhaps four, K^+ binding sites in their pore (Neyton and Miller, 1988). One way that these sites could be formed by the backbone carbonyl oxygens of only a few residues per subunit is illustrated in Fig. 6 *e*.

The models have also been influenced by results of mutagenesis experiments on blockade of the pore by TEA at both the extracellular and intracellular entrances of the narrow portions of the pore. Results of these experiments are illustrated in Fig. 6 *d*. The model was constructed so that the green residues, where mutations affect blockade by extracellular TEA (De Biasi et al., 1994; Hartmann et al., 1991; MacKinnon and Yellen, 1990), are accessible from the outside and so that the cyan residues, where mutations alter blockade by TEA from the inside (Choi et al., 1993; Hartmann et al., 1991; Yellen et al., 1991), are accessible from the inside. P20 residues from the four subunits were positioned near the extracellular entrance so that when tyrosine or phenylalanine are present, their aromatic groups can form much of the TEA binding site (Heginbotham and MacKinnon, 1993). The mutation (DeBiasi et al., 1994) of residue P9 to histidine is especially interesting because it alters blockade by extracellular TEA, pH, zinc, and histidine reagents even though it is near the middle of the P segment and only two residues away from the first position where mutations alter the blockade by intracellular TEA. This result supports our contention (Durell and Guy, 1992; Guy, 1988, 1990; Guy and Conti, 1990; Guy and

less well conserved residues around the conserved core. Color code: green = retinal group, yellow = no substitutions, orange = one substitution, red = two substitutions, magenta = three substitutions, blue = four or more substitutions. Coordinates obtained from Protein Data Bank, model in Henderson et al. (1990).

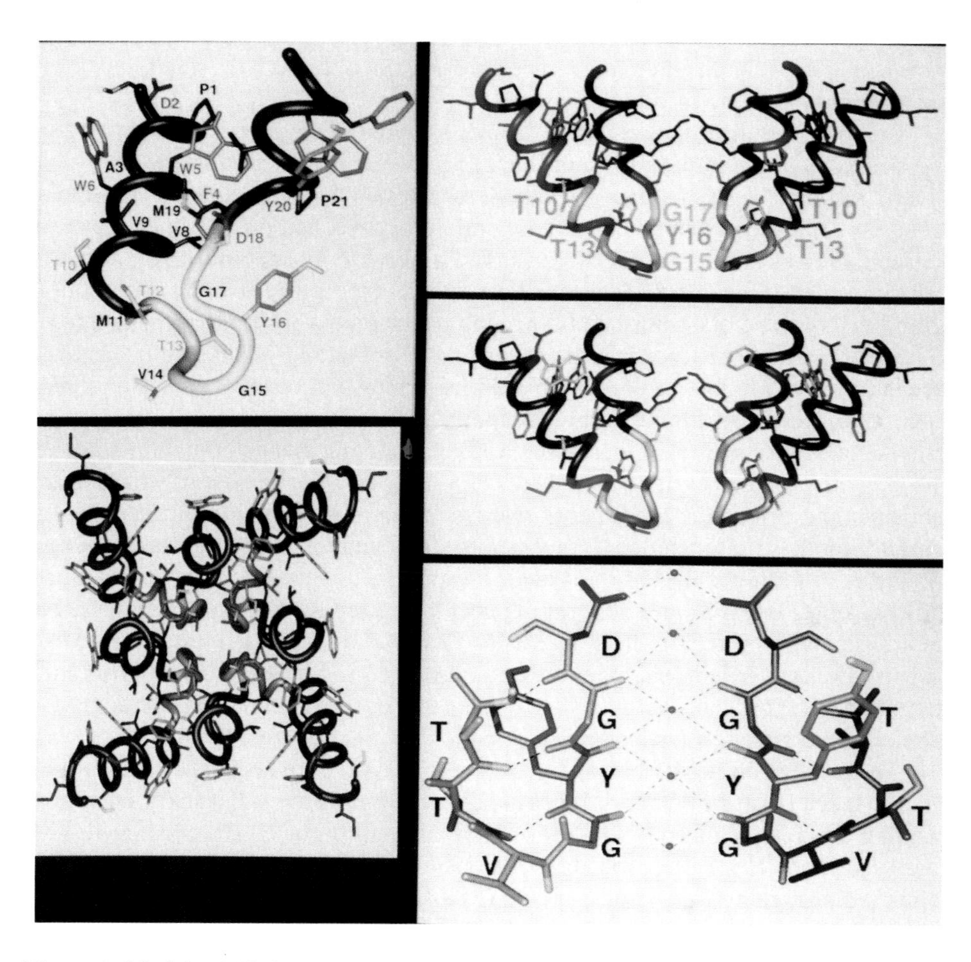

Figure 6. Models of *Shaker* K^+ channel P segments. (*A*) Monomer. View is from inside the channel looking toward the lipid, top is outside the cell. Black tubes are backbones of α helices; white tubes are random coil backbone. Side chain color code: gray = alkyl carbons, purple = aromatic carbons, pink = uncharged oxygens, red = negatively charged oxygens, light blue = uncharged nitrogens, yellow = sulfurs. (*B*) Assembly of four P segments to form outer vestibule and ion selective region of the channel; view form outside the cell. (*C*) Cross-section of outer vestibule and ion selective region showing two P segments. View is the same as in *A* except monomers are rotated by 90° about the pore's axis. Residues are colored coded as in Fig. 5 according to how well they are conserved among distantly related K^+ channels. (*D*) Side view of P segments illustrating residues where mutations alter TEA binding from outside (MacKinnon and Yellen, 1990; Hartmann et al., 1991; DeBiasi et al., 1993) (*green*) and inside (Yellen et al., 1991; Hartmann et al., 1991; and Choi et al., 1993) (*cyan*) the cell. (*E*) Side view of putative random coil segment illustrating how several K^+ binding sites may be formed by only a few residues. All backbone atoms are shown. Color code: cyan = potassium ions, pink = uncharged oxygens of backbone or side chain, red = negatively charged side chain oxygens, light blue = uncharged nitrogens, red dashed lines = salt bridges, black dashed lines = hydrogen bonds. The lining may be quite flexible and have this conformations only when a K^+ is at each site.

Seetharamulu, 1986) that the P segments do not span the entire transmembrane region, but rather form a narrow ion selective region in the outer half of the transmembrane region, and that a larger pore is formed through the inner half of the transmembrane region by other segments.

Comparisons of voltage-gated K^+ channel sequences reveal that most insertions and deletions occur in putative loop regions between helices and in the NH_2- and COOH-terminal segments that precede S1 and follow S6. The putative transmembrane S1–S3 helices exhibit a pattern of "unilateral conservation" in which the more hydrophobic helical face is poorly conserved and the opposite, more polar face, is well conserved (Guy, 1988). The presence of this type of pattern has been interpreted as indicative of a transmembrane alpha helix that is positioned with its poorly conserved hydrophobic face in contact with lipid and its more polar well conserved face in contact with other protein segments or forming part of the pore's lining (Guy, 1988, 1990; Rees, DeAntonio, and Eisenberg, 1989). These unilateral conservation patterns may thus be used in three-dimensional modeling in much the same way that one uses the edge pieces of a jigsaw puzzle; i.e., they allow one to identify the surface segments and orient these segments relative to the lipid-protein interface. In our tentative models that include all transmembrane segments, there are three concentric layers of proteins around the axis of the pore; S1–S3 and to a lesser extent S5 form the outer layer, S4 and S6 form the next layer, and for the extracellular half of the transmembrane region the P segments form the inner layer. In modeling Mg^{2+} blocked inward rectifying channels the outer layer is formed by two helices from each subunit that are probably homologous to S5 and S6 of the other channel proteins and the inner layers are formed by the P segments.

The relationship described above between the extent of sequence conservation and the spacial distribution of residues is valid only for proteins that share a common function which is localized to a specific region of the protein; e.g., ion selectivity (K^+ channels), light absorbtion (bacteriorhodopsin), or oxygen binding (globins). Although voltage-gated K^+ channels are more closely related to voltage-gated Na^+ and Ca^{2+} than they are to some other K^+ channel families, the P regions are poorly conserved when K^+ channel sequences are compared to those of Na^+ and Ca^{2+} channels. Likewise; plant and *Eag* K^+ channels are more closely related to cation-selective cyclic nucleotide-gated channels than to some other types of K^+ channels. In this case, the first part of P, which in our more recent models forms an alpha helix, is well conserved between the K^+ and cyclic nucleotide-gated channels, whereas there is a deletion of one or two residues and no sequence identity in the latter portion, which is postulated to determine selectivity.

The information obtained by comparing distantly related sequences in the approach described above relies primarily upon identifying those positions that are well conserved in the multiple sequence alignment. Additional information for three-dimensional modeling can be obtained by comparing interactions between residues that are not well conserved. A basic assumption of this approach is that the general folding pattern is essentially the same for those portions of the proteins that can be aligned unambiguously and that do not have insertions and deletions in the alignment. In developing the models, we attempt to optimize favorable residue-residue interactions (e.g., disulfide bridges, salt bridges, hydrogen bonds between side chains, aromatic-aromatic interactions), residue-water interactions (polar side chains on the water accessible surface), and residue-lipid interaction (hydrophobic residues exposed to the alkyl chain portion of the membrane, positively charged residues at the cytoplasmic lipid head group region, aromatic side chains at the extracellular head group region). These interactions should be satisfied equally well for portions of all homologous proteins that have the same general folding patterns. Thus, we favor models for which this postulate is valid. Compensatory substitutions may be especially informative in this regard; e.g., a hydrophobic-hydrophobic interaction in one protein may be replaced by a salt-bridge interaction in an homologous protein.

Summary

Molecular modeling and mutagenesis analysis of voltage-gated channels have succeeded in identifying much of the topology of the proteins and in identifying which sequential segments are involved in functional mechanisms such as activation gating, inactivation gating, ion selectivity, and ligand binding. Efforts are currently underway to use these methods to model the protein structure and functional mechanisms more precisely. The experimental and theoretical efforts are dependent to a considerable extent upon information obtained by comparing homologous sequences. Although the fine details of models developed in this manner are unlikely to be as correct as models developed from x-ray crystallography and NMR, they still may contribute substantially to our understanding of the structure and function of these important proteins.

References

Anderson, J. A., S. S. Huprikar, L. V. Kochian, W. J. Lucas, and R. F. Gaber. 1992. Functional expression of a probable *Arabidopsis thaliana* potassium channel in *Saccharomyces cerevisiae. Proceedings of the National Academy of Sciences, USA.* 89:3736–3740.

Armstrong, C. 1981. Sodium channels and gating currents. *Physiological Review.* 61:644–683.

Atkinson, N. S., G. A. Robertson, and B. Ganetzky. 1991. A component of calcium-activated potassium channels encoded by the *Drosophila slo* locus. *Science.* 253:551–555.

Backx, P., D. Yue, J. Lawrence, E. Marban, and G. Tomaselli. 1992. Molecular localization of an ion-binding site within the pore of mammalian sodium channels. *Science.* 257:248–251.

Choi, K. L., C. Mossman, J. Aubie, and G. Yellen. 1993. The internal quaternary ammonium receptor site of Shaker potassium channels. *Neuron.* 10:533–541.

Chua, K., J. Tytgat, E. Liman, and P. Hess. 1992. Membrane topology of RCK1 K-channels. *Biophysical Journal.* 61:289*a*. (Abstr.)

Cowan, S. W., T. Schirmer, G. Rummel, M. Steiert, R. Ghosh, R. A. Pauptit, J. N. Jansonius, and J. P. Rosenbusch. 1992. Crystal structures explain functional properties of two *E. Coli* porins. *Nature.* 358:727–733.

DeBiasi M., J. A. Drewe, G. E. Kirsch, and A. M. Brown. 1994. Histidine substitution identifies a surface position and confers C_s^+ selectivity on a K^+ pore. *Biophysical Journal.* 65:1235–1242.

Durell, S. R., and H. R. Guy. 1992. Atomic scale structure and functional models of voltage-gated potassium channels. *Biophysical Journal.* 62:238–250.

Guy, H. R. 1990. Models of voltage- and transmitter-activated channels based on their amino acid sequences. *In* Monovalent Cations in Biological Systems. C. A. Pasternak, editor. CRC Press/Boca Raton, Florida. 31–58.

Guy, H. R. 1988. A model relating the sodium channel's structure to its function. *In* Molecular Biology of Ionic Channels: Current Topics in Membrane Transport; Vol. 33. W. S. Agnew, T. Claudio, and F. J. Sigworth, editors. Academic Press, San Diego. 289–308.

Guy, H. R., and F. Conti. 1990. Pursuing the structure and function of voltage-gated channels. *Trends in Neuroscience.* 13:201–206.

Guy, H. R., and P. Seetharamulu. 1986. Molecular model of the action potential sodium channel. *Proceedings National Academy Sciences, USA.* 83:508–512.

Guy, H. R., S. R. Durell, J. Warmke, R. Drysdale, and B. Ganetzky. 1991. Similarities in amino acid sequences of *Drosophila eag* and cyclic nucleotide gated channels. *Science.* 254:730.

Hartmann, H. A., G. E. Kirsch, J. A. Drewe, M. Taglialatela, R. H. Joho, and A. M. Brown. 1991. Exchange of conduction pathways between two related K^+ channels. *Science.* 251:942–944.

Heginbotham, L., T. Abramson, and R. MacKinnon. 1992. A functional connection between the pores of distantly related ion channels as revealed by mutant K^+ channels. *Science.* 258:1152–1155.

Heginbotham, L., and R. MacKinnon. 1992. The aromatic binding site for tetraethylammonium ion on potassium channels. *Neuron.* 8:483–491.

Heginbotham, L., Z. Lu, T. Abramson, and R. MacKinnon. 1994. Mutation in the K^+ channel signature sequence. *Biophysical Journal.* 66:1–7.

Heinemann, S. H., H. Terlau, W. Stühmer, K. Imoto, and S. Numa. 1992. Calcium channel characteristics conferred on the sodium channel by singel mutations. *Nature.* 356:441–443.

Heinemann, S. H., H. Terlau, and K. Imoto. 1992. Molecular basis for pharmacological differences between brain and cardiac sodium channels. *Pflügers Archiv.* 422:90–92.

Henderson, R., J. M. Baldwin, T. A. Ceska, F. Zemlin, E. Beckmann, and K. H. Downing. 1990. Model for the structure of bacteriorhodopsin based on high-resolution electron cryo-microscopy. *Journal Molecular Biology.* 213:899–929.

Hille, B. 1975. The receptor for tetrodotoxin and saxitoxin: a structural hypothesis. *Biophysical Journal* 15:615–619.

Hille B. 1977. Local anesthetics: Hydrophilic and hydrophobic pathways for the drug-receptor reaction. *Journal General Physiology.* 69:497–515.

Ho, K., C. G. Nichols, W. J. Lederer, J. Lytton, P. M. Vassilev, M. V. Kanazirska, and S. C. Hebert. 1993. Cloning and expression of an inwardly rectifying ATP-regulated potassium channel. *Nature.* 262:31–38.

Kaupp, U. B., T. Niidome, T. Tanabe, S. Terada, W. Bonigk, W. Stühmer, N. J. Cook, K. Kangawa, H. Matsuo, T. Hirose, T. Miyata, and S. Numa. 1989. Primary structure and functional expression from complementary DNA of the rod photoreceptor cyclic GMP-gated channel. *Nature.* 342:762–766.

Kubo, Y., T. J. Baldwin, Y. N. Jan, and L. Y. Jan. 1993. Primary structure and functional expression of a mouse inward rectifier potassium channel. *Nature.* 362:127–132.

Liman, E. R., J. Tytgat, and P. Hess. 1992. Subunit stoichiometry of a mammalian K^+ channel determined by construction of multimeric cDNA's. *Neuron.* 9:861–871.

MacKinnon, R. 1991. Determination of the subunit stoichiometry of a voltage-activated potassium channel. *Nature.* 350:232–234.

MacKinnon, R., and G. Yellen. 1990. Mutations affecting TEA blockade and ion permeation in voltage-activated K^+ channels. *Science.* 250:276–279.

Milkman, R., and M. McKane Bridges. 1993. An *E. coli* homologue of eukaryotic potassium channels. Genebank entry ECOKCH.

Neyton, J., and C. Miller. 1988. Discrete Ba^{2+} block as a probe of ion occupancy and pore structure in the high-conductance Ca^{2+}-activated K^+ channel. *Journal of General Physiology.* 92:569–586.

Noda, M., S. Shimizu, T. Tanabe, T. Takai, T. Kayano, T. Ikeda, H. Nakayama, Y. Kanaoka, N. Minamino, K. Kangawa, H. Matsuo, M. A. Raftery, T. Hirose, M. Notake, S. Inayama, H. Hayashida, T. Miyata, and S. Numa. 1984. Primary structure of *Electrophorus electricus* sodium channel deduced from cDNA sequence. *Nature.* 312:121–127.

Noda, M., T. Ikeda, T. Kayano, H. Suzuki, H. Takeshima, M. Kurasaki, H. Takahashi, and S. Numa. 1986. Existence of distinct sodium channel messenger RNAs in rat brain. *Nature.* 320:188–192.

Noda, M., H. Suzuki, S. Numa, and W. Stühmer. 1989. A single point mutation confers tetrodotoxin and saxitoxin insensitivity on the sodium channel II. *FEBS Letters.* 259:213–216.

Overington, J., D. Donnelly, M. S. Johnson, A. Sali, and T. L. Blundell. 1992. Environment-specific amino acid substitution tables: tertiary templates and prediction of protein folds. *Protein Science.* 1:216–226.

Papazian, D. M., L. C. Timpe, Y. N. Jan, and L. Y. Jan. 1991. Alteration of voltage-dependence of *Shaker* potassium channel by mutations in the S4 sequence. *Nature.* 349:305–310.

Patton, D. E., J. W. West, W. A. Catterall, and A. L. Goldin. 1993. A region critical for sodium channel inactivation functions as an inactivation gate in a potassium channel. 11th International Biophysics Congress. 133.

Peled, H., and Y. Shai. 1993. Membrane interaction and self-assembly within phospholipid membranes of synthetic segments corresponding to the H-5 region of the *Shaker* K^+ channel. *Biochemistry.* 32:7879–7885.

Pusch, M., M. Noda, and F. Conti. 1991. Gating currents of a sodium-channel mutant with very low conductance. *Biophysical Journal.* 59:25*a.* (Abstr.)

Rees D. C., L. DeAntonio, and D. Eisenberg. 1989. Hydrophobic organization of membrane proteins. *Science.* 245:510–586.

Satin, J., J. W. Kyle, M. Chen, P. Bell, L. L. Cribbs, H. A. Fozzard, and R. B. Rogart. 1992. A mutant of TTX-resistant cardiac sodium channels with TTX-sensitive properties. *Science.* 256:1202–1205.

Shen, N. V., X. Chen, M. M. Boyer, and P. J. Pfaffinger. 1993. Deletion analysis of K^+ channel assembly. *Neuron.* 11:67–76.

Stühmer, W., F. Conti, H. Suzuki, X. Wang, M. Noda, N. Yahagi, H. Kubo, and S. Numa. 1989*a*. Structural parts involved in activation and inactivation of the sodium channel. *Nature.* 339:597–603.

Stühmer, W., J. P. Rupperberg, K. H. Schroter, B. Sakmann, M. Stocker, K. P. Giese, A. Perschke, A. Baumann, and O. Pongs. 1989*b*. Molecular basis of functional diversity of voltage-gated potassium channels in mammalian brain. *EMBO Journal.* 8:3235–3244.

Tanabe, T., H. Takeshima, A. Mikami, V. Flockerzi, H. Takahashi, K. Kangawa, M. Kojima, H. Matsuo, T. Hirose, and S. Numa. 1987. Primary structure of the receptor for calcium channel blockers from skeletal muscle. *Nature.* 328:313–318.

Tempel, B. L., D. M. Papazian, T. L. Schwarz, Y. N. Jan, and L. Y. Jan. 1987. Sequence of a probable potassium channel component encoded at *Shaker* locus of *Drosophila. Science.* 237:770–775.

Terlau, H., S. H. Heinemann, W. Stühmer, M. Pusch, F. Conti, H. Imoto, and S. Numa. 1991. Mapping the site of tetrodotoxin and saxitoxin of sodium channel II. *FEBS Letters.* 293:93–96.

Vassilev, P. M., T. Scheuer, and W. A. Catterall. 1988. Identification of an intracellular peptide segment involved in sodium channel inactivation. *Science.* 241:1658–1664.

Warmke, J., R. Drysdale, and B. Ganetzky. 1991. A distinct potassium channel polypeptide encoded by the *Drosophila eag* locus. *Science.* 252:1560–1562.

Weiss, M. S., U. Abele, J. Weckesser, W. Welte, E. Schiltz, and G. E. Schulz. 1991. Molecular architecture and electrostatic properties of a bacterial porin. *Science.* 254:1627–1630.

Yellen, G., M. E. Jurman, T. Abramson, and R. MacKinnon. 1991. Mutations affecting internal TEA blockade identify the probable pore-forming region of a K^+ channel. *Science.* 251:939–942.

Zagotta, W. N., T. Hoshi, and R. W. Aldrich. 1990. Restoration of inactivation in mutants of *Shaker* potassium channels by a peptide derived from ShB. *Science.* 250:568–571.

Molecular Evolution of K^+ Channels in Primitive Eukaryotes

Timothy Jegla* and Lawrence Salkoff*‡

**Department of Anatomy and Neurobiology, and ‡Department of Genetics, Washington University School of Medicine, St. Louis, Missouri 63110*

Summary

Cnidarians and ciliate protozoans represent evolutionary interesting phylogenetic groups for the study of K^+ channel evolution. Cnidaria is a primitive metazoan phylum consisting of simple diploblast organisms which have few tissue types such as jellyfish, hydra, sea anemones, and corals. Their divergence from the rest of the metazoan line may predate the radiation of the major triploblast phyla by several hundred million years (Morris, 1993). Cnidarians are the most primitive metazoans to have an organized nervous system. Thus, comparing K^+ channels cloned from cnidarians to those cloned from more advanced metazoans may reveal which types of K^+ channel are most fundamental to electrical excitability in the nervous system. In contrast, channels in ciliate protozoans such as *Paramecium* may not have been designed to send electrical signals between cells, but simply to control the behavior, such as an avoidance reaction, of a single cell. Hence, comparing cloned *Paramecium* K^+ channels to K^+ channels cloned from cnidarians and other metazoans may reveal which types of K^+ channel are most fundamental to electrical excitability in eukaryotes, and which K^+ channels are specialized for neuronal signaling.

Potassium channels are involved in a diversity of tasks and are universally present in eukaryotes. K^+ channels set the resting membrane potentials of most metazoan and protozoan cells and are fundamental components of membrane electrical activity in virtually all eukaryotic systems. These channels control the shape, duration and frequency of metazoan action potentials and are known to participate in the action potentials of protozoans, fungi and plants as well (Hille, 1992). Voltage-clamp recordings have shown that a various assortment of voltage-gated K^+ channels as well as Ca^{2+}-activated K^+ channels are widespread in eukaryotes (Hille, 1992). Thus, K^+ channels appear to be crucial to behavioral responses in all classes of eukaryotes, including locomotion in metazoans and protozoans, and rapid growth responses and cell shape changes in plants.

K^+ channel diversity is by far the greatest in metazoans, which have made a strong commitment to electrically excitable cellular networks. There is an apparent need for a great diversity of K^+ channel subtypes in these metazoans. Over 50 K^+ channel sequences from many distinct gene families have been reported so far, and all but two (both from plants) have been found in triploblast metazoans. The complex needs of neuronal integration and neuromuscular transmission in triploblasts require exquisite control of cellular excitability. This is in large part achieved by an extensive and diverse set of K^+ channels. The two closely related plant K^+ channels, *AKT1* and *KAT1* (Anderson, Huprikar, Kochian, Lucas, and Gaber, 1992; Sentenac, Bonneaud, Minet, Lacroute, Salmon, Gaymard, and Grignon, 1992), share only the most basic structural features with triploblast K^+ channels, a likely consequence of the very different roles K^+ channels may play in plants.

The molecular basis of K^+ currents in the more primitive diploblast metazoans including such animals as jellyfish (which have simple nervous systems) and in the electrically excitable single cells of the protozoan groups, has remained a mystery. Will the most primitive metazoans share K^+ channel types with the triploblast metazoans, reflecting a set of excitable molecules universal to all metazoan life forms? Will the most metazoans share K^+ channel types with protozoans, reflecting a common origin of electrical excitability? Discovering the molecular identity of K^+ channels in these organisms may help us relate the emergence of specific K^+ channel types to key points in the evolution of electrical excitability. To this end, we have studied the evolutionary origins of voltage-gated K^+ channels by isolating K^+ channel genes in the diploblast cnidarians (jellyfish) and ciliate protozoans (*Paramecium*). What follows is a brief description of the diversity of triploblast voltage-gated K^+ channels and their position within the larger family of K^+ channel genes, followed by a description of the K^+ channels that we have found in jellyfish and *Paramecium* and a discussion of what these findings might be telling us about the evolution of voltage-gated K^+ channels.

Voltage-gated K^+ Channels in Higher Metazoa: *Shaker, Shab, Shal,* and *Shaw*

Molecular studies have shown that an extended gene family is responsible for much of the biophysical diversity of voltage-gated K^+ channels in triploblast metazoans. Four structurally related subfamilies of voltage-gated K^+ channels (*Shaker, Shab, Shal,* and *Shaw*) were originally isolated as single genes in *Drosophila* (Butler, Wei, Baker, and Salkoff, 1989). *Shaker* and *Shal* encode transient A-type K^+ currents that differ in voltage-sensitivity and kinetics, whereas *Shab* and *Shaw* encode delayed rectifier K^+ currents that also differ in biophysical properties (Fig. 1). Each of these four *Drosophila* K^+ channel genes is represented by a multiple gene subfamily in *Mus* (Fig. 2). The *Shab* and *Shal* subfamilies share the greatest level of sequence and functional conservation between *Drosophila* and *Mus,* whereas the *Shaker* and *Shaw* subfamilies show relatively more variability in their biophysical properties (Salkoff, Baker, Butler, Covarrubias, Pak, and Wei, 1992). *Shaker, Shaw,* and *Shab* homologs have also been found in *Aplysia* (Pfaffinger, Furukawa, Zhao, Dugan, and Kandel, 1991; Zhao and Kandel, 1991), a *Shaker* homolog has been found in Annelids (Johansen, Wei, Salkoff, and Johansen, 1990) and a *Shaw* homolog has been found in *C. elegans* (Wei, Jegla, and Salkoff, 1991). These genes represent distinct lineages that must have evolved from a common ancestral gene before the divergence of the major triploblast metazoan groups (Salkoff et al., 1992). *Shaker, Shab, Shal,* and *Shaw* genes may have been conserved amongst these diverse triploblast metazoans to serve shared essential functions, such as action potential repolarization and the generation of patterned neuronal output. *Shaker, Shab, Shal,* and *Shaw* may represent an essential set of voltage-gated K^+ channels necessary for electrical signaling in neuromuscular systems and for neuronal integration. Thus, exploring the phylogenetic distribution (and therefore evolution) of these voltage-gated K^+ channel

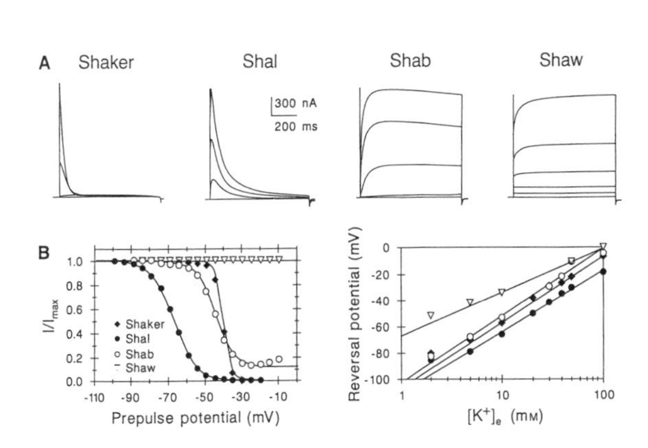

Figure 1. Channels encoded by *Shaker, Shal, Shab,* and *Shaw* express currents that cover a wide range of biophysical properties. (*A*) Outward currents recorded in response to voltage steps ranging from −80 to +20 mV in 20-mV steps from a holding potential of −90 mV. Records were leak subtracted assuming a linear leak current. (*B*) Comparison of (*left*) steady state inactivation and (*right*) reversal potentials for *Shaker, Shal, Shab,* and *Shaw* currents. (Wei et al., 1990; Salkoff et al., 1992).

subfamilies in primitive metazoans and other electrically excitable eukaryotes is valuable in understanding their roles in membrane excitability.

Voltage-gated K^+ Channels Are Part of a Large Ion Channel Gene Family

The extended family of voltage-gated K^+ channels includes other ion channel genes which have a similar structural motif. *Shaker, Shab, Shal,* and *Shaw* channels share a structural motif that includes six transmembrane domains, S1–S6 (Fig. 3). Each

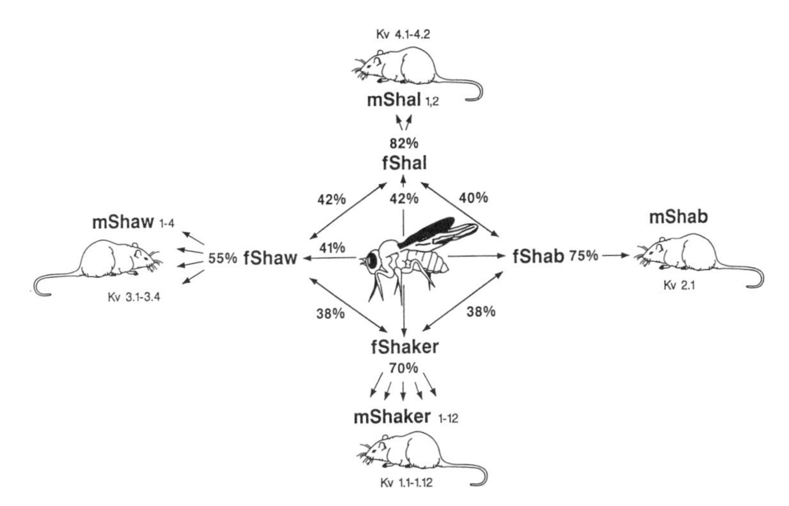

Figure 2. Conservation of voltage gated K^+ channels. *Shaker, Shal, Shab,* and *Shaw* each represent a separate conserved gene subfamily. Numbers represent percentage of identity in the membrane-spanning regions S1–S6. The pairwise comparison of the proteins encoded by the *Drosophila Shaker, Shal, Shab,* and *Shaw* genes (designated by the prefix *f*) shows that percentage identity between them averages 40%. However, comparison of each of the *Drosophila* channel proteins with its mammalian homolog (designated by the prefix *m*) reveals a much higher level of identity. The mammalian K^+ channels within each subfamily share high identity, averaging greater than 70%. As in *Drosophila,* a comparison of mammalian K^+ channels between subfamilies shows an identity of ~40%. (Salkoff et al., 1992).

contains a string of positive charges in the S4 region that serves as the channel's voltage sensor (Papazian, Timple, Jan, and Jan, 1991; Liman, Hess, Weaver, and Koren, 1991; Logothetis, Kammen, Lindpainter, Bisbas, and Nadal-Ginard, 1993) and a highly conserved membrane-embedded loop between S5 and S6 which forms the highly K^+ selective pore of the ion channel (Hartmann, Kirsch, Drewe, Taglialatella, Joho, and Brown, 1991; Yool and Schwarz, 1991; Yellen, Jurman, Abramson, and MacKinnon, 1991). Ca^{2+}-activated K^+ channels from *Drosophila* (*Slo;* Atkinson, Robertson, and Ganetzky, 1991) and *Mus* (*mSlo;* Butler et al., 1993) and the two hyperpolarization-dependent K^+ channels from plants also share this structural motif. Additionally, cation-selective cyclic nucleotide-gated channels from the bovine retina (Kaupp et al., 1989) and catfish olfactory organ (Goulding, Ngai, Kramer, Colicos, Axel, Sugelbaum, and Chess, 1992) contain the six transmembrane domain motif, but show key differences in the ion-selective pore region. This extended family of K^+ channel genes shares a common evolutionary origin with the voltage-gated Na^+ and Ca^{2+} channel gene families. Voltage-gated Na^+ and Ca^{2+} channels contain four tandem repeats of the S1–S6 motif found the K^+ channel gene family. These new channel structures presumably arose from two intragenic duplications of the S1–S6

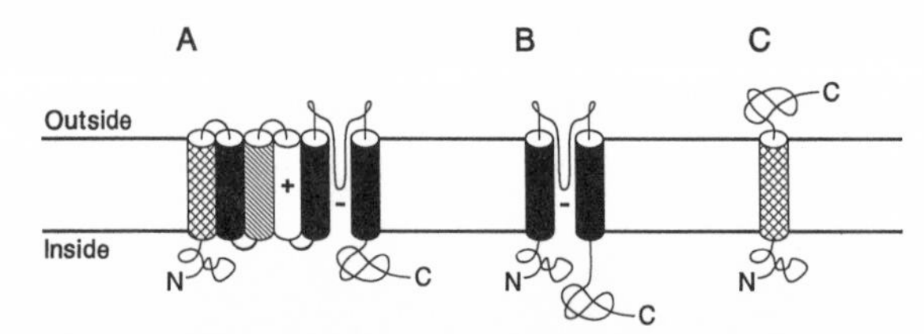

Figure 3. Topology of three classes of K^+ channel. (*A*) The structure of the voltage-gated K^+ channels *Shaker, Shab, Shal,* and *Shaw* consists of six transmembrane domains (S1 to S6) with two features of particular interest: a string of positive charges in S4 which likely forms the channel's voltage sensor (marked with a +) and a membrane-embedded loop between S5 and S6 which forms the K^+-selective pore of the channel (marked with a −). (*B*) A second family of inward rectifying K^+ channels (*ROMK1, IRK1,* and *GIRK1*) contains only two transmembrane domains surrounding the pore loop. (*C*) The *minK* K^+ channel consists of only a single transmembrane domain.

motif in an ancestral gene (Hille, 1992). Evidence suggests that members of the K^+ channel gene family function as tetramers (Agnew, 1988; Timpe, Schwarz, Tempel, Papazian, Jan, and Jan, 1988) and that voltage-gated Na^+ and Ca^{2+} channels function as "pseudotetramers," a name denoting the fact that the four linked homologous domains of these channels are similar but not identical (Noda et al., 1984). In all cases, the functional channels contain 24 membrane spanning domains.

Other K^+ Channel Genes

The complete diversity of K^+ channels is not explained by the previously mentioned K^+ channel genes. A second family consisting of three inwardly rectifying K^+ channels from *Mus,* called *IRK1* (Kubo et al., 1993*a*), *ROMK1* (Ho et al., 1993) and GIRK1 (which is G-protein-coupled, Kubo, Baldwin, Jan, and Jan, 1993*b*) does not contain the S1–S6 motif. These three genes show strong homology to the voltage-gated K^+ channels in the K^+-selective pore region, but lack an S4 voltage sensor and contain only two transmembrane domains (Fig. 3). *IRK1, ROMK1,* and *GIRK1* show little or no intrinsic voltage sensitivity which is not surprising in channels lacking an S4 voltage sensor. A third type of mammalian K^+ channel characterized by a single transmembrane domain, *minK,* was originally cloned from rat (Takumi, Onkubu, and Nakanishi, 1988). *MinK* appears to encode a very slow voltage-gated K^+ current. Although these two classes of K^+ channels have so far only been reported in mammals, they too may represent gene families conserved in other triploblast metazoans.

Cnidarians and Ciliate Protozoans Represent Evolutionarily Interesting Phylogenetic Groups for the Study of K^+ Channel Evolution

It is likely that much of the molecular evolution of neurons occurred in simple diploblast cnidarians such as the jellyfish. As the most primitive metazoans to contain nervous systems, cnidarians must contain the basic set of molecules that allow for the production and function of an electrically excitable neuronal network. Many of the voltage-gated K^+ channel current types characterized in higher metazoans have in fact been observed in cnidarians by physiological techniques (Anderson, 1989). Any K^+ channels conserved between cnidarians and higher metazoans are likely to be essential to electrical signaling in all metazoans.

Ciliate protozoans are unicellular organisms that use membrane excitation to control swimming behavior and are only distantly related to metazoans. Ciliates represent a very early branch point in the evolution of eukaryotes that predates the advent of multicellularity (Wainright, Hinkle, Sogin, and Stickel, 1993): Many even have a unique genetic code (Hanyu, Kuchino, Nishimura, and Beier, 1986; King, Maley, Ling, Kanabrocki, and Kung, 1991). Physiological analysis of the ciliate protozoan *Paramecium tetraurelia* has revealed that its action potentials are both Ca^{2+} and Na^+ dependent, and that it has a broad array of K^+ currents (Saimi and Kung, 1987). Two of these K^+ currents are gated by calmodulin (Preston, Saimi, Amberger, and Kung, 1990) and have not been observed in other phylogenetic groups. Mutant analysis has shown that these K^+ currents are critical for producing proper swimming patterns in *Paramecium* (Preston et al., 1990). Ciliates contain all the K^+ channels necessary to allow a single cell to use membrane excitation to convert environmental stimuli into quick behavioural responses. However, they will lack any K^+ channels specifically designed to control electrical transmission between cells.

Because of their interesting positions in evolution and the physiological evidence for voltage-gated K^+ channels in each, we decided to use cnidarians and *Paramecium* as model organisms for studying the evolutionary origins of the voltage-gated K^+ channel gene subfamilies conserved in triploblast metazoans, *Shaker, Shal, Shab,* and *Shaw.* We reasoned that we should find homologs of these channels in cnidarians if, indeed, they are fundamental to electrical signaling in nervous systems, and that we might also find homologs in *Paramecium* if these channels are fundamental to membrane excitability.

Diploblasts and Triploblasts Share a Similar Set of K^+ Channels

We found that cnidarians do contain homologs of voltage-gated K^+ channels found in higher metazoans. We used degenerate primers based on pore and S6 sequences conserved between the four *Drosophila* voltage-gated K^+ channels (*Shaker, Shab, Shal,* and *Shaw*) to amplify fragments of six K^+ channels from the genomic DNA of the jellyfish, *Polyorchis penicillatus.* These fragments were then used as high stringency probes on both cDNA and genomic libraries from *Polyorchis* to obtain larger

K^+ channel clones. The search has produced clones of two distinct jellyfish *Shaker* homologs, *jShak1* and *jShak2,* and a clone of a jellyfish *Shal* homolog, *jShal* (Fig. 4). In addition to these clones, sequence homology suggests that two other amplified fragments, for which larger clones have not been found, may well represent jellyfish *Shab* and *Shaw* homologs.

The S1 through S6 regions of *JShak1* and *jShak2* share ~50% amino acid identity to each other, and share similar identity to the *Drosophila* and *Mus* homologs of *Shaker.* This level of conservation is lower than the 70% amino acid identity that is typically shared between *Drosophila* and *Mus* homologs of *Shaker* (Salkoff et al., 1992), but is significantly higher than the 40% level of conservation these jellyfish homologs of *Shaker* share with *Drosophila Shab, Shaw,* and *Shal. JShak1* also has *Shaker*-like features in two functional domains of the amino terminal cytoplasmic region: a region encoding the "inactivation ball" of the *Drosophila ShakerB* splice variant (Zagotta, Hoshi, and Aldrich, 1990) and a region thought to be involved in

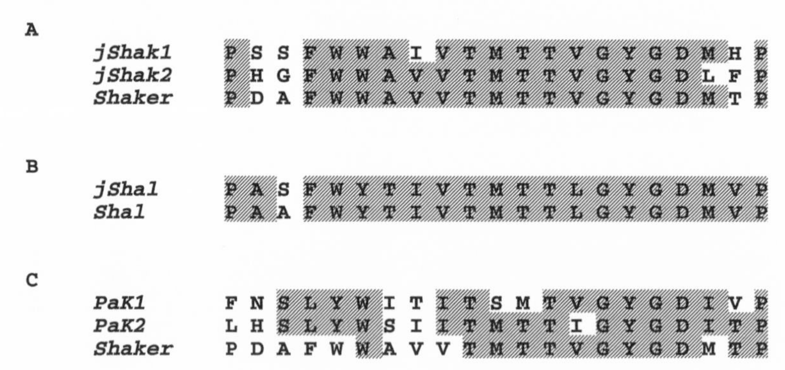

Figure 4. Alignments of jellyfish and *Paramecium* K^+ channel sequences to *Drosophila* K^+ channels in the K^+-selective pore region. (*A*) Alignment of the jellyfish *Shaker* homologs *jShak1* and *jShak2* to *Drosophila Shaker.* (*B*) Alignment of the jellyfish *Shal* homolog *jShal* to *Drosophila Shal.* (*C*) Alignment of the novel *Paramecium* K^+ channels *Pak1* and *Pak2* to *Drosophila Shaker.*

the tetramerization of *Shaker* subunits into functional channels (Li, Jan, and Jan, 1992; Shen, Chen, Boyer, and Pfaffinger, 1993). The *JShak2* cDNA is truncated just 5′ of S1, so it is not clear if it also contains these features.

JShal shows a greater level of conservation to *Drosophila* and *Mus Shal* homologs (over 70% amino acid identity from S1 to S6, and nearly as much conservation at the amino and carboxyl termini). This higher level of conservation is consistent with the observation that *Shal* is the most highly conserved subfamily of voltage-gated K^+ channels in higher metazoans (Pak, Baker, Covarrubias, Butler, Ratcliffe, and Salkoff, 1991; Salkoff et al., 1992). Remarkably, the position of an intron in the gene region encoding the pore of *jShal* and *Drosophila Shal* is identically conserved, suggesting the conservation of gene structure extending back, perhaps, a billion years. Thus, *Shaker, Shal* (and likely *Shab* and *Shaw*) appear to have been highly conserved as distinct lineages from the very beginnings of metazoan evolution. This evidence suggests that many of the genetic determinants of membrane electrical properties may be common to virtually all nervous systems.

Paramecium tetraurelia **Contains a Novel Family of K^+ Channels, but Appears to Lack Homologs of the Metazoan Voltage-gated K^+ Channels**

Our attempts to isolate homologs of *Shaker, Shal, Shab,* and *Shaw* from *Paramecium tetraurelia* were unsuccessful. Instead, fragments of five channels representing a new and highly divergent family of K^+ channels were isolated. Using these fragments, complete genomic clones of two K^+ channels were found. The encoded channel subunits, *PaK1* and *PaK2,* share ~34% amino acid identity to each other in the core (S1 to S6) region. *PaK1* and *PaK2,* share much lower amino acid identity with the metazoan K^+ channels, *Shaker, Shal, Shab,* and *Shaw* except for the highly conserved

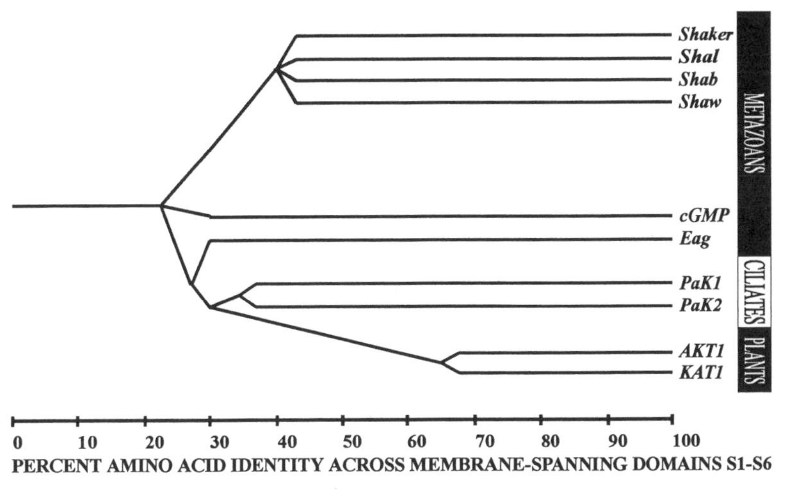

Figure 5. Possible evolutionary tree of the metazoan voltage-gated K^+ channels and related ion channels containing a single S1–S6 motif. Note that the branch containing the voltage-gated K^+ channels *Shaker, Shal, Shab,* and *Shaw* is strictly metazoan, whereas a second more diverse branch of the tree is much more widely spread in eukaryotes. This second branch includes the *Paramecium* K^+ channels *PaK1* and *PaK2,* as well as the metazoan cyclic nucleotide gated channels.

K^+-selective pore region (Fig. 4). The string of positive charges in the S4 region that forms the voltage sensor of metazoan voltage-gated K^+ channels is poorly conserved in *PaK1* and *PaK2,* suggesting the possibility that *PaK1* and *PaK2* are not purely voltage-gated channels. In fact, these *Paramecium* clones show somewhat greater conservation to metazoan cyclic nucleotide-gated channels and to the hyperpolarization-dependent K^+ channels from plants (26–30%) than they do to *Shaker, Shal, Shab,* and *Shaw* (19–23%). *Paramecium* also contains calmodulin-gated K^+ channels that may be unique to ciliates (Saimi and Kung, 1987). It is not clear whether this novel K^+ channel gene family represented by *PaK1* and *PaK2* will encode the entire range of K^+ currents seen in *Paramecium,* but the low sequence homology within this family suggests a diversity of functional properties greater than that seen among the metazoan voltage-gated K^+ channels.

The Voltage-gated K^+ Channels *Shaker, Shal, Shab,* and *Shaw* May Be Specialized for the Signaling Needs of Metazoans, and May Not Be Found Outside the Metazoan Line

Our amplification screen for homologs of *Shaker, Shal, Shab,* and *Shaw* has been universally successful in finding one or more homologs for these genes in all metazoans on which it has been tried. Thus, it has been used successfully on jellyfish, nematode, annelids, *Mus* and *Drosophila.* It is significant then that this screen has consistently failed to amplify homologs of *Shaker, Shal, Shab,* or *Shaw* from organisms outside the metazoan line. We have not found homologs of these channels after repeated screening attempts on plants (*Arabidopsis thaliana*), yeast (*S. cerevisiae*), flagellate protozoans (*Chlamydomonas*) and ciliate protozoans (*Paramecium*). This is a strong suggestion (but not proof) that *Shaker, Shal, Shab,* and *Shaw* are strictly metazoan and may represent a set of voltage-gated K^+ channels specifically designed for the propagation of electrical signals in metazoan nervous systems. It will be interesting to see if more primitive metazoans such as sponges, which do not contain nervous systems but can propagate electrical signals from cell to cell, contain homologs of these voltage-gated K^+ channels.

In contrast to the *Shaker* K^+ channel family, the K^+ channel gene family that includes the metazoan cyclic nucleotide-gated channels, plant hyperpolarization-dependent K^+ channels and the *Paramecium* channels appears to be much more widespread in eukaryotes and thus may be far older (Fig. 5). Because of this phylogenetic distribution, the origins of this loose group of channels must have predated the advent of multicellularity, and probably electrical excitability. Their ancestral functions were likely related to the maintenance of resting potentials and volume regulation, having later been altered to serve a wider variety of functions in a diverse group of organisms. The metazoan voltage-gated K^+ channels may well have evolved from an ancestral channel that looked very much like members of this second group of channels.

References

Agnew, W. S. 1988. A rosetta stone for K^+ channels. *Nature.* 331:114–115.

Anderson, J. A., S. S. Huprikar, L. V. Kochian, W. J. Lucas, and R. F. Gaber. 1992. Functional expression of a probable *Arabidopsis thaliana* potassium channel in *Saccharomyces cerevisiae. Proceedings of the National Academy of Sciences, USA.* 89:3736–3740.

Anderson, P. A. V. 1989. Ionic currents of the *Scyphozoa. In* Evolution of the First Nervous Systems. NATO ASI Series A, Life Science. Vol. 188. Plenum Publishing Corp., NY. 267–280.

Atkinson, N. S., G. A. Robertson, and B. Ganetzky. 1991. A component of calcium-activated potassium channels encoded by the *Drosophila slo* locus. *Science.* 253:551–555.

Butler, A., A. Wei, K. Baker, and L. Salkoff. 1989. A family of putative potassium channel genes in *Drosophila. Science.* 243:943–947.

Goulding, E. H., J. Ngai, R. H. Kramer, S. Colicos, R. Axel, S. A. Siegelbaum, and A. Chess. 1992. Molecular cloning and single-channel properties of the cyclic nucleotide-gated channel from catfish olfactory neurons. *Neuron.* 8:45–58.

Hanyu, N., Y. Kuchino, S. Nishimura, and H. Beier. 1986. Dramatic events in ciliate evolution: alteration of UAA and UAG termination codons to glutamine codons due to anticodon mutations in *Tetrahymena* $tRNA^{Gln}$. *EMBO Journal* 5:1307–1311.

Hartmann, H. A., G. E. Kirsch, J. A. Drewe, M. Taglialatela, R. H. Joho, and A. M. Brown. 1991. Exchange of conduction pathways between two related K^+ channels. *Science.* 251:942–944.

Hille, B. 1992. Ionic Channels of Excitable Membranes. Second edition. Sinauer Associates, Inc., Sunderland, MA.

Ho, K., C. G. Nichols, W. J. Lederer, J. Lytton, P. M. Vassilev, M. V. Kanazirska, and S. C. Hebert. 1993. Cloning and expressing of an inwardly rectifying ATP-regulated potassium channel. *Nature.* 362:31–38.

Johansen, K., A. Wei, L. Salkoff, and J. Johansen. 1990. A leech gene sequence homologous to *Drosophila* and mammalian *Shaker* K^+ channels. *Journal of Cell Biology.* 111:60*a.* (Abstr.)

Kaupp, B. U., T. Niidome, T. Tanabe, S. Terada, W. Bonigk, W. Suhmer, N. Cook, K. Kangawa, H. Matsuo, T. Hirose, and S. Numa. 1989. Primary structure and functional expression from complementary DNA of the rod photoreceptor cyclic GMP-gated channel. *Nature.* 342:762–766.

King, J. A., M. E. Maley, K.-Y. Ling, J. A. Kanabrocki, and C. Kung. 1991. Efficient expression of the *Paramecium* calmodulin gene in *Escherichia coli* after four TAA-to-CAA changes through a series of polymerase chain reactions. *Journal of Protozoology.* 38:441–447.

Kubo, Y., T. J. Baldwin, Y. N. Jan, and L. Y. Jan. 1993*a.* Primary structure and functional expression of a mouse inward rectifier potassium channel. *Nature.* 362:127–132.

Kubo, Y., E. Reuveny, P. A. Slesinger, Y. N. Jan, and L. Y. Jan. 1993*b.* Primary structure and functional expression of a rat G-protein-coupled muscarinic potassium channel. *Nature.* 364:802–806.

Li, M., Y. N. Jan, and L. Y. Jan. 1992. Specification of subunit assembly by the hydrophilic amino-terminal domain of the *Shaker* potassium channel. *Science.* 235:1225–1230.

Liman, E. R., P. Hess, F. Weaver, and G. Koren. 1991. Voltage-sensing residues in the S4 region of a mammalian K^+ channel. *Nature.* 353:752–756.

Logothetis, D. E., B. F. Kammen, K. Lindpainter, D. Bisbas, and B. Nadal-Ginard. 1993. Gating charge differences between two voltage-gated K^+ channels are due to the specific charge content of their respective S4 regions. *Neuron.* 10:1121–1129.

Morris, S. C. 1993. The fossil record and the early evolution of the Metazoa. *Nature.* 361:219–225.

Noda, M., S. Shimizu, T. Tanabe, T. Takai, T. Kayano, T. Ikeda, H. Takahashi, H. Nakayama, Y. Kanaoka, N. Minamino, K. Kangawa, H. Matsuo, M. A. Raftery, T. Hirose, S. Inayama, H. Hayashida, T. Miyata, and S. Numa. 1984. Primary structure of *electrophorus electricus* sodium channel deduced from cDNA sequence. *Nature.* 312:121–127.

Pak, M. D., K. Baker, M. Covarrubias, M., A. Butler, A. Ratcliffe, and L. Salkoff. 1991. *MShal,* a subfamily of A-type K^+ channel cloned from mammalian brain. *Proceedings of the National Academy of Sciences, USA.* 88:4386–4390.

Papazian, D. M., L. C. Timple, Y. N. Jan, and L. Y. Jan. 1991. Alteration of voltage-dependence of *Shaker* potassium channel by mutations in the S4 sequence. *Nature.* 349:305–310.

Pfaffinger, P. J., Y. Furukawa, B. Zhao, D. Dugan, and E. R. Kandel. 1991. Cloning of expression of an *Aplysia* K^+ channel and comparison with native *Aplysia* K^+ currents. *Journal of Neuroscience.* 11:918–927.

Preston, R. R., Y. Saimi, E. Amberger, and C. Kung. 1990. Interactions between mutants with defects in two Ca^{2+}-dependent K^+ currents of *Paramecium tetraurelia*. *Journal of Membrane Biology*. 115:61–69.

Saimi, Y., and C. Kung. 1987. Behavioral genetics of *Paramecium*. *Annual Review of Genetics*. 21:47–65.

Salkoff, L., K. Baker, A. Butler, M. Covarrubias, M. Pak, and A. Wei. 1992. An essential 'set' of K^+ channels conserved in flies, mice and humans. *Trends in Neuroscience*. 15:161–166.

Sentenac, H., N. Bonneaud, M. Minet, F. Lacroute, J.-M. Salmon, F. Gaymard, and C. Grignon. 1992. Cloning and expression in yeast of a plant potassium ion transport system. *Science*. 256:663–665.

Shen, N. V., X. Chen, M. M. Boyer, and P. J. Pfaffinger. 1993. Deletion analysis of K^+ channel assembly. *Neuron*. 11:67–76.

Takumi, T., H. Onkubo, and S. Nakanishi. 1988. Cloning of a membrane protein that induces a slow voltage-gated potassium current. *Science*. 242:1042–1045.

Timpe, L. C., T. L. Schwarz, B. L. Tempel, D. M. Papazian, Y. N. Jan, and L. Y. Jan. 1988. Expression of functional potassium channels from *Shaker* cDNA in *Xenopus* oocytes. *Nature*. 331:143–145.

Wainwright, P. O., G. Hinkle, M. L. Sogin, and S. K. Stickel. 1993. Monophyletic origins of the Metazoa: and evolutionary link with fungi. *Science*. 260:340–341.

Wei, A., M. Covarrubias, A. Butler, K. Baker, M. D. Pak, and L. Salkoff. 1990. K^+ current diversity is produced by an extended gene family conserved in *Drosophila* and mouse. *Science*. 248:599–603.

Wei, A., T. Jegla, and L. Salkoff. 1991. A *C. elegans* potassium channel gene with homology to *Drosophila* Shaw. *Society for Neuroscience Abstracts*. 17:1281(S10.10).

Yellen, G., M. E. Jurman, T. Abramson, and R. Mackinnon. 1991. Mutations affecting internal TEA blockade identify the probable pore-forming region of a K^+ channel. *Science*. 251:939–942.

Yool, A. J., and T. L. Schwarz. 1991. Alteration of ionic selectivity of a K^+ channel by mutation of the H5 region. *Nature*. 349:700–704.

Zagotta, W. N., T. Hoshi, and R. W. Aldrich. 1990. Restoration of inactivation in mutants of *Shaker* potassium channels by a peptide derived from *ShB*. *Science*. 250:568–571.

Zhao, B., and E. R. Kandel. 1991. Molecular cloning of a new K^+ channel in *Aplysia*. *Society of Neuroscience Abstracts*. 17:1281(S10.11).

The Connexins and Their Family Tree

M. V. L. Bennett, X. Zheng, and M. L. Sogin

Marine Biological Laboratory, Woods Hole, Massachusetts 02543, and Albert Einstein College of Medicine, Bronx, New York 10461

Introduction

The connexins are the proteins that form gap junctions. Cloning of connexin cDNAs has revealed twelve distinct isoforms in mammals, each encoded by a separate gene. All are capable of forming homomeric channels, generally thought to be comprised of a hexamer in each of the apposed membranes (cf. Bennett, Barrio, Bargiello, Spray, Hertzberg, and Saez, 1991). Particular connexins are found in many or few tissues; single cells can express more than one connexin. Heteromeric hemichannels, i.e., those containing more than one connexin type, may occur for some combinations (Kordel, Nicholson, and Harris, 1993); heterotypic junctions can form between some but not all combinations of homomeric hemichannels (e.g., Werner, Rabadan, Levine, and Dahl, 1993). The functional differences among the connexins are being actively explored (e.g., Hertzberg and Johnson, 1988; Hall, Zampighi, and Davis, 1992).

Comparison of connexin sequences can generate initial hypotheses about structure-function relations that are testable by site-directed mutagenesis and domain replacement (see Rubin, Verselis, Bennett, and Bargiello, 1992; Bennett, Rubin, Bargiello, and Verselis, 1993; Verselis, Ginther, and Bargiello, 1994). Conserved sequences and structural motifs are presumed to be functional. Variable regions are presumed either to be neutral or to confer functional differences. A neutral difference in the genes' coding regions may be accompanied by a functional difference in a regulatory sequence (cf. Li and Noll, 1994). Thus, a difference in gene regulation may be the only functional difference between two homologs that differ in amino acid sequence. In the available genomic sequences of connexins there is no indication of alternative splicing in the coding region, unlike the situation for K^+ channels and GluR1-4 and NMDA receptors (Jan and Jan, 1992; Wisden and Seeburg, 1993; Durand, Bennett, and Zukin, 1993; Hollmann, Boulter, Maron, Beasley, Sullivan, Pecht, and Heinemann, 1993). However, an intron has been found in the 5′ noncoding region of several connexins, which may provide for different pathways of transcriptional and posttranscriptional regulation (Henneman, Kozjek, Dahl, Nicholson, and Willecke, 1992; Willecke, Henneman, Dahl, Jungbluth, and Heynkes, 1991*a*). The noncoding exon provides for a further functional difference between connexins unrelated to differences in the coding sequences.

Phylogenetic trees of gene families from the same and different species can delineate the number of gene duplications and sometimes the relative timing of these events. The structure of the connexin family tree is a major subject of this paper. Within a species, members of a gene family are evolutionary paralogues (genes that have evolved in parallel) and they are generally derived from ancestral gene duplications. In different species, corresponding members of a gene family are evolutionary orthologues. Whether particular family members are orthologous or paralogous can be inferred from molecular phylogenies (with varying degrees of confidence, see below). Orthologous genes should cluster together near the terminal branches of the tree forming parallel subtrees of speciation. Each subtree describes evolutionary relationships after the duplication giving rise to the paralogues and should be congruent with organism phylogenies inferred from other genotypic, phenotypic, and paleontological evidence. Branching patterns that define relations between the subtrees can provide clues about the number and order of duplication events that led to the gene families.

Sequence Comparisons

We used all of the connexin sequences available in GenBank, but omitted from the final tree several rat sequences that were essentially redundant with included mouse sequences. Drs. D. Paul (Harvard University, Cambridge, MA), E. Beyer (Washington University, St. Louis, MO), K. Willecke (Universität Bonn, Germany) and their colleagues kindly allowed us access to several of their sequences before publication. 12 distinct rodent connexins have been identified, six in mouse and rat, four in mouse only and two in rat only (Table I). It is likely that all 12 occur in both groups (or that one or more have been lost), since the connexins so far found in only one of these rodents are too distinct to have arisen after rat and mouse lines diverged. Sequences of a number of orthologues of the rodent connexins are known from other mammals, including dog, cattle, and human. Several orthologues are known in chicken and *Xenopus.* The connexins are named by their molecular weight in kilodaltons, using tenths if necessary to differentiate connexins of similar sizes. Thus, connexin 32 or Cx32 has a predicted molecular weight of ~32 kD, similarly for Cx31 and Cx31.1. Where necessary the species name precedes the connexin name. The nucleic acid sequences were translated into their putative amino acid sequences and aligned using CLUSTAL. The alignments (available upon request) were further adjusted using GDE2.0 (Genetic Data Environment 2.0) to ensure the juxtapostioning of coding regions for functionally equivalent amino acid sequences.

The connexin molecules can be divided into nine principal domains: four membrane-spanning regions (M1-M4), two extracellular loops (E1, E2), and in the cytoplasm the NH_2 terminus (NT), the COOH terminus (CT) and a cytoplasmic loop (CL) connecting M2 and M3 (Fig. 1). The general topology is supported by antibody and protease studies (cf. Bennett, Barrio, Bargiello, Spray, Hertzberg, and Saez, 1991).

There are two highly variable and two relatively conserved regions. The two conserved regions are NT-M1-E1-M2 and M3-E2-M4 extending slightly into CL and CT. The inter-domain boundaries are generally well conserved and readily identified. The ends of M3 are not clear but the alignment in this region is unequivocal. (The M2 of Cx45 may be longer by three amino acids). NT is more variable than the M and E domains, but has several conserved and evenly spaced hydrophobic residues that may be involved in membrane insertion, because the molecules lack a signal sequence. The variable regions are a major part of CL and most of CT shortly after M4. These differ in length as well as in sequence; CL ranges from 37 amino acids in Cx31.1 to 70 amino acids in Cx45; CT varies from 18 amino acids in Cx26 to 211 in Cx50 (and 268 in chicken Cx56). The greater variability in length of the COOH terminus (CT) may in part result from point mutations inserting or removing a stop

TABLE I
Rodent Connexins Aligned

```
             1                         --NT--><--------M1--------><--E1- 50
 musCx26     M.DWGTLQSI LGGVNKHSTS IGKIWLTVLF IFRIMILVVA AKEVWGDEQA  49
 musCx32     M.NWTGLYTL LSGVNRHSTA IGRVWLSVIF IFRIMVLVVA AESVWGDEKS  49
 musCx303    M.NWGFLQGI LSGVNKYSTA LGRIWLSVVF IFRVLVYVVA AEEVWDDDQK  49
 musCx31     M.DWKKLQDL LSGVNQYSTA FGRIWLSVVF VFRVLVYVVA AERVWGDEQK  49
 musCx311    M.NWSVFEGL LSGVNKYSTA FGRIWLSLVF VFRVLVYLVT AERVWGDDQK  49
Iden. I      M.-W------ L-GVN--ST- -G--WL---F -FR-----V- A--VW-D---
 ratCx33     MSDWSALHQL LEKVQPYSTA GGKVWIKVLF IFRILLLGTA IESAWSDEQF  50
 musCx43     MGDWSALGKL LDKVQAYSTA GGKVWLSVLF IFRILLLGTA VESAWGDEQS  50
 musCx37     MGDWGFLEKL LDQVQEHSTV VGKIWLTVLF IFRILILGLA GESVWGDEQS  50
 ratCx46     MGDWSFLGRL LENAQEHSTV IGKVWLTVLF IFRILVLGAA AEEVWGDEQS  50
 musCx40     MGDWSFLGEF LEEVHKHSTV IGKVWLTVLF IFRMLVLGTA AESSWGDEQA  50
 musCx45     M.SWSFLTRL LEEIHNHSTF VGKIWLTVLI VFRIVLTAVG GESIYYDEQS  49
 musCx50     MGDWSFLGNI LEEVNEHSTV IGRVWLTVLF IFRILILGTA AEFVWGDEQS  50
Iden. II     M--W--L--- L------ST- -G--W--VL- -FR------- -E----DEQ-
Iden. I+II   M--W------ L------ST- -G--W----- -FR------- ------D---
                                   --NT--><--------M1--------><--E1--

             51                           --E1--><--------M2-------->-->100
 musCx26     DFVCNTLQPG CKNVCYDHHF PISHIRLWAL QLIMVSTPAL LVAMHVAYRR  99
 musCx32     SFICNTLQPG CNSVCYDHFF PISHVRLWSL QLILVSTPAL LVAMHVAHQQ  99
 musCx303    DFICNTKQPG CPNVCYDEFF PVSHVRLWAL QLILVTCPSL LVVMHVAYRE  99
 musCx31     DFDCNTRQPG CTNVCYDNFF PISNIRLWAL QLIFVTCPSM LVILHVAYRE  99
 musCx311    DFDCNTRQPG CTNVCYDEFF PVSHVRLWAL QLILVTCPSL LVVMHVAYRK  99
Iden. I      -F-CNT-QPG C--VCYD--F P-S--RLW-L QLI-V--P-- LV--HVA---
 ratCx33     EFHCNTQQPG CENVCYDQAF PISHVRLWVL QVIFVSVPTL LHLAHVYYVI  100
 musCx43     AFRCNTQQPG CENVCYDKSF PISHVRFWVL QIIFVSVPTL LYLAHVFYVM  100
 musCx37     DFECNTAQPG CTNVCYDQAF PISHIRYWVL QFLFVSTPTL IYLGHVIYLS  100
 ratCx46     DFTCNTQQPG CENVCYDRAF PISHIRFWAL QIIFVSTPTL IYLGHVLHIV  100
 musCx40     DFRCDTIQPG CQNVCYDQAF PISHIRYWVL QIIFVSTPSL VYMGHAMHTV  100
 musCx45     KFVCNTEQPG CENVCYDAFA PLSHVRFWVF QIILVATPSV MYLGYAIHKI  99
 musCx50     DFVCNTQQPG CENVCYDEAF PISHIRLWVL QIIFVSTPSL MYVGHAVHHV  100
Iden. II     -F-C-T-QPG C-NVCYD--- P-SH-R-W-- Q---V--P-- ----------
Iden. I+II   -F-C-T-QPG C--VCYD--- P-S--R-W-- Q---V--P-- ----------
                                      --E1--><--------M2-------->--><--E2--

             101                                              150
 musCx26     HEKKR..... .......... ......KFMK GEIKNEFKDI EEIKTQ....  124
 musCx32     HIEKK..... .......... .......MLR LEGHGDPLHL EEVKRH....  123
 musCx303    ERERK..... .......... ......HRLK HGPNAPALYS NLSKKR....  124
 musCx31     ERERK..... .......... ......HRQK HGEQCAKLYS HPGKKH....  124
 musCx311    AREKK..... .......... ......YQEK IGEGY..LYP NPGKKR....  122
Iden.I       -----..... .......... ......---- ---------- ---K--....
 ratCx33     RQNEK..... .......... ......LKKQ EEEELKVAHF NGASGERRL.  128
 musCx43     RKEEK..... .......... ......LNKK .EEELKVAQT DGVNVEMHL.  127
 musCx37     RREER..... .......... ......L.RQ KEGELRALPS KDLHVERAL.  127
 ratCx46     RMEEK..... .......... ......KKER EEELLRRDNP QHGRGREPM.  128
 musCx40     RMQEK..... .......... ......QKLR DAEKAKEAHR TGAY...EY.  125
 musCx45     AKMEHGEADK KAARSKPYAM RWKQHRALEE TEEDHEEDPM MYPEMELES.  148
 musCx50     RMEEK..... .......... ......RKDR EAEELCQQSR SNGGERVPIA  129
Iden. II     ---E------ ---------- ---------- ---------- ----------
Iden. I+II   ---------- ---------- ---------- ---------- ----------
```

Groups I and II and identities are indicated. Interdomain boundaries are shown above and below the sequences. Conserved regions in the CT domains are shown in bold face.

codon; however the short conserved sequences in the middle and near the end of the COOH terminus (see below) indicate that other mechanisms must also be operative.

Many of the conserved regions of the connexins have been noted before (cf. Bennett et al., 1991). In M3 the conserved acidic, basic and polar groups that are separated by several nonpolar residues are thought to form the channel lining. The

TABLE I
(Continued)

```
            151                         --CL--><---------M3--------><--E2--
  musCx26   .......... ........KV RIE.GSLWWT YTTSIFFRVI FEAVFMYVFY  155
  musCx32   .......... ........KV HIS.GTLWWT YVISVVFRLL FEAVFMYVFY  154
  musCx303  .......... .......... ....GGLWWT YLLSLIFKAA VDSGFLYIFH  150
  musCx31   .......... .......... ....GGLWWT YLFSLIFKLI IELVFLYVLH  150
  musCx311  .......... .......... ....GGLWWT YVFSLSFKAT IDIIFLYLFH  148
Iden. I     .......... ........-- ---.G-LWWT Y--S--F--- --F-Y-----
  ratCx33   ..QKHTGKHI KCGSKEHGNR KMR.GRLLLT YMASIFFKSV FEVAFLLIQW  175
  musCx43   ..KQIEIKKF KYGIEEHGKV KMR.GGLLRT YIISILFKSV FEVAFLLIQW  174
  musCx37   ..AAIEHQMA KISVAEDGRL RIR.GALMGT YVVSVLCKSV LEAGFLYGQW  174
  ratCx46   ..RTGSPRDP PL.RDDRGKV RIA.GALLRT YVFNIIFKTL FEVGFIAGQY  174
  musCx40   ..PVAEKAEL SCWKEVDGKI VLQ.GTLLNT YVCTILIRTT MEVAFIVGQY  172
  musCx45   ..EKENKEQS QPKPKHDGRR RIREDGLMKI YVLQLLARTV FEVGFLIGQY  196
  musCx50   PDQASIRKSS SS.SKGTKKF RLE.GTLLRT YVCHIIFKTL FEVGFIVGHY  177
Iden. II    ---------- ---------- ------L--- Y--------- -E--F---Q-
Iden. I+II  ---------- ---------- ------L--- Y--------- ----F-----
                                --CL--><---------M3--------><--E2--

            201                                    --E2--><---------M4----
  musCx26   IMYNGFFMQR LVKCNAW.PC PNTVDCFISR PTEKTVFTVF MISVSGICIL  204
  musCx32   LLYPGYAMVR LVKCEAF.PC PNTVDCFVSR PTEKTVFTVF MLAASGICII  203
  musCx303  CIYKDYDMPR VVAC.SVTPC PHTVDCYIAR PTEKKVFTYF MVVTAAICIL  199
  musCx31   TLWHGFTMPR LVQCASIVPC PNTVDCYIAR PTEKKVFTYF MVGASAVCII  200
  musCx311  AFYPRYTLPS MVKCHA.EPC PNTVDCFIAK PSEKNIFIVF MVVTAVICIL  197
Iden. I     ---------- -V-C----PC P-TVDC---- P-EK--F--- M------CI-
  ratCx33   YLY.GFTLSA VYICE.QSPC PHRVDCFLSR PTEKTIFILF MLVVSMVSFV  223
  musCx43   YIY.GFSLSA VYTCK.RDPC PHQVDCFLSR PTEKTIFIIF MLVVSLVSLA  222
  musCx37   RLY.GWTMEP VFVCQ.RAPC PHIVDCYVSR PTEKTIFIIF MLVVGVISLV  222
  ratCx46   FLY.GFQLQP LYRCD.RWPC PNTVDCFISR PTEKTIFVIF MLAVACASLV  222
  musCx40   LLY.GIFLDT LHVCR.RSPC PHPVNCYVSR PTEKNVFIVF MMAVAGLSLF  220
  musCx45   FLY.GFQVHP FYVCS.RLPC PHKIDCFISR PTEKTIFLLI MYGVTGLCLI  244
  musCx50   FLY.GFRILP LYRCS.RWPC PNVVDCFVSR PTEKTIFILF MLSVAFVSLF  225
Iden. II    --Y.G----- ---C----PC P----C--SR PTEK--F--- M--V------
Iden. I+II  ---------- ---C----PC P----C---- P-EK--F--- M---------
                                                   --E2--><---------M4----

            ---><--CT--                                            300
  musCx26   LNITELCYLF VRYCSGKS.K RPV                                   226
  musCx32   LNVAEVVYLI IRACARRA.Q RRSNPPSRKG SGFGHRLSPE YKQNEINKLL  252
  musCx303  LNLSEVVYLV GKRCMEVF.R PRR....... RKASRRHQLP DTCPPYVIS.  240
  musCx31   LTICEICYLI FHRIMRGI.S KGKSTKSISS PKSSSRASTC RCHHKLLESG  249
  musCx311  LNLVELIYLV IKRCSECA.Q LRRPPTA.HA KNDPNWANSP SKEKDFLSCD  245
Iden. I     L--------- --------.- ---------- ---------- ----------
  ratCx33   LNVIELFYVL FKAIKNHL.G NEKEEVYC.. ........NP VEL.QKPSCV  261
  musCx43   LNIIELFYVF FKGVKDRV.K GRSDPYHA.. ........TT GPLSPSKDCG  261
  musCx37   LNLLELVHLL CRCVSREI.K ARRDH.DA.. ........RP AQGSASDPYP  260
  ratCx46   LNMLEIYHLG WKKLKQGV.T NHFNPDASEV RHKPLDPLSE AANSGPPSVS  271
  musCx40   LSLAELYHLG WKKIRQRF.G KSRQGVD... ........KH QLPGPPTSLV  258
  musCx45   LNIWEMLHLG FGTIRDSLNS KRRELDDPGA YNYPFTWNTP SAPPGYNIAV  294
  musCx50   LNIMEMSHLG MKGIRSAF.K RPVEQPLGEI AEKSLH..SI AVSSIQKAKG  272
Iden. II    L--------- ---------- ---------- ---------- ----------
Iden. I+II  L--------- ---------- ---------- ---------- ----------
            ---><--CT--
```

conserved phenylalanines may operate in gating by occluding the channel when the M3 helix tilts or rotates. The extracellular loops, E1 and E2, are highly conserved, particularly compared to the extracellular or cytoplasmic domains of many other integral membrane proteins; this conservation presumably reflects their role in interconnecting hemichannels. The most striking conserved features of E1 and E2 are the three cysteines in both loops and the PCP motif in E2 which would be

TABLE I
(Continued)

	301				350	
musCx32	SEQDGSLKDI	LRRSPGTGAG	LAEKSDRCSA	C		283
musCx303	.KGGHPQDES	VILTKAGMAT	VDAGVYP			266
musCx31	DPEADPASEK	LQASAPSLTP	I			270
musCx311	LIFLGSDAHP	PLLPDRPRAH	VKKTIL			271
Iden. I	----------	----------	------...	....	....	
ratCx33	SSSAVLTTIC	SS..DQVVPV	GLSSFYM			286
musCx43	SPKYAYFNGC	SSPTAPLSPM	SPPGYKLVTG	DRNNSSCRNY	N.........	302
musCx37	EQVFFYLPMG	EGPSSPPCP.	TYNG......	....	....	283
ratCx46	IGL.......	PPYYTHPACP	TVQGKATGFP	GAPLLPADFT	VVTLNDA...	311
musCx40	QSLTPPPDFN	QCLKNSSGEK	FFSDFSN...	....	....	285
musCx45	KPDQIQYTEL	SNA.......	....	....	....	307
musCx50	YQLLEEEKIV	SHYFPLTEVG	MVETSPLSAK	PFSQFEEKIG	TGPLADMSRY	322
Iden. II	----------	----------	----------	----------	----------	
Iden. I+II	----------	----------	----------	----------	----------	

	351				400	
musCx43	....	....KQA**SEQ**	**NWANYSAEQ**N	RMGQAGSTIS	NSHAQPFDFP	338
musCx37	....	....LSS**TEQ**	**NWANLTTEE**R	LTSSRPPPFV	N.........	310
ratCx46	.QGRGHPVKH	CNGHHLT**TEQ**	**NWASLGAEP**Q	TPASKPSSAA	SSPHGRKGLT	360
musCx40	....	....NMG**SRK**	**NPDALATGE**V	PNQEQIPGEG	FIHMHYSQKP	321
musCx45	....	....KIA**YKQ**	**NKANIAQEQ**Q	YGSHEEHLPA	DLETLQREIR	343
musCx50	YQETLPSYAQ	VGVQEVE**REE**	**PPIEEAVEP**E	VGEKKQEAEK	VAPEGQETVA	372

	401				450	
musCx43	DDSQNAKKVA	AGHELQPLA.	....	....	....IVDQRP	363
musCx37	....	....	....	....	....TAPQGG	316
ratCx46	DSSGSSLEES	ALVVTPEGEQ	ALATTVEMH.	....	SPPLVLLDPE	399
musCx40	....	....	....	..EYASGASA	GHRLPQGYHS	339
musCx45	MAQERLDLAI	QAYH......	....	HQNNPHGPRE	KKAKVGSKSG	377
musCx50	VPDRERVETP	GVGKEDEKEE	LQAEKVTKQG	LSAEKAPSLC	PELTTDDNRP	422

	451		
musCx43	SSRASSR**ASS**	**RPRPDDL**EI	382
musCx37	RKSPSRP**NSS**	**ASKKQYV**	333
musCx46	RSSKS..**SSG**	**RARPGDL**AI	416
musCx40	DKRRLSK**ASS**	**KARSDDL**SV	358
musCx45	GSNKSSI**SSK**	**SGDGKTS**VWI	396
musCx50	PLSRLSK**ASS**	**RARSDDL**TI	441

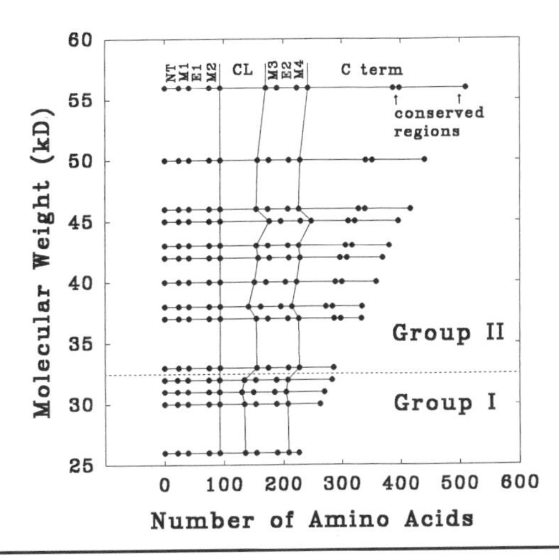

Figure 1. Domain organization of the connexins. Each horizontal line represents a connexin with location on the abscissa indicating the predicted molecular weight and length indicating number of amino acids. The interdomain boundaries and the conserved region in the central region of the COOH terminus are indicated by filled circles. The connexins included are mouse Cx26, *Xenopus* Cx30, mouse Cx31, mouse Cx32, rat Cx33, mouse Cx37, *Xenopus* Cx38, mouse Cx40, chicken Cx42, mouse Cx43, mouse Cx45, rat Cx46, mouse Cx50 and chicken Cx56.

expected to prevent helical structure in this region. In E1 15 of 35 residues are identical in all, 11 of 12 consensus is present for 6 more sites. E2 is less well conserved; identity for 8 of 37, 11 of 12 consensus in 4 more. E2 is more divergent in its NH_2-terminal half.

There are two short regions of similarity in the COOH terminus in the larger connexins (Fig. 1, Tables I and II). The function of the more NH_2-terminal sequence is unknown. The sequences near the COOH-terminal ends contain Ser and basic amino acids and are likely to be phosphorylation sites, as has been shown directly for (rat) Cx43 (Saez, Nairn, Czernik, Spray, Hertzberg, Greengard, and Bennett, 1993). (An unrelated phosphorylation site for A kinase is found in Cx32 [Saez, Nairn, Czernik, Spray, and Hertzberg, 1990]).

TABLE II
Group II C-terminal Homologies

xenCx38	-GGHNWSRIQMEQ-	284	-HQTSSKQQYV-	333
musCx37	-TEQNWANLTTEE-	298	-NSSASKKQYV-	334
musCx40	-SRKNPDALATGE-	300	-ASSKARSDDL-	356
dogCx40	-SQQNTDNLATEQ-	299	"	355
chiCx42	-SQQNTANFATER-	308	"	367
mamCx43	-SEQNWANYSAEQ-	317	-ASSRPRPDDL-	380
musCx45	-YKQNKANIAQEQ-	322	-SSKSGDGKTS-	393
ratCx46	-TEQNWASLGAEP-	339	-SSGRARPGDL-	414
musCx50	-REEPPIEEAVEP-	351	-ASSRARSDDL-	438
	divergent		conserved	
musCx40	-SRKNPDALATGE-	300	-ASSKARSDDL-	356
dogCx40	-SQQNTDNLATEQ-	299	-ASSKARSDDL-	355
Identities	-S----D-LAT--		-ASSKARSDDL-	
	conserved		divergent	
musCx37	-TEQNWANLTTEE-	298	-NSSASKKQYV-	334
ratCx46	-TEQNWASLGAEP-	339	-SSGRARPGDL-	414
Identities	-TEQNWA-L--E--		--S---------	

Sequences from the two conserved regions found in most Group II connexins are shown. These regions have diverged at different rates in different lineages; either region can be more divergent. The numbers refer to the last amino acid in the sequences.

Sequence comparisons have been useful in analysis of voltage dependence of the conductance of gap junctions formed of Cx26 and Cx32 (Rubin et al., 1992; Bennett et al., 1993; Verselis et al., 1994). These two connexins each form gap junctions closed by transjunctional voltage, with equal sensitivity for voltages of either sign. The symmetry is ascribable to there being two oppositely oriented hemichannels, one in each membrane, one closing for either polarity of voltage. The two amino acids, KE (LysGlu), at the beginning of E1 in Cx26 are unique for this position among the connexins. Mutation of these residues to the consensus sequence for this position, ES (GluSer), increased the steepness of the conductance/voltage relation. The converse change in Cx32 increased the steepness of its conductance/voltage relation. The properties of heterotypic wild type/mutant channels revealed that the polarity of sensitivity of the hemichannels was opposite in the two cases, that is the mutant hemichannel properties occurred with the mutant side positive in the case of Cx26 and with the mutant side negative in the case of Cx32. Changing domains showed that the NT-M1 domains determined the polarity of sensitivity (Verselis et al., 1994). The only site in the NT-M1 domains in which the two differ in charge is the third amino acid which is Glu in Cx26 and Asp in Cx32. Exchange of these amino acids in each case reversed the polarity of voltage sensitivity. The conclusion was that all three amino acids are part of the voltage sensor. This study is illustrative of how sequence comparisons can guide structure function analyses.

It has recently been reported that X-linked Charcot-Marie-Tooth disease is associated with mutations in the Cx32 gene (Bergoffen, Scherer, Wang, Scott, Bone, Paul, Chen, Lensch, Chance, and Fischback, 1993; Fairweather, Bell, Cochrane, Chelly, Wang, Mostacciuolo, Monaco, and Haites, 1994). 15 different lineages with distinct mutations are known. Several of the mutations are frame shifts that would be certain to prevent the gene product from forming junctions. Other point mutations in M3 and E1 and E2 are of highly conserved residues; it appears likely that these mutant proteins also would not generate functional gap junctions. These predictions can be and are being tested by site directed mutagenesis. Two additional lineages show no mutations in the coding sequence; disruption of the promoter region or regulatory sites may produce the phenotype in these cases.

The Tree(s)

We restricted our phylogenetic analyses to the two major conserved regions in connexin genes (encoding positions 1–105 and 173–251 in Table I). Because of higher rates of substitution leading to frequent but unseen reversals, only the first and second positions of codons that could be unambiguously aligned were included in the sequence comparisons. For phylogenetic inferences based upon distance matrix methods, sequence similarities were converted to evolutionary distances expressed in nucleotide changes/site by the method of Jukes and Cantor (1969). Trees were constructed using a modification of the least squares distance matrix methods (Elwood, Olsen, and Sogin, 1985). Trees based upon the principles of maximum parsimony were inferred using heuristic search options in the computer program PAUP3.0 (Swofford, 1991). Initial tree topologies were obtained by addition of randomly selected taxa using ten repetitions for each round of parsimony analysis. To assess the fraction of sites that support elements in our distance and parsimony analyses, we employed bootstrap procedures (Felsenstein, 1985).

The distance matrix tree in Fig. 2 describes evolutionary relationships between orthologous and paralogous members of the connexin gene family. It is largely congruent with trees generated using either the first or the second conserved region alone, and it is consistent with bootstrap analyses based upon parsimony. The number of bootstrap replicates displaying corresponding topological elements in distance matrix and parsimony procedures (in parentheses) is indicated in the figure. Values below 50 percent are not included in the figure. Rather than interpreting bootstrap values as measures of statistical confidence in phylogenetic tree reconstructions, they should be regarded as measures of relative confidence between topological elements in the tree. Bootstrap values are known to be influenced by the number of changes that define internal branches in phylogenetic trees (Hillis and Bull, 1993), and are often influenced by use of different taxa or genes in molecular reconstructions (Liepe, Gunderson, Nerad, and Sogin, 1993).

Although the tree is unrooted, separation into groups I and II as indicated appears reasonable. Branch points that represent duplications are indicated by black ovals. Branch points that represent speciation are unmarked. One branch is questionable in nature (Cx37-Cx38, see below). The second and third duplications within the group I lineage appear well resolved by the distance method. The later node is less probable according to parsimony analysis, but this uncertainty may be an artifact of the long branch going to the Cx30.3, Cx31, Cx31.1 group. Relations among Cx30.3, Cx31, and Cx31.3 are uncertain, although they are a group and their genes are on the same chromosome (Schwarz, Chang, Hennemann, Dahl, Lalley, and Willecke, 1992). The duplications giving rise to this group may be relatively recent, and intergene transfer may be responsible for the ambiguity in the tree in this region. There is some conservation in the CL domain in these connexins, but inclusion of this

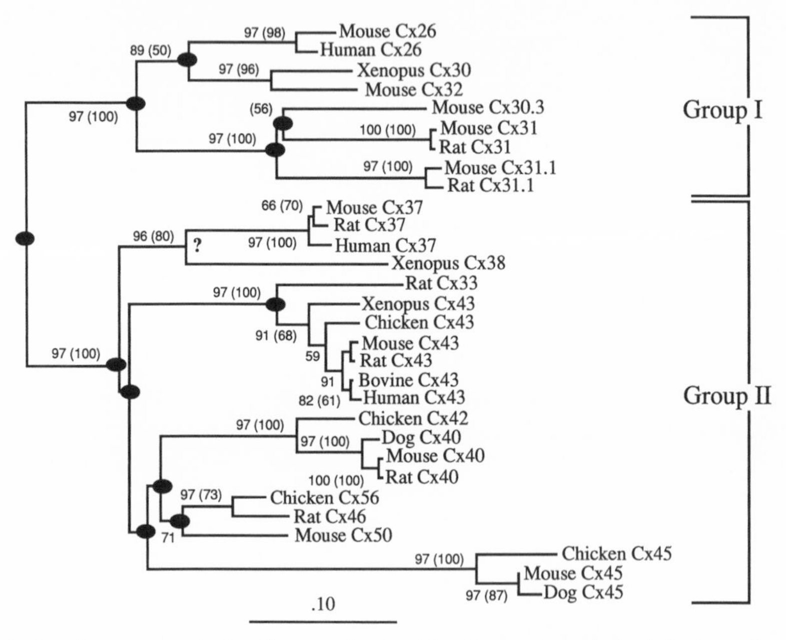

Figure 2. Our best connexin family tree obtained by distance methods. The branch points at filled ovals represent gene duplications. The branch points without ovals represent speciation. One branch point indicated by a question mark is ambiguous (see text). The numbers on the horizontal lines represent the number of bootstrap replicates of 100 trials displaying corresponding topological elements in distance matrix and (*in parentheses*) parsimony procedures. The calibration represents a difference of 0.1 out of a possible 1.0, i.e., 10%.

information in the tree construction did not clarify the relations between them. Their COOH termini are very divergent and are not informative in this respect. In group II, the affiliations of Cx43s and Cx33 and of Cx37s and Cx38 are well resolved. Deep branches involving these groups, Cx50 and the presumably orthologous Cx40, Cx42 group, the Cx45 group and the Cx46, Cx56 group are ambiguous. Resolution of these branches may be possible when sequences from lower forms become available.

The connexin trees are consistent with vertebrate evolution in that mammalian orthologues are closer to each other than to their chicken orthologues (see Cx43s, Cx45s and the Cx40, Cx42 group). Similarly, mammalian Cx43s are more like each other than they are like *Xenopus* Cx43 and mammalian Cx32s are more like each other than they are like *Xenopus* Cx30. The rate of divergence differs among connexins. Cx43s of mammals and *Xenopus* are less divergent than are the orthologues, *Xenopus* Cx30 and the mammalian Cx32s. *Xenopus* Cx38 and mouse Cx37 are more closely related to each other than to any other connexin but considerably more divergent than Cx30 and Cx32. These differences in rates of divergence are true of the less conserved regions of the sequences as well as those used to make the tree. Comparing mouse and *Xenopus* CL and CT domains are 85 and 78% identical for Cx43; 68 and 36% identical for Cx30, Cx32, and 32 and 22% identical for Cx37, Cx38. Although Cx37 and Cx38 are functionally related by their pronounced voltage dependence, their very disparate tissue distribution and CT sequences suggest that they are distinct connexins (Willecke, Heynkes, Dahl, Stutenkemper, Henneman, Jungbluth Suchyna, and Nicholson, 1991*b*). There may be a Cx38 orthologue in mammals and/or a Cx37 orthologue in *Xenopus*.

As noted above in relation to the E1 and E2 domains, different conserved regions of the connexin molecules have diverged at different rates. Differences in rates of change are also evident in the two regions of conservation in the COOH-terminal domain, either of which can be more divergent. In Cx43s, both regions are completely conserved in all the sequences, those of *Xenopus,* chicken and mammal, again indicating the relative stability of this gene. In the group II connexins the distal COOH-terminal sequences that contain putative phosphorylation sites give two clusters, the Cx37, Cx38 group and the others with Cx45 as an outlier. The more central conserved region shows closer affiliation between Cx37, Cx43, and Cx46 than between Cx37 and Cx38. In the Cx40, Cx42 orthologous group the dog and chicken orthologues are more closely related in this region than the dog and rodent orthologues, and in rodent Cx40 and Cx50 this region could only be identified on the basis of the overall alignment. It may be that the function of this region, what ever it is, has been lost in *Xenopus* Cx38 and mouse Cx40 and Cx50.

The large distance between groups I and II compared to the distance between mammalian and amphibian orthologues suggests that divergence between groups I and II took place early in or before vertebrate divergence. The divergences represented by the deeper branches also are likely to have been very early. There have been no sequences obtained from earlier diverging forms. Gap junctions do occur in most coelenterates (Mackie, 1990) (and in mesozoa, which may be even more primitive, Revel, 1988), and antibodies to liver gap junctions were reported to cross react with *Hydra* junctions (Frazer, Green, Boder, and Gilula, 1987). Thus, it is likely that connexins appear very early in metazoan evolution. Sequences from earlier diverging forms may help in tree refinement and may also reveal precursor molecules in unicellular organisms that in general lack gap junctions.

Summary

The connexins, gap junction forming proteins, are encoded by a gene family. Sequence comparisons reveal regions of conservation with functional implications for voltage dependence of junctional conductance, junction formation and regulation by phosphorylation. The best connexin tree shows that most gene duplications giving rise to the family occurred early in or before vertebrate divergence. The topology of most deep branches of the tree is uncertain. Evolutionary rates vary for different paralogous connexin genes.

Acknowledgements

This work was supported in part by grants from the National Institutes of Health NS-07512 and HD-04248 to M. V. L. Bennett and GM-32964 to M. L. Sogin. M. V. L. Bennett is the Sylvia and Robert S. Olnick Professor of Neuroscience.

References

Bennett, M. V. L., L. Barrio, T. A. Bargiello, D. C. Spray, E. Hertzberg, and J. C. Saez. 1991. Gap junctions: new tools, new answers, new questions. *Neuron.* 6:305–320.

Bennett, M. V. L., J. B. Rubin, T. A. Bargiello, and V. K. Verselis. 1993. Structure-function studies of voltage sensitivity of connexins, the family of gap junction forming proteins. *Japanese Journal of Physiology.* 43:301–310.

Bergoffen, J., S. S. Scherer, S. Wang, M. O. Scott, L. J. Bone, D. L. Paul, K. Chen, M. W. Lensch, P. F. Chance, and K. H. Fischbeck. 1993. Connexin mutations in X-linked Charcot-Marie-Tooth disease. *Science.* 262:2039–242.

Durand, G. M., M. V. L. Bennett, and R. S. Zukin. 1993. Splice variants of the *N*-methyl-D-aspartate receptor NR1 identify domains involved in regulation by polyamines and protein kinase C. *Proceedings of the National Academy of Sciences, USA.* 90:6731–6735.

Elwood, H. J., G. J. Olsen, and M. L. Sogin. 1985. The small-subunit Ribosomal RNA gene sequences from the hypotrichous ciliates *Oxytricha nova* and *Stylonychia pustulata. Molecular Biology and Evolution.* 2:399–410.

Fairweather, N., C. Bell, S. Cochrane, J. Chelly, S. Wang, M. L. Mostacciuolo, A. P. Monaco, and N. E. Haites. 1994. Mutations in the connexin 32 gene in X-linked dominant Charcot-Marie-Tooth disease (CMTX1). *Human Molecular Genetics.* 3:29–34.

Felsenstein, J. 1985. Confidence limits on phylogenies: An approach using the bootstrap. *Evolution.* 39:783–791.

Fraser, S. E., C. R. Green, H. R. Bode, and N. B. Gilula. 1987. Selective disruption of gap junctional communication interferes with a patterning process in hydra. *Science.* 237:49–55.

Hall, J. E., G. A. Zampighi, and R. M. Davis, editors. 1993. Gap junctions. *Progress in Cell Research.* 3:1–333.

Hennemann, H., G. Kozjek, E. Dahl, B. Nicholson, and K. Willecke. 1992. Molecular cloning of mouse connexins26 and -32: similar genomic organization but distinct promoter sequences of two gap junction genes. *European Journal of Cell Biology.* 58:81–89.

Hertzberg, E. L., and R. Johnson, editors. 1988. Gap Junctions. Alan R. Liss, Inc., NY. 541 pp.

Hillis, D. M., and J. J. Bull. 1993. An empirical test of bootstrapping as a method for assessing confidence in phylogenetic analysis. *Systematic Biology.* 42:182–192.

Hollmann, M., J. Boulter, C. Maron, L. Beasley, J. Sullivan, G. Pecht, and S. Heinemann. 1993. Zinc potentiates agonist-induced currents at certain splice variants of the NMDA receptor. *Neuron.* 10:943–954.

Jan, L. Y., and Y. N. Jan. 1992. Structural elements involved in specific K^+ channel functions. *Annual Review of Physiology.* 54:537–555.

Jukes, T. H., and C. R. Cantor. 1969. Evolution of protein molecules. *In* Manual of Protein Metabolism. H. N. Munro, editor. Academic Press, NY. 21–32.

Kordel, M., B. J. Nicholson, and A. L. Harris. 1993. Crosslinking studies of rat and mouse liver connexins in purified form and in plasma membranes. *Biophysical Journal.* 64:A92. (Abstr.)

Li, X., and M. Noll. 1994. Evolution of distinct developmental functions of three *Drosophilia* genes by acquisition of different *cis*-regulatory regions. *Nature.* 367:83–87.

Leipe, D. D., J. H. Gunderson, T. A. Nerad, and M. L. Sogin. 1993. Small subunit ribosomal RNA of *Hexamita inflata* and the quest for the first branch in the eukaryotic tree. *Molecular and Biochemical Parasitology.* 59:41–48.

Mackie, G. O. 1990. Evolution of cnidarian giant axons. *In* Evolution of the First Nervous Systems. P. A. V. Anderson, editor. Plenum Publishing Corp., NY. 395–407.

Revel, J-P. 1988. The oldest multicellular animal and its junctions. *In* Gap Junctions. E. L. Hertzberg and R. Johnson, editors. Alan R. Liss, Inc., NY. 135–149.

Rubin, J. B., V. K. Verselis, M. V. L. Bennett, and T. A. Bargiello. 1992. Molecular analysis of voltage dependence of heterotypic gap junctions formed by connexins 26 and 32. *Biophysical Journal.* 62:197–207.

Saez, J. C., A. C. Nairn, A. J. Czernik, D. C. Spray, E. L. Hertzberg, P. Greengard, and M. V. L. Bennett. 1990. Phosphorylation of connexin 32, the hepatocyte gap junction protein, by cAMP-dependent protein kinase, protein kinase C and Ca^{2+}/calmodulin-dependent protein kinase II. *European Journal of Biochemistry.* 192:263–273.

Saez, J. C., A. C. Nairn, A. J. Czernik, D. C. Spray, and E. Hertzberg. 1993. Rat connexin43: regulation by phosphorylation in heart. *Progress in Cell Research.* 3:275–281.

Schwartz, H. J., Y. S. Chang, H. Hennemann, E. Dahl, P. A. Lalley, and K. Willecke. 1992. Chromosomal assignments of mouse connexin genes, coding for gap junctional proteins, by somatic cell hybridization. *Somatic Cell and Molecular Genetics.* 18:351–359.

Swofford, D. L. 1991. Phylogenetic Analysis Using Parsimony (PAUP), version 3.0s. Illinois Natural History Survey, Champaign, Il.

Verselis, V.K., C.S. Ginther, and T.A. Bargiello. 1994. Opposite voltage gating polarities of two closely related connexims. *Nature.* 368:348–351.

Werner, R., C. Rabadan-Diehl, E. Levine, and G. Dahl. 1993. Affinities between connexins. *Progress in Cell Research,* 3:21–24.

Willecke, K., H. Hennemann, E. Dahl, S. Jungbluth, and R. Heynkes. 1991*a.* The diversity of connexin genes encoding gap junctional proteins. *European Journal of Cell Biology.* 56:1–7.

Willecke, K., R. Heynkes, E. Dahl, R. Stutenkemper, H. Hennemann, S. Jungbluth, T. Suchyna, and B. J. Nicholson. 1991*b.* Mouse connexin37: cloning and functional expression of a gap junction gene highly expressed in lung. *Journal of Cell Biology.* 114:1049–1057.

Wisden, W., and P. H. Seeburg. 1993. Mammalian ionotropic glutamate receptors. *Current Opinion in Neurobiology.* 3:291–298.

Relationships of G-Protein–coupled Receptors. A Survey with the Photoreceptor Opsin Subfamily

Meredithe L. Applebury

Visual Sciences Center, The University of Chicago, Chicago, Illinois 60637

Relationships among the G-protein–coupled receptors were evaluated using several distance matrices with the neighbor-joining method of Saitou and Nei (1987). The relationships generated vary depending upon alignment, length, or region of sequence compared, and the distance matrix used to score similarity. To provide a statistical level of confidence, bootstrap resampling was applied to the analysis of a selection of G-protein–coupled receptors and the subfamily photoreceptor opsins. A general consensus indicates that the opsins behave as a discrete subfamily among the superfamily of G-protein–coupled receptors. Their relationship to other subfamilies remains unresolved. Within the opsin subfamily, the retinochromelike opsins segregate as a discrete group, but are more closely related to the invertebrate than vertebrate opsins. Among vertebrate opsins, the long wavelength cone opsins, the blue/violet opsins, and the rod opsins (including a class of green cone opsins) form distinct subgroups, but their relationships to one another remain unresolved. For this superfamily of receptors, the confidence levels for many branch pairings are low. The application of methods complimentary to those used in this preliminary study will be necessary to resolve questions about appropriate pairing and evolutionary relationships.

Introduction

The G-protein–coupled receptors form one of the largest families of proteins whose primary sequences have been assessed. These receptors detect extracellular ligands, or light, and transduce the binding information by coupling to intracellular trimeric GTP-binding proteins. The receptor-G-protein pathways may signal intracellular changes in second messengers, protein phosphorylations, or ion channel conductances. Each receptor is a single polypeptide composed of seven relatively hydrophobic transmembrane regions. Each polypeptide folds to form an integral membrane protein that specifies a ligand binding pocket accessible from the extracellular space; a set of intracellular loops that link the membrane spanning regions bind intracellular proteins for signaling and signal regulation. Members of the family are expressed in most eucaryotic cell types and are found in organisms ranging from *Dictyostelium* to man; curiously, no examples have been identified yet in plants. With the advent of molecular cloning and recent polymerase chain reaction (PCR) gene amplification, the number of receptors in the superfamily now exceeds 200 sequences. Over 1,000 receptor types are estimated to be expressed in humans; the subfamilies of olfactory or gustatory receptors may alone exceed 1,000 types.

Visual inspection of protein sequences, alignment, and observation of strong identities have generally been used to classify the subfamilies within the superfamily (Applebury and Hargrave, 1986; Probst, Snyder, Schuster, Brosius, and Sealfon, 1992). As the database has grown, a more extended evaluation of the relationships within the subfamilies and interrelationships among the subfamilies has become possible. To better understand the function of these receptors, investigators have sought structural features that provide ligand specificity, interaction with specific types of G-proteins (Wilkie, 1994), regulation by receptor kinases (Inglese, Freedman, Koch, and Lefkowitz, 1993) and regulation by arrestins (Wilson and Applebury, 1993). If valid groups can be established, structural features associated with specific sequences may better predict physiological function and aid in deriving the evolution of subtypes.

Although the database is large, the majority of G-protein–coupled receptor sequences are of mammalian origin; the proteins are often nearly identical or highly similar. Thus, in fact, the number of sequences that distinguishes any group is limited. The most extensive dataset available is the subfamily of photoreceptor opsins. Receptors from many species, both invertebrate and vertebrate, have been cloned and sequenced. Predictions of evolutionary relationships in this family have been given by Applebury (1987*a,b*) Goldsmith (1990), Yokoyama and Yokoyama (1990), Okano, Kojima, Fukada, Shichida, and Yoshizawa (1992), and Johnson, Grant, Zankel, Boehm, Merbs, Nathans, and Nakanishi, (1993). The following work was undertaken to explore the relationships among the opsins using the current larger database. In addition, the relationships of the opsins to other members of the superfamily were explored to question whether opsins form a unique family of their own or whether they may be more closely related to some other receptor types, for example, other sensory receptors such as the olfactory or gustatory receptors. The assessment of the opsins identifies several problems in predicting relationships in this family of proteins and provides a guideline for assessment of other subfamilies among the extensive G-protein–coupled receptor superfamily.

Methods

Databases

The sources of amino acid sequences for visual pigment opsins, including related retinaldehyde-binding retinochrome proteins, and selected G-protein–coupled receptors are listed in Tables I and II. A selection of 27 G-protein–coupled receptors, which represent different receptor subtypes, was selected from human sequences for consistency; as many invertebrate sequences were included as could be found. Sequences, such as those for metabotrophic glutamate and prostaglandin EP1 receptors that were difficult to align were dropped in the course of analysis. Most amino acid sequences were accessed through the ENTREZ database obtained from the National Center for Biotechnology Information (NCBI; Bethesda, MD). The accession number is given for reference. For data entered from the literature, references are provided. The murine and canine sequences are unpublished data available from the author's laboratory. The identity of most receptors is given by its ligand specificity; thus, the general ligand type is provided for the G-protein–coupled receptors and the characteristic wavelengths of absorption maxima are provided for

TABLE I
Database for Opsin Sequences

File	Phylum*	Common name	Organism	Wave length‡	Opsin	Amino acids	Accession number
Visual pigments							
1 ancP625	C	American chameleon	*Anolis carolinensis*	625/A2	red cone	369	*a*
2 bovP498	C	cow	*Bos taurus*	498	rod	348	*b*
3 calP488	Ar	blowfly	*Calliphora erythrocephala*	488	Rh1-6	371	gbJO5596
4 canP498	C	dog	*Canis familiaris*	498	rod	348	*b*
5 cavG101	C	cavefish (blind)	*Astyamax fasciatus*	G	green cone	353t	gbM38619
6 cavG103	C	cavefish (blind)	*Astyamax fasciatus*	G(554)	green cone	355t	spP22330
7 cavR007	C	cavefish (blind)	*Astyamax fasciatus*	R(596)	red cone	342t	gbM38652
8 chkP415	C	chicken	*Gallus gallus*	415	violet cone	347	gbM92039
9 chkP455	C	chicken	*Gallus gallus*	455	blue cone	361	gbM92037
10 chkP503	C	chicken	*Gallus gallus*	503	rod	351	sp22320
11 chkP508	C	chicken	*Gallus gallus*	508	green cone	355	gbM92038
12 chkP571	C	chicken	*Gallus gallus*	571	red cone	362	gbM62903
13 crfP535	Ar	crayfish	*Procambarus clarkii*	535	rhabdomere	376	bbs122965
14 cgRHO	C	Ch. hampster	*Cricetulus griseus*	498§	rod	348	emblX61084
15 drmgRH1	Ar	fruit fly	*Drosophila melanogaster*	480	Rh1-6	372	pir1107195A
16 drmgRH2	Ar	fruit fly	*Drosophila melanogaster*	420	Rh2 ocelli	381	gbM12896
17 drmgRH3	Ar	fruit fly	*Drosophila melanogaster*	335/350	Rh3 (7)	383	gbM17718
18 drmgRH4	Ar	fruit fly	*Drosophila melanogaster*	335/350	Rh4 (7)	378	gbM17730
19 drpsRH1	Ar	fruit fly	*Drosophila pseudoobscura*	480§	Rh1-6	374	emblX65877
20 drpsRH2	Ar	fruit fly	*Drosophila pseudoobscura*	420§	Rh2 ocelli	381	emblX65878
21 drpsRH3	Ar	fruit fly	*Drosophila pseudoobscura*	335/350§	Rh3 (7)	382	emblX65879
22 drpsRH4	Ar	fruit fly	*Drosophila pseudoobscura*	335/350§	Rh4 (7)	380	emblX65880
23 drviRH4	Ar	fruit fly	*Drosophila virilis*	335/350§	Rh4 (7)	383	bgM77281
24 frgP502	C	frog leopard	*Rana pipiens*	502	rod	354	pirS27231
25 gecP467	C	gecko tokay	*Gekko gekko*	467	transrod	355	gbM92035
26 gecP521	C	gecko tokay	*Gekko gekko*	521	transrod	365	gbM92036
27 gofP441	C	goldfish	*Carassius auratus*	441	blue cone	351	pirE45229
28 gofP492	C	goldfish	*Carassius auratus*	492	rod	354	pirC45229
29 gofP505	C	goldfish	*Carassius auratus*	505	green 1 cone	349	pirA45229
30 gofP511	C	goldfish	*Carassius auratus*	511	green 2 cone	349	pirB45229
31 gofP525	C	goldfish	*Carassius auratus*	525	red cone	357	pirD45229
32 hcrP520	Ar	horseshoe crab	*Limulus polyphemus*	520	lateral eye	376	gbL03781
33 hcrP530	Ar	horseshoe crab	*Limulus polyphemus*	530/360	median ocelli	376	gbL03782
34 humP426	C	human	*Homo sapiens*	426	blue cone	348	gbM13299
35 humP498	C	human	*Homo sapiens*	498	rod	348	gbK02281
36 humP530	C	human	*Homo sapiens*	530	green cone	364	gbK03495
37 humP557	C	human	*Homo sapiens*	557	red cone	364	gbM13305
38 lmpP497	C	lamprey	*Lampetra japonica*	497	rod short	353	gbM63632
39 murBLUE	C	mouse	*Mus musculus*	nd	blue cone	346	*b*
40 murP498	C	mouse	*Mus musculus*	498	rod	348	gbM36699
41 murP505	C	mouse	*Mus musculus*	505	green cone	359	*b*
42 octP475	M	octopus	*Paroctopus defleini*	475	rhabdomere	455	emblX07797
43 oviP498	C	sheep	*Ovis aries*	498	rod	348	spP02700
44 sgbP501	C	sand goby	*Pomatoschistus minutus*	501	rod	352	bbs113148
45 sqLFP493	M	squid	*Loligo forbesi*	493	rhabdomere	452	gbX56788
46 sqTPP480	M	squid	*Todarodes Pacificus*	480	rhabdomere	448	bbs124052
47 ssOPS	Ar	p. mantis	*Sphodromantis*	nd	rhabdomere	376	gbX71665
48 xenP502	C	frog	*Xenopus laevis*	502	rod	354	gbL07770
49 zefP360	C	zebra fish	*Brachydanio rerio*	360	UV	354	gbL11014
Retinochromes							
1 bovRPER	C	cow	*Bos taurus*	nd	retinochrome	291	*c*
2 squidRC	M	squid	*Todarodes pacificus*	490	retinochrome	301	spP23820

*Phyla are abbreviated: C, Chordata, Ar, Arthopoda, M, Mollusca.

‡Absorption maximum of visual pigments formed upon binding A1 or A2 retinaldehydes are taken from Lythgoe (1972), Hardie (1985), Britt et al. (1993), or original references cited with accession numbers.

§Wavelength based on similarity to like-pigments. (t) Incomplete sequence truncated at COOH terminus.

(*a*) Kawamura and Yokoyama (1993).

(*b*) Sequences available from the author's laboratory.

(*c*) Jiang et al. (1993).

the visual pigments (formed from opsins that covalently bind A1 or A2 retinaldehydes).

Analyses

Two procedures for calculating relatedness of protein sequences were used. The first method provided an initial survey for the G-protein–coupled receptors that include opsins and a survey for relations among the opsins themselves. The work was carried

TABLE II
Database for G-Protein–Coupled Receptors

File	Phylum*	Common name	Organism	Ligand	Receptor	Amino acids	Accession number
1 dictcAMP	Ac	slime mold	*Dictyostelium discoideum*	cAMP	cAMP	392	gb M21824
2 drmg5HT1	Ar	fruit fly	*Drosophila melanogaster*	serotonin	5-HT-1	564	gb M55533
3 drmgMUS_M1	Ar	fruit fly	*Drosophila melanogaster*	acetylcholine	ACM1	722	gb M27495
4 drmgOCT	Ar	fruit fly	*Drosophila melanogaster*	octopamine	tyramine	601	gb M60789
5 drmgTACH	Ar	fruit fly	*Drosophila melanogaster*	tachykinin	tachykinin NKD	504	gb M77168
6 hamβADR1	C	hamster	*Mesocricetus auratus*	epinephrine	β-adrenergic 1	417	embl X03804
7 h5HT1A	C	human	*Homo sapiens*	serotonin	5-HT-1A	422	gb M28269
8 hADEN_A1	C	human	*Homo sapiens*	adenosine	adenosine A1	326	sp P30542
9 hADRα1A	C	human	*Homo sapiens*	epinephrine	α-adrenergic 1A	501	gb M76446
10 hADRβ1	C	human	*Homo sapiens*	epinephrine	β-adrenergic 1	477	gb J03019
11 h1L8_1	C	human	*Homo sapiens*	interleukin-8	interleukin-8	350	gb M68932
12 hBK_2	C	human	*Homo sapiens*	bradykinin	bradykinin 2	364	gb M88714
13 hCCK_A	C	human	*Homo sapiens*	cholecystokinin	CCK-A	428	gb L13605
14 hCCK_B	C	human	*Homo sapiens*	gastrin	CCK-B	447	gb L08112
15 hDOPA_DIA	C	human	*Homo sapiens*	dopamine	dopamine D1A	446	embl X55760
16 hMUS_M1	C	human	*Homo sapiens*	acetylcholine	muscarinic M1	460	embl X15263
17 hNK_1	C	human	*Homo sapiens*	substance P	neurokinin-1	407	gb M76675
18 hNK_2	C	human	*Homo sapiens*	substance K	neurokinin-2	398	gb M57414
19 hNK_3	C	human	*Homo sapiens*	neuromedin	neurokinin-3	465	gb M89473
20 hOLFPO7J	C	human	*Homo sapiens*	unknown	olfactory	320	embl X64995
21 hOXYR	C	human	*Homo sapiens*	oxytocin	oxytocin	389	embl X64878
22 hSST_1A	C	human	*Homo sapiens*	somatostatin	somatostatin 1A	391	gb M81829
23 hTSHR	C	human	*Homo sapiens*	thyrotropin	thyrotropin SH	764	gb M31774
24 mPGEP_3	C	mouse	*Mus musculus*	prostaglandin	prostaglandin EP3	365	gbD10204
25 mTRHR	C	mouse	*Mus musculus*	thyrotropin	thyrotropin RH	393	gb M94384
26 rGUST27	C	rat	*Rattus norvegicus*	unknown	gustatory	312	gbD12820
27 rOLF_F3	C	rat	*Rattus norvegicus*	unknown	olfactory	333	gb M64376

*Phyla are abbreviated: Ac, Acrasiomycota, Ar, Arthropoda, C, Chordata.

out with a package of programs written by D. Kaznadzey for Imagene (Imagene, Chicago, IL). The programs provide very rapid assessment of large databases, use dynamic alignment, comparison matrix scoring, and pairwise grouping according to the neighbor-joining method of Saitou and Nei (1987). The alignment was carried out according to Myers and Miller (1988) based on the methods of Gotoh (1982) and Needleman and Wunsch (1970). Gap initiation parameters (gip) and gap extension penalties (gep) were adjusted to maximize the similarity scores for alignment. Four scoring matrices were tested: the mdm78 matrix of (Fig. 85, Schwartz and Dayhoff, 1978), the mstge structural and genetic matrix of Feng, Johnson, and Doolittle (1985), a genetic code matrix (Smith, Waterman, and Fitch, 1981), and a unitary matrix (Doolittle, 1981) based on scoring 1 for identities. The utility of these matrices have been evaluated previously by Feng et al. (1985).

In the second procedure, a specific alignment of sequences was used with programs for analysis of molecular sequence data provided in the PHYLIP package, Version 3.5c, assembled by Felsenstein (1989). The sequences were initially aligned

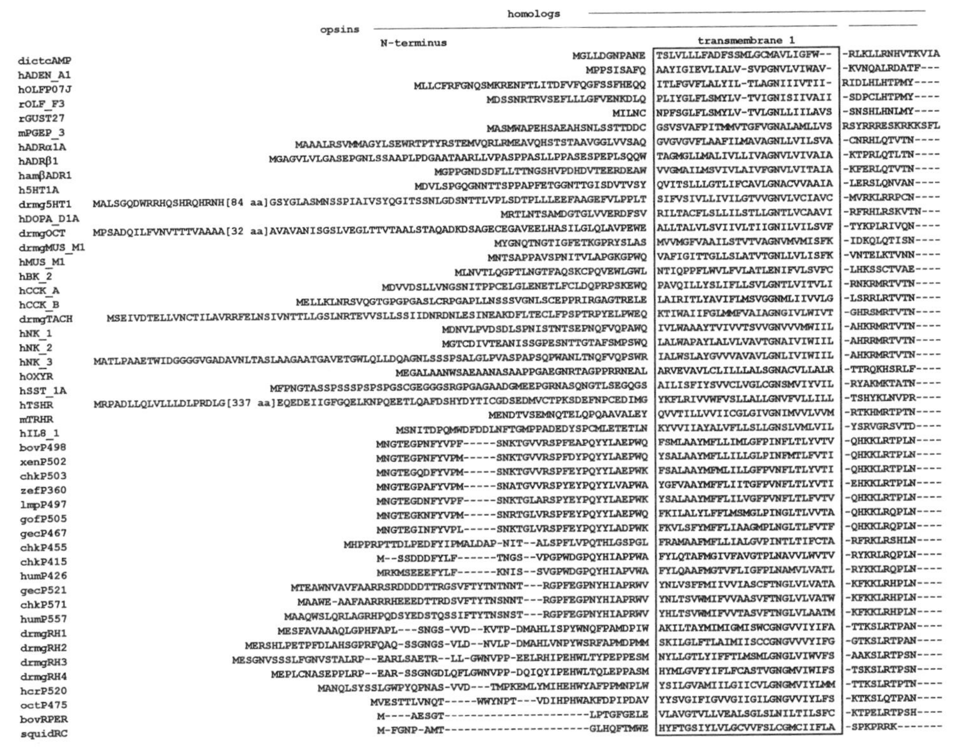

Figure 1. Amino acid sequences and alignment of G-protein–coupled receptor sequences used in the distant matrix evaluation of pairwise relationships. Alignment was achieved with McCAW and some regions were adjusted to improve alignment locally. For several members, long inserts of sequence noted by the number of amino acids (aa) are omitted. The regions of sequences used for distance comparisons are noted by the lines above the data. The overlying line labeled homologs indicates regions of amino acid sequence used for evaluations presented in Fig. 2. The line labeled opsins indicates the regions used to generate relationships in Fig. 3.

with McCAW, a program for multiple alignment construction (Schuler, Altschul, and Lipman, 1991), obtained from NCBI. Further minor local alignment was carried out by eye. The alignments of the receptors and selected opsins are shown in Fig. 1. The portions of the sequences used for calculations are indicated by lines above the sequences presented (Fig. 1). The distance matrices were calculated with PRODIST which uses the Dayhoff PAM 001 matrix (Fig. 82, Dayhoff, Schwartz, and Orcutt, 1978). Pairwise grouping was performed with NEIGHBOR using the neighbor-joining method of Saitou and Nei (1987) and trees drawn with DRAWTREE. The order of sequence entry was assessed with calculations using different entry orders and by conducting randomized order of entry. No effect of order was observed. To provide indices of significance, the method of bootstrap resampling (Felsenstein, 1985, 1988) was applied to generate multiple ($n = 100$) randomly resampled data sets from the initial specified alignment. For each set, distance matrices were calculated with PRODIST and pairwise grouping generated with NEIGHBOR. A consensus tree for the 100 sets was generated with CONSENSE. The later program produces values indicating the number of times pairs appeared in the 100 trees calculated. Values below 50% are not reliable indicators of paired relationships; the program generates branches of the tree in which the consensus is less than 50% simply to fill out the tree with all sequences being assessed. No consideration of branch lengths was made in this work.

	transmembrane 2		transmembrane 3		transmembrane 4	
homologs						
opsins						
dictcAMP	CFCATSFCKDFPSTILTLTNTA-	-------------VNGGFPCY	LYAIVITYGSFACWLWTLCLAI	SIYMLIVKREPEPERFEK-	---YYYLLCWGLPLISTIVMLAKN	--------------
hADEN_A1	CFIVSLAVADVAVGALVIPLAIL	----INIGPQTY-----FHTCL	MVACPVLILTQSSILALLAIAV	DRYLRVKIPLRYKMVVTPR	RAAVAIAGCWILSFVVGLTPMFGW	-NNLSAVERAWAANG
hOLFP07J	FFLSMLSTSETVYTLVILPRMLS	-----SLVGMSQP--MSLAGCA	TQMFFFVTFGITNCFLLTAMGY	DRYVAICNPLRYMVIMNKR	LRIQLVLGACSIGLIVAITQVTSV	---------FRLPFC
rOLF_F3	FFLSNLSFVDICFISTTVPKMLV	-----NIQTQNNV--ITYAGCI	TQIYFFLLFVELDNFLLTIMAY	DRYVAICHPMHYTVIMNYK	LCGFLVLVSWIVSVLHALFQSLMM	--------LALPFCT
rGUST27	FFLSNLSFVDICFISTTIPKMLV	-----NIHSQTKD--ISYIECL	SQVYFLTTFGGMDNFLLTLMAC	DRYVAICHPLNYTVIMNLQ	LCALLILMFWLIMFCVSLIHVLLM	--------NELNFSR
mPGEP_3	LCIGWLALTDLVGQLLTSPVVIL	---VYLSQRRWEQLDPSGRLCT	FFGLTMTVFGLSSLLVASAMAV	ERALAIRAPHWYASHMKTR	-ATPVLLGVWLSVLAFALLPVLGV	------------GRY
hADRα1A	YFIVNLAVADLLLSATVLPFSAT	-----MEVLGFWA--FGRAFCD	VWAAVDVLCCTASILSLCTISV	DRYVGVRHSLKYPAIMTER	KAAAILALLWVVALVVSVGPLLGW	-------------KE
hADRβ1	LFIMSLASADLVMGLLVVPFGAT	-----IVVWGRWE--YGSFFCE	LWTSVDVLCVTASIETLCVIAL	DRYLAITSPFRYQSLLTRA	RARGLVCTVWAISALVSFLPILMH	------------WWR
hamβADR1	YFITSLACADLVMGLAVVPFGAS	-----HILMKMWN--FGNFWCE	FWTSIDVLCVTASIETLCVIAV	DRYIAITSPFKYQSLLTKN	KARMVILMVWIVSGLTSFLPIQMH	------------WYR
h5HT1A	YLIGSLAVTDLMVSVLVLPMAAL	-----YQVLNKWT--LGQVTCD	LFIALDVLCCTSSILHLCAIAL	DRYWAITDPIDYVNKRTPR	RAAALISLTWLIGFLISIPPMLGW	------------RTP
drmg5HT1	YLLVSLALSDLCVALLVMPMALL	-----YEVLEKWN--FGPLLCD	IWVSFDVLCCTASILNLCAISV	DRYLAITKPLEYGVKRTPR	RMMLCVGIVWLAAACISLPPLLIL	-----------GNEH
hDOPA_D1A	FFVISLAVSDLLVAVLVMPWKAV	-----AEIAGFWP--FGSF-CN	IWVAFDIMCSTASILNLCVISV	DRYWAISSPFRYERKMTPK	AAFILISVAWTLSVLISFIPVQLS	WHKAKPTSPSDGNAT
drmgOCT	FFIVSLAVADLTVALLVLPFNVA	-----YSILGRWE--FGIHLCK	LWLTCDVLCCTSSILNLCAIAL	DRYWAITDPINYAQKRTVG	RVLLLISGVWLLSLLISSPPLIGW	-----------NDWP
drmgMUS_M1	YFLFSLAIADFAIGTISMPLFAV	-----TTILGYWP--LGPIVCD	TWLALDYLASNASVLNLLIINF	DRYFSVTRPLTYRAKGTTN	R-AAVMIGAWGISLLLWPPWIYSW	----------PYIEG
hMUS_M1	YFLLSLACADLIIGTFSMNLYTT	-----YLLMGHWA--LGTLACD	LWLALDYVASNASVMNLLLISF	DRYFSVTRPLSYRAKRTPR	RAALMIGLAWLVSFVLWAPAILFW	----------QYLVG
hBK_2	IYLGNLAAADLILACGLPFWAIT	-----ISNNFDWL--FGETLCR	VVNAIISMNLYSSICFLMLVSI	DRYLALVKTMSMGRMRGVR	WAKLYSLVIWGCTLLLSSPMLVFR	---------TMKEYS
hCCK_A	IFLLSLAVSDLMLCLFCMPFNLI	-----PNLLKDFI--FGSAVCK	TTTYFMGTSVSVSTFNLVAISL	ERYGAICKPLQSRVWQTKS	HALKVIAATWCLSFTIMTPYPIYS	--------NLVPFTK
hCCK_B	AFLLSLAVSDLLLAVACMPFTLL	-----PNLMGTFI--FGTVICK	AVSYLMGVSVSVSTLSLVAIAL	ERYSAICRPLQARVWQTRS	HAARVIVATWLLSGLLMVPYPVYT	------------VVQ
drmgTACH	YFLLNLSIADLLMSSLNCVFNFI	-----FMLNSDWP--FGSIYCT	INNFVANVTVSTSVFTLVAISF	DRYIAIVDPLKRRTSRRK-	-VRIILVLIWALSCVLSAPCLLYS	--------SIMTKQY
hNK_1	YFLVNLAFAEASMAAFNTVVNFT	-----YAVHNEWY--YGLFYCK	FHNFFPIAACFASIYSMTAVAF	DRYMAIIHPLQPRLSAT--	ATKVVICVIWVLALLLAFPQGYYS	-------------TT
hNK_2	YFIVNLALADLCMAAFNAAFNFV	-----YASHNIWY--FGRAFCY	FQNLFPITAMFVSIYSMTAIAA	DRYMAIVHPFQPRLSAP--	STKAVIAGIWLVALALASPQCFYS	-------------TV
hNK_3	YFLVNLAFSDASMAAFNTLVNFI	-----YALHSEWY--FGANYCR	FQNFFPITAVFASIYSMTAIAV	DRYMAIIDPLKPRLSAT--	ATKIVIGSIWILAFLLAFPQCLYS	-------------KT
hOXYR	FFMKHLSIADLVVAVFQVLPQLL	-----WDITFRFY--GPDLLCR	LVKYLQVVGMFASTYLLLLMSL	DRCLAICQPLRSLRRR---	TDRLAVLATWLGCLVASAPQVHIF	------------SLR
hSST_1A	IYILNLAIADELLMLSVPFLVTS	------TLLRHWP--FGALLCR	LVLSVDAVNMFTSIYCLTVLSV	DRYVAVVHPIKAARYRRPT	VAKVVNLGVWVLSLLVILPIVVFS	------------RTA
hTSHR	FLMCNLAFADFCMGMYLLLIASV	DLYTHSEYYNHAIDWQTGPGCN	TAGFFTVFASELSVYTLTVITL	ERWYAITFAMRLDRKIRLR	HACAIMVGGWVCCFLLALLPLVGI	------------SSY
mTRHR	CYLVSLAVADLMVLVAAGLPNIT	D----SIY-GSWV--YGYVGCL	CITYLQYLGINASSCSITAFTI	ERYIAICHPIKAQFLCTFS	RAKKIIIFVWAFTSIYCMLWFFLL	--------DLNISTY
hIL8_1	VYLLNLALADLLFALTLPIWAAS	-------KVNGWI--FGTFLCK	VVSLLKEVNFYSGILLLACISV	DRYLAIVHATRTLTQKRH-	LVKFVCLGCWGLSMNLSLPFFLFR	------------QAY
bovP498	YILLNLAVADLFMVFGGFTTTLY	-----TSLHGYFV--FGPTGCN	LEGFFATLGGEIALWSLVVLAI	ERYVVVCKPMSNFRFGEN-	HAIMGVAFTWVMALACAAPPLVGW	------------SRY
xenP502	YILLNLVFANHFMVLCGFTVTMY	-----TSMHGYFI--FGPTGCY	IEGFFATLGGEVALWSLVVLAV	ERYIVVCKPMANFRFGEN-	HAIMGVAFTWIMALSCAAPPLFGW	------------SRY
chkP503	YILLNLVVADLFMVFGGFTTTMY	-----TSMNGYFV--FGVTGCY	IEGFFATLGGEIALWSLVVLAV	ERYVVVCKPMSNFRFGEN-	HAIMGVAFSWIMAMACAAPPLFGW	------------SRY
zefP360	YILLNLAIADLFMVFGGFTTTMY	-----TSLHGYFV--FGRLGCN	LEGFFATLGGEMGLKSLVVLAI	ERWMVVCKPVSNFRFGEN-	HAIMGVAFTWVMACSCAVPPLVGW	------------SRY
lmpP497	YILLNLAMANLFMVLFGFTVTMY	-----TSMNGYFV--FGPTMCS	IEGFFATLGGEVALWSLVVLAI	ERYIVICKPMGNFRFGNT-	HAIMGVAFTWIMALACAAPPLVGW	------------SRY
gofP505	FILVNLAVAGTIMVCFGFTVTFY	-----TAINGYFV--LGPTGCA	VEGFMATLGGEVALWSLVVLAI	ERYIVVCKPMGSFKFSSS-	HAFAGIAFTWVMALACAAPPLFGW	------------SRY
gecP467	YILVNLAAANLVTVCCGFTVTFY	-----ASWYAYFV--FGPIGCA	IEGFFATIGGQVALWSLVVLAI	ERYIVICKPMGNFRFSAT-	HAIMGIAFTWFMALACAGPPLFGW	------------SRF
chkP455	YILVNLALANLLVILVGSTTACY	-----SFSQMYFA--LGPTACK	IEGFAATLGGMVSLWSLAVVAF	ERFLVICKPLGNFTFRGS-	HAVLGCVATWVLGFVASAPPLFGW	------------SRY
chkP415	YILVNISASGFVSCVLSVFVVFV	-----ASARGYFV--FGKRVCE	LEAFVGTHGGLVTGWSLAFLAF	ERYIVICKPFGNFRFSSR-	HALLVVVATWLIGVGVGLPPFFGW	------------SRY
humP426	YILVNVSFGGFLLCIFSVFPVFV	-----ASCNGYFV--FGRHVCA	LEGFLGTVAGLVTGWSLAFLAF	ERYIVICKPFGNFRFSSK-	HALTVVLATWTIGIGVSIPPFFGW	------------SRF
gecP521	WILVNLAFVDLVETLVASTISVF	-----NQIFGYFI--LGHPLCV	IEGYVVSSCGITGLWSLAIISW	ERWFVVCKPFGNIKFDSK-	LAIIGIVFSWVWAWGWSAPPIFGW	------------SRY
chkP571	WILVNLAVADLGETVIASTISVI	-----NQISGYFI--LGHPMCV	VEGYTVSACGITALWSLAIISW	ERWFVVCKPFGNIKFDGK-	LAVAGILFSWLWSCAWTAPPIFGW	------------SRY
humP557	WILVNLAVADLAETVIASTISIV	-----NQVSGYFV--LGHPMCV	LEGYTVSLCGITGLWSLAIISW	ERWLVVCKPFGNVRFDAK-	LAIVGIAFSWIWSAVWTAPPIFGW	------------SRY
drmgRH1	LLVINLAISDFGIMIT-NTPMMG	----INLYFETWV--LGPMMCD	IYAGLGSAFGCSSIWSMCMISL	DRYQVIVKGMAGRPMTIP-	LALGKIAYIWFMSSIWCLAPAFGW	------------SRY
drmgRH2	LLVLNLAFSDFCMMAS-QSPVMI	----INFYYETWV--LGPLWCD	IYAGCGSLFGCVSIWSMCMIAF	DRYNVIVKGINGTPMTIK-	TSIMKILFIWMMAVFWTVMPLIGW	------------SAY
drmgRH3	ILVINLAFCDFMMMV--KTPIFI	----YNSFHQGYA--LGHLGCQ	IFGIIGSYTGIAAGATNAFIAY	DRFNVITRPMEGK-MTHG-	KAIAMIIFIYMYATPWVVACYTET	-----------WGRF
drmgRH4	MFVLNLAVFDLIMCL--KAPIF-	-----NSFHRGFAIYLGNTWCQ	IFASIGSYSGIGAGMTNAAIGY	DRYNVITKPMN-RNMTFT-	KAVIMNIIIWLYCTPWVVLPLTQF	-----------WDRF
hcrp520	LLVVNLAFSDFCMMAFM-MPTMT	----SNCFAETW-I-LGPFMCE	VYGMAGSLFGCASIWSMVMITL	DRYNVIVRGMAAAPLTHK-	KATLLLLFVWIWSGGWTILPFFGW	------------SRY
octP475	MFIINLAMSDLSFSAINGFPLKT	----ISAFMKKW-I-FGKVACQ	LYGLLGGIFGFMSINTMAMISI	DRYNVIGRPMAASKKMSHR	RAFLMIIFVWMWSIVWSVGPVFNW	------------GAY
bovRPER	LLVLSLALADSGISLNAL--VAA	----TSSLLRRW-PY-GSEGCQ	AHGFQGFVTALASICSSAAVAW	GRYHHFCTRSRLDW---N-	TAVSLVFFVWLSSAFWAALPLLGW	------------GHY
squidRC	YAILIHVLITA-MAVNGGDPAHA	----SSSIVGRW-LY-GSVGCQ	LMGFWGFFGGMSHIWMLFAFAM	ERYMAVCHREFYQQ-MPSV	YYSIIVGLMYTFGTFWATMPLLGW	------------ASY

Figure 1. (Continued)

Results and Discussion

The principle criterion by which receptors are assigned to the G-protein–coupled receptor superfamily is the presence of seven transmembrane membrane spanning regions in a single polypeptide chain. As indicated in Fig. 1 and observed in the sequences aligned by Probst et al. (1992), the seven hydrophobic helical domains are relatively uniform throughout the family. Some selected residues that are highly conserved anchor the alignment. For example, the Asn(N) residue in transmembrane 1, the cys(C) residues in loop 2–3 and loop 3–4, and the AsnPro(NP) residues in transmembrane 7 provide guidelines for alignment and illustrating similarity.

However, no residue is invariant in the greater superfamily. In contrast to the uniformity of the transmembrane domains, the amino- and carboxyl-termini, as well as the intervening extracellular and intracellular loops are highly variable (Fig. 1). The overall configuration of the receptors makes it difficult to align the more distant members of the superfamily and poses the problem of which subregions should be chosen for comparison. In general, the subgroups of the superfamily show ~15–25% identity to one another. The invertebrate opsins are related to vertebrate opsins by 25–35% identity, the rod and cone opsins by 40–50% identity, the rodlike pigments by >80% identity, and the mammalian rod pigments by >93% identity (Applebury, 1987*a;* Okano et al., 1992).

```
homologs    ______________________________________________________________________          ________________________________
opsins      ______________________________________________________________________          ________________________________
                                              transmembrane 5                                                                       transmembrane 6
dictcAMP    TVQFVGNWCWIG-------------   | VSFTGYRFGLFYGPFLFIWAISAV | LVGLTSRYTYVVIHNGVSD------------------NKEKHLTYQF | KLINYIIVFLVCWVFAVVNRIVNGL
hADEN_A1    SMGEPVIKCEFE--------KVISME  | YMVYFNFFVWVLPPLLLMVLIYLE | VFYLIRKQLNKKVSASSGDPQK---------------YYGKELKIAK | SLALILFLFALSWLPLHILNCITLF
hOLFP07J    ARKVPHFFCDIRPVMKLSCIDTTVNE  | ILTLIISVLVLVVPMGLVFISYVL | IISTIL-------------------------------KIASVEGRKK | AFATCASHLTVVIVHYSCASIAYLK
rOLF_F3     HLEIPHYFCEPNQVIQLTCSDAFLND  | LVIYFTLVLLATVPLAGIFYSYFK | IVSSIC-------------------------------AISSVHGKYK | AFSTCASHLSVVSLFYCTGLGVYLS
rGUST27     GTEIPHFFCELAQVLKVANSDTHINN  | VFMYVVTSLLGLIPMTGILMSYSQ | IASSLL-------------------------------KMSSSVSKYK | AFSTCGSHLCVVSLFYGSATIVYFC
mPGEP_3     SVQWPGTWCFIS-------TGPAGNE  | TDPAREPGSVAFASAFACLGLLAL | VVTFACNLATIKALVSRCRAKAAVSQSSAQ--------WGRITTETAI | QLMGIMCVLSVCWSPLLIMMLKMIF
hADRα1A     PVPPDERFCGIT-----------EEA  | GYAVFSSVCSFYLPMAVIVVMYCR | VYVVARSTTRSLEAGVKRERGKASEVVLRIH[31 aa]KFSREKKAAK | TLAIVVGVFVLCWFPFFFVLPLGSL
hADRβ1      AESDEARRCYND----PKCCDFVTNR  | AYAIASSVVSFYVPLCIMAFVYLR | VFREAQKQVKKIDSCERRFLGGPARPPSPSP[37 aa]VALREQKALK | TLGIIMGVFTLCWLPFFLANVVKAF
hamβADR1    ATHQKAIDCYHK----ETCCDFFTNQ  | AYAIASSIVSFYVPLVVMVFVYSR | VFQVAKRQLQKIDKSEGRFHSPNLGQVEQDG[11 aa]FCLKEHKALK | TLGIIMGTFTLCWLPFFIVNIVHVI
h5HT1A      EDRSDPDACTIS-----------KDH  | GYTIYSTFGAFYIPLLLMLVLYGR | IFRAARFRIRKTVKKVEKTGADTRHGASPAP[87 aa]ALARERKTVK | TLGIIMGTFILCWLPFFIVALVLPF
drmg5HT1    EDEEGQPICTVC-----------QNF  | AYQIYATLGSFYIPLSVMLFVYYQ | IFRAARRIVLEEKRAQTHLQQALNGTGSPSA[69 aa]QLAKEKKAST | TLGIIMSAFTVCWLPFFILALIRPF
hDOPA_D1A   SLAETIDNCDSS-----------LSR  | TYAISSSVISFYIPVAIMIVTYTR | IYRIAQKQIRRIAALERAAVHAKNCQTTTGN[15 aa]SFKRETKVLK | TLSVIMGVFVCCWLPFFILNCILPF
drmgOCT     DEFTSATPCELT-----------SQR  | GYVIYSSLGSFFIPLAIMTIVYIE | IFVATRRRLRERARANKLNTIALKSTELEPM[197aa]SLSKERRAAR | TLGIIMGVFVICWLPFFLMYVILPF
drmgMUS_M1  KRTVPKDECYIQ--------FIETNQ  | YITFGTALAAFYFPVTIMCFLYWR | IWRETKKRQKDLPNLQAGKKDSSKRSNSSDE[384aa]EKRQESKAAK | TLSAILLSFIITWTPYNILVLIKPL
hMUS_M1     ERTMLAGQCYIQ---------FLSQP  | IITFGTAMAAFYLPVTVMCTLYWR | IYRETENRARELAALQGSETPGKGGGSSSSS[114aa]SLVKEKKAAR | TLSAILLAFILTWTPYNIMVLVSTF
hBK_2       DEGHNVTACVIS---YPS---LIWEV  | FTNMLLNVVGFLLPLSVITFCTMQ | IMQVLRNNEMQKFK-----------------------EIQTERRATV | LVLVVLLLFIICWLPFQISTFLDTL
hCCK_A      NNNQTANMCRFL------LPNDVMQQ  | SWHTFLLLILFLIPGIVMMVAYGL | ISLELYQGIKFEASQKKSAKERKPSTTSSGK[41 aa]NLMAKKRVIR | MLIVIVVLFFLCWMPIFSANAWRAY
hCCK_B      PVGPRVLQCVHR---WPS---ARVRQ  | TWSVLLLLLLFFIPGVVMAVAYGL | ISRELYLGLRFDGDSDSDSQSRVRNQGGLPG[52 aa]KLLAKKRVVR | MLLVIVVLFFLCWLPVYSANTWRAF
drmgTACH    YNGKSRTVCFMM-WPDGRYPTSMADY  | AYNLIILVLTTGIPMIVMLICYSL | MGRVPGGSRSIGENTDRQME-----------------SMKSKRKVVR | MFIAIVSIFAICWLPYHLFFIYAYH
hNK_1       ETMPSRVVCMIE-WPEH--PNKIYEK  | VYHICVTVLIYFLPLLVIGYAYTI | VGITLWASEIPGDSSDRYHE-----------------QVSAKRKVVK | MMIVVVCTFAICWLPFHIFFLLPYI
hNK_2       TMDQGATKCVVA-WPED--SGGKTLL  | LYHLVVIALIYFLPLAVMFVAYSV | IGLTLWRRAVPGHQAHGANLR----------------HLQAKKKFVK | TMVLVVLTFAICWLPYHLYFILGSF
hNK_3       KVMPGRTLCFVQ-WPEG----PKQHF  | TYHIIVIILVYCFPLLIMGITYTI | VGITLWGGEIPGDTCDKYHE-----------------QLKAKRKVVK | MMIIVVMTFAICWLPYHIYFILTAI
hOXYR       EVADGVFDCWAV------FIQPWGPK  | AYITWITLAVYIVPVIVLATCYGL | ISFKIWQNLRLKTAAAAAAEAPEGAAAGDGG[12 aa]ISKAKIRTVK | MTFIIVLAFIVCWTPFFFVQMWSVW
hSST_1A     ANSDGTVACNML----MPEPAQRWLV  | GFVLYTFLMGFLLPVGAICLCYVL | IIAKMRMVALKAGWQQ---------------------RKRSERKITL | MVMMVVMVFVICWMPFYVVQLVNVF
hTSHR       ---AKVSICLPM------DTETPLAL  | AYIVFVLTLNIVAFVIVCCCHVKI | YITVRNPQYNP--------------------------GDKDTKIAKR | MAVLIFTDF-ICMAPISFYALSAIL
mTRHR       K-NAVVVSCGYK-------ISRNYYS  | PIYLMDFGVFYVVPMILATVLYGF | IARILFLNPIPSDPKENSKMWKNDSIHQNKN[12 aa]TVSSRKQVTK | MLAVVVILFALLWMPYRTLVVVNSF
hIL8_1      HPNNSSPVCYEV----LGNDTAKWRM  | VLRILPHTFGFIVPLFVMLFCYGF | TLRTLFKA-----------------------------HMGQKHRAMR | VIFAVVLIFLLCWLPYNLVLLADTL
bovP498     IPEGMQCSCGID---YYTPHEETNNE  | SFVIYMFVVHFIIPLIVIFFCYGQ | LVFTVKEAAAQQQESAT--------------------TQKAEKEVTR | MVIIMVIAFLICWLPYAGVAFYIFT
xenP502     IPEGMQCSCGVD---YYTLKPEVNNE  | SFVIYMFIVHFTIPLIVIFFCYGR | LLCTVKEAAAQQQESLT--------------------TQKAEKEVTR | MVVIMVVFFLICWVPYAYVAFYIFT
chkP503     IPEGMQCSCGID---YYTLKPEINNE  | SFVIYMFVVHFMIPLAVIFFCYGN | LVCTVKEAAAQQQESAT--------------------TQKAEKEVTR | MVIIMVIAFLICWVPYASVAFYIFT
zefP360     IPEGMQCSCGVD---YYTRTPGVNNE  | SFVIYMFIVHFFIPLIVIFFCYGR | LVCTVKEAARQQQESET--------------------TQRAEREVTR | MVIIMVIAFLICWLPYAGVAWYIFT
lmpP497     IPEGMQCSCGPD---YYTLNPNFNNE  | SYVVYMFVVHFLVPFVIIFFCYGR | LLCTVKEAAAAQQESAS--------------------TQKAEKEVTR | MVVLMVIGFLVCWVPYASVAFYIFT
gofP505     IPEGMQCSCGPD---YYTLNPDYNNE  | SYVIYMFVCHFILPVAVIFFTYGR | LVCTVKAAAAQQQDSAS--------------------TQKAEREVTK | MVILMVFGFLIAWTPYATVAAWIFF
gecP467     IPEGMQCSCGPD---YYTLNPDFHNE  | SYVIYMFIVHFTVPMVVIFFSYGR | LVCKVREAAAQQQESAT--------------------TQKAEKEVTR | MVILMVLGFLLAWTPYAATAIWIFT
chkP455     IPEGLQCSCGPD---WYTTDNKWHNE  | SYVLFLFTFCFGVPLAIIVFSYGR | LLITLRAVARQQEQSAT--------------------TQKADREVTK | MVVVMVLGFLVCWAPYTAFALWVVT
chkP415     MPEGLQCSCGPD---WYTVGTKYRSE  | YYTWFLFIFCFIVPLSLIIFSYSQ | LLSALRAVAAQQQESAT--------------------TQKAEREVSR | MVVVMVGSFCLCYVPYAALAMYMVN
humP426     IPEGLQCSCGPD---WYTVGTKYRSE  | SYTWFLFIFCFIVPLSLICFSYTQ | LLRALKAVAAQQQESAT--------------------TQKAEREVSR | MVVVMVGSFCVLYVPYAAFAMYMVN
gecP521     WPHGLKTSCGPD---VFSGSVELGCQ  | SFMLTLMITCCFLPLFIIIVCYLQ | VWMAIRAVAAQQKESES--------------------TQKAEREVSR | MVVVMIVAFCICWGPYASFVSFAAA
chkP571     WPHGLKTSCGPD---VFSGSSDPGVQ  | SYMVVLMVTCCFFPLAIIILCYLQ | VWLAIRAVAAQQKESES--------------------TQKAEKEVSR | MVVVMIVAYCFCWGPYTFFACFAAA
humP557     WPHGLKTSCGPD---VFSGSSYPGVQ  | SYMIVLMVTCCIIPLAIIMLCYLQ | VWLAIRAVAKQQKESES--------------------TQKAEKEVTR | MVVVMIFAYCVCWGPYTFFACFAAA
drmgRH1     VPEGNLTSCGID---YLE--RDWNPR  | SYLIFYSIFVYYIPLFLICYSYWF | IIAAVSAHEKAMREQAKKMNVKSLRSSEDA--------EKSAEGKLAK | VALVTITLWFMAWTPYLVINCMGLF
drmgRH2     VPEGNLTACSID---YMT--RMWNPR  | SYLITYSLFVYYTPLFLICYSYWF | IIAAVAAHEKAMREQAKKMNVKSLRSSEDC--------DKSAEGKLAK | VALTTISLWFMAWTPYLVICYFGLF
drmgRH3     VPEGYLTSCTFD---YLT--DNFDTR  | LFVACIFFFSFVCPTTMITYYYSQ | IVGHVFSHEKALRDQAKKMNVESLRSNVDKN-------KETAEIRIAK | AAITICFLFFCSWTPYGVMSLIGAF
drmgRH4     VPEGYLTSCSFD---YLS--DNFDTR  | LFVGTIFFFSFVCPTLMILYYYSQ | IVGHVFSHEKALREQAKKMNVESLRSNVDKS-------KETAEIRIAK | AAITICFLFFVSWTPYGVMSLIGAF
hcrP520     VPEGNLTSCTVD---YLT--KDWSSA  | SYVVIYGLAVYFLPLITMIYCYFF | IVHAVAEHEKQLREQAKKMNVASLRANADQQ-------KQSAECRLAK | VAMMTVGLWFMAWTPYLIISWAGVF
octP475     VPEGILTSCSFD---YLS--TDPSTR  | SFILCMYFCGFMLPIIIIAFCYFN | IVMSVSNHEKEMAAMAKRLNAKELRKAQA---------GASAEMKLAK | ISMVIITQFMLSWSPYAIIALLAQF
bovRPER     DYEPLGTCCTLD---Y-S-RGDRNFT  | SFLFTMAFFNFLLPLFITVVSYRL | MEQ----------------------------------KLGKTSRPPV | NTVLPARTLLLGWGPYALLYLYATI
squidRC     GLEVHGTSCTIN---Y-S-VSDESYQ  | SYVFFLAIFSFIFPMVSGWYAISK | AWSGLSAIPDAEKEKDKD-------------------ILSEEQLTAL | AGAFILISLISWSGFGYVAIYSALT
```

Figure 1. (Continued)

In this work, an exploratory assessment of relationships among the G-protein–coupled receptors and opsins, and the opsins themselves was carried out first. The method of dynamic alignment (Myers and Miller, 1988) was used with four different comparison matrices for scoring. Relationships among the proteins were produced by the pairwise grouping method of neighbor-joining (Saitou and Nei, 1987). These algorithms work well on large databases and can be carried out relatively rapidly. 51 opsins (Table I) and some 40 related homolog receptors (27 presented in Table II) were evaluated together and as subgroups. The sequences were made to be the same relative length by truncating the NH_2 and COOH termini to the full length of the opsins; in the case of muscarinic and octopamine receptors, the loop 5–6 was shortened as well. In the opsin subfamily, different regions (NH_2 termini, central, COOH termini) of the sequences were independently analyzed for comparison of trees generated.

Some consistent general relationships were observed that were independent of length of sequence examined, comparison matrix used, and alignment forced by changing gip and gap parameters. The opsins, the biogenic amine receptors (including muscarinic), the neurokinin receptors, the gustatory/olfactory receptors, the cholecystokinin/gastrin receptors, and bradykinin/somatostatin receptors formed independent groups in the larger superfamily. Other receptors such as TSH, TRH, oxytocin varied in their pairings. The opsins themselves segregate as retino-chrome-

```
homologs
opsins
                                 transmembrane 7                                              C-terminus
dictcAMP     NMFPPALNILHTYL----- SVSHGFWASVTFIYNNPLMW---- RYFGAKILTVFTFFGYFTDVQKKLEKNKNNNNPSPYSSSRGTSGKTMGGHPTG[64 aa]STSTNGQGNN* 392
hADEN_A1     PSCHKPSI----------- LTYIAIFLTHGNSAMNPIVYA--- -FRIQKFRVTFLKIWNDHFRCQPAPPIDEDLPEERPDD* 326
hOLFP07J     PKSENTREH---------- DQLISVTYTVITPLLNPVVYTLRN KEVKDALCRAVGGKFS* 320
rOLF_F3      SAANNSSQA---------- SATASVMYTVVTPMVNPFIYSLRN KDVKSVLKKTLCEEVIRSPPSLLHFFLVLCHLPCFIFCY* 333
rGUST27      SSVLHSTHKK--------- -MIASLMYTVISPMLNPFIYSLRN KDVKGALGKLFIRVASCPLWSKDFRPKFILKPERQSL* 312
mPGEP_3      NQMSVEQCKTQMGKEKECN SFLIAVRLASLNQILDPWVYLLLR KILLRKFCQIRDHTNYASSSTSLPCPGSSALMWSDQLER* 365
hADRα1A      FPQLKPSEG---------- VFKVIFWLGYFNSCVNPLIYPCSS REFKRAFLRLLRCQCRRRRRRRPLWRVYGHHWRASTSGLRQDCAPSSGDAPPG[74 aa]PSAQRWRLCP* 501
hADRβ1       HRELVPDR----------- LFVFFNWLGYANSAFNPIIY-CRS PDFRKAFQGLLcCARRAARRRHATHGDRPRASGCLARPGPPPSPGAASDDDDD[34 aa]RPGFASESKV* 477
hamβADR1     QDNLIPKE----------- VYILLNWLGYVNSAFNPLIY-CRS PDFRIAFQELLCLRRSSSKAYGNGYSSNSNGKTDYMGEASGCQLGQEKESERL[26 aa]RNCSTNDSPL* 418
h5HT1A       CESSCHMPTL--------- LGAIINWLGYSNSLLNPVIYAYFN KDFQNAFKKIIKCKFCRQ* 422
drmg5HT1     ETMHVPAS----------- LSSLFLWLGYANSLLNPIIYATLN RDFRKPFQEILYFRCSSLNTMMRENYYQDQYGEPPSQRVMLGDERHGARESFL* 564
hDOPA_D1A    CGSGETQPFCIDSN----- TFDVFVWFGWANSSLNPIIYA-FN ADFRKAFSTLLGCYRLCPATNNAIETVSINNNGAAMFSSHHEPRGSISKECNL[49 aa]PITQNGQHPT* 446
drmgOCT      CQTCCPTNK---------- FKNFITWLGYINSGLNPVIYTIFN LDYRRAFKRLLGLN* 601
drmgMUS_M1   -TTCSDCIPTE-------- LWDFFYALCYINSTINPMSYALCN ATFRRTYVRILTCKWHTRNREGMVRGVYN* 722
hMUS_M1      CKDCVPET----------- LWELGYWLCYVNSTINPMCYALCN KAFRDTFRLLLLCRWDKRRWRKIPKRPGSVHRTPSRQC* 460
hBK_2        HRLGILSSCQDERIIDV-- ITQIASFMAYSNSCLNPLVYVIVG KRFRKKSWEVYQGVCQKGGCRSEPIQMENSMGTLRTSISVERQIHKLQDWAGSRQ*  364
hCCK_A       DTASAERRLSGT------- PISFILLLSYTSSCVNPIIYCFMN KRFRLGFMATFPCCPNPGPPGARGEVGEEEEGGTTGASLSRFSYSHMSASVPPQ* 428
hCCK_B       DGPGAHRALSGA------- PISFIHLLSYASACVNPLVYCFMH RRFRQACLETCARCCPRPPRARPRALPDEDPPTPSIASLSRLSYTTISTLGPG* 447
drmgTACH     NNQVASTKYVQH------- MYLGFYWLAMSNAMVNPLIYYWMN KRFRMYFQRIICCCCVGLTRHRFDSPKSRLTNKNSSNRHTRAETKSQWKRSTM[74 aa]NPVELSPKQM* 504
hNK_1        NPDLYLKKFIQQ------- VYLAIMWLAMSSTMYNPIIYCCLN DRFRLGFKHAFRCCPFISAGDYEGLEMKSTRYLQTQGSVYKVSRLETTISTVV[35 aa]SFSFSSNVLS* 407
hNK_2        QEDIYCHKFIQQ------- VYLALFWLAMSSTMYNPIIYCCLN HRFRSGFRLAFRCCPWVTPTKEDKLELTPTTSLSTRVNRCHTKETLFMAGDTA[24 aa]LAPTKTHVEI* 398
hNK_3        YQQLNRWKYIQQ------- VYLASFWLAMSSTMYNPIIYCCLN KRFRAGFKRAFRWCPFIKVSSYDELELKTTRFHPNRQSSMYTVTRMESMTVVF[42 aa]SPYTSVDEYS* 465
hOXYR        DANAPKEAS---------- AFIIVMLLASLNSCCNPWIYMLFT GHLFHELVQRFLCCSASYLKGRRLGETSASKKSNSSSFVLSHRSSSQRSCSQPSTA* 389
hSST_1A      AEQDDAT------------ VSQLSVILGYANSCANPILYGFLS DNFKRSFQRILCLSWMDNAAEEPVDYYATALKSRAYSVEDFQPENLESGGVFRNGTCTSRITTL* 391
hTSHR        NKPLITVSN---------- SKILLVLFYPLNSCANPFLYAIFT KAFQRDVFILLSKFGICKRQAQAYRGQRVPPKNSTDIQVQKVTHDMRQGLHNM[29 aa]ISEEYMQTVL* 764
mTRHR        LSSPFQENW---------- FLLFCRICIYLNSAINPVIYNLMS QKFRAAFRKLCNCKQKPTEKAANYSVALNYSVIKESDRFSTELEDITVTDTYVSTTKVSFDDTCLASEN* 393
hIL8_1       MRTQVIQETCERR-NNIGR ALDATEILGFLHSCLNPIIYAFIG QNFRHGFLKILAMHGLVSKEFLARHRVTSYTSSSVNVSSNL* 350
bovP498      HQGSDFGPI---------- FMTIPAFFAKTSAVYNPVIYIMMN KQFRNCMVTTLCCGKNPLG-DDE---ASTT-VSKTE-----TSQVAPA* 348
xenP502      HQGSNFGPV---------- FMTVPAFFAKSSAIYNPVIYIVLN KQFRNCLITTLCCGKNPFG-DEDG--SSAA-TSKTEASSVSSSQVSPA* 354
chkP503      NQGSDFGPI---------- FMTIPAFFAKSSAIYNPVIYIVMN KQFRNCMITTLCCGKNPLG-DEDT---SAG---KTETSSVSTSQVSPA* 351
zefP360      HQGSEFGPV---------- FMTLPAFFAKTSAVYNPCIYICMN KQFRHCMITTLCCGKNPFE-EEEG--ASTT-ASKTEASSVSSSSVSPA* 354
lmpP497      HQGSDFGAT---------- FMTLPAFFAKSSALYNPVIYILMN KQFRNCMITTLCCGKNPLG-DDESG-AST---SKTEVSSVSTSPVSPA* 353
gofP505      NKGADFSAK---------- FMAIPAFFSKSSALYNPVIYVLLN KQFRNCMLTTIFCGKNPLG-DDE---SSTVSTSKTEVSS-----VSPA* 349
gecP467      NRGAAFSVT---------- FMTIPAFFSKSSSIYNPIIYVLLN KQFRNCMVTTICCGKNPFG-DEDV--SSSVSQSKTEVSSVSSSQVAPA* 355
chkP455      HRGRSFEVG---------- LASIPSVFSKSSTVYNPVIYVLMN KQFRSCMLKLLFCGRSPFG-DDED--VSGSSQA-TQVSSVSSSHVAPA* 361
chkP415      NRDHGLDLR---------- LVTIPAFFSKSACVYNPIIYCFMN KQFRACIMETVC-GK-PLT-DDSD--ASTSAQ-RTEVSSVSSSQVGPT* 347
humP426      NRNHGLDLR---------- LVTIPSFFSKSACIYNPIIYCFMN KQFQACIMKMVC-GK-AMT-DESD--TCSSQ--KTEVSTVSSTQVGPN* 348
gecP521      NPGYAFHPL---------- AAALPAYFAKSATIYNPVIYVFMN RQFRNCIMQL-----FGKKVDD-G--SEASTTSRTEVSSVSNSSVAPA* 365
chkP571      NPGYAFHPL---------- AAALPAYFAKSATIYNPIIYVFMN RQFRNCILQL-----FGKKVDD-G--SEV-STSRTEVSSVSNSSVSPA* 362
humP557      NPGYAFHPL---------- MAALPAYFAKSATIYNPVIYVFMN RQFRNCILQL-----FGKKVDD-G--SELSSASKTEVSSV--SSVSPA* 364
drmgRH1      KFE-GLTPL---------- NTIWGACFAKSAACYNPIVYGISH PKYRLALKEKCPCCVFGKV-DDGK--SSDAQSQAT-----ASEAESKA* 373
drmgRH2      KID-GLTPL---------- TTIWGATFAKTSAVYNPIVYGISH PKYRIVLKEKCPMCVFGNT-DEPKPDAPASD---TE---TTSEADSKA* 381
drmgRH3      GDKTLLTPG---------- ATMIPACACKMVACIDPFVYAISH PRYRMELQKRCPWLALNEKAPESSAVASTS---TTQE----PQQTTAA* 383
drmgRH4      GDKSLLTQG---------- ATMIPACTCKLVACIDPFVYAISH PRYRLELQKRCPWLGVNEKSGE----ISSAQSTTTQE----QQQTTAA* 378
hcrP520      SSGTRLTPL---------- ATIWGSVFAKANSCYNPIVYGISH PRYKAALYQRFPSLACG--SGESGSDVKSEASATTT--MEEKPKIPEA* 376
octP475      GPAEWVTPY---------- AAELPVLFAKASAIHNPIVYSVSH PKFREAIQTTFPWLLTCCQFDEKECEDANDAEEEVVASERGGESRDAAQMKE[73 aa]QGVDNQAYQA* 455
bovRPER      ADATSISPK---------- LQMVPALIAKAVPTVNAMNYALGS EMVHRGIWQCLSPQRREHSREQ-------------------------* 291
squidRC      HGGAQLSHL---------- RGHVPPIMSKTGCALFPLLIFLLT ARSLPKSDTKKP-----------------------------------* 301
```

Figure 1. (Continued)

like opsins, mollusk (squid and octopus), arthropod, and chordate groups. Among the latter vertebrate opsins, general clustering of the rodlike pigments (including a set of rodlike green cone pigments), the blue and violet cone pigments, and the long wavelength red and green cone pigments is observed. These branches are consistent with earlier evaluations (Applebury, 1987*a,b;* Goldsmith, 1990).

Variation in the trees produced is observed, however, among and within these groups. Alignment set by adjusting gip and gap affected the relationships; although the values were optimized to obtain maximal identities and minimal gaps, visual inspection showed that some pairs might be better aligned in local regions. Further truncation of the NH_2 and COOH termini, for which alignment is most difficult, changed the pairwise branching of groups, as did the selection of subregions of the opsin sequences in rod opsin assessments. The four comparison matrices produced different branching relationships. Variation was seen in branching between major receptor groups. The blue/violet opsins were sometimes paired with the red/green opsins and sometimes with the rod opsins. Among the nearly identical mammalian rod opsins, which have 93–95% identities, variance in pairs was also observed. The variance indicates that this family of proteins does not behave consistently under analysis with distance matrix methods, in contrast to the tyrosine kinases or globins analyzed by Feng et al. (1985).

To fix some parameters in assessment, a specific alignment was adopted and the relationships were reevaluated with the PHYLIP package of programs assembled by Felsenstein (1989). A group of 27 representative G-protein–coupled receptors and 21 selected opsins were chosen to evaluate the relationships between the opsins and the larger receptor family. The opsins were selected to represent the opsin groups generally observed in the initial survey. The database and alignment is shown in Fig. 1. The overlying line labeled homologs in Fig. 1 indicates the regions of sequence used for analysis. The NH_2 and COOH termini were truncated, loop 5–6 was restricted to its NH_2 and COOH ends, and some insertions were represented by one or two residues rather than the whole insertion. The pairings were evaluated with the PAM-250(001) matrix (Dayhoff et al., 1978) available in the PRODIST program. This matrix has been shown to be appropriate for evaluating sequences of 20–30% similarity (Dayhoff et al., 1978; Feng et al., 1985). To obtain some index of significance for the branching, the bootstrap resampling method of Felsenstein (1985, 1988) was applied and $n = 100$ sample sets were generated from the primary data and their consensus tabulated.

The consensus relationships produced for the homolog receptors and subset of opsins is given in Fig. 2. The values at the nodes indicate the number of times the sequences were paired in the 100 trees produced. As indices of confidence, values of greater than 60 mark pairs that are reliably consistent; values <60 are observed for the same branches that proved variable in the initial survey. For the given alignment (Fig. 1) and adoption of bootstrap values of >60 as a statistical index of confidence, the following relationships can be suggested from the analysis presented in Fig. 2: The opsins form a subgroup of receptors unto themselves. Other well-formed groups include the biogenic amine receptors and muscarinic acetylcholine receptors (AmineR), the gustatory and olfactory group (GusR and OlfR), and the neurokinin group (NKR). However, the pairing between any of these groups or other individual receptors whose pairings show low confidence values remains unresolved, as indicated by numbers of <60 given in parentheses. The *Dictyostelium* cAMP receptor does serve as a reliable outgroup for the superfamily. Any one of the G-protein–coupled receptors would appear to serve as an appropriate outgroup for a more detailed analysis of the opsins.

Within the well resolved groups, several subpairings show lack of statistical significance. To resolve some of these issues, the opsin subfamily was evaluated by itself. A larger database (49 opsins and 2 retinochromes) was assessed. The sequences were truncated to the length of mammalian rod opsins (348 amino acids) as shown in Fig. 1 by the overlying line labeled opsins. Some minor refinements in the NH_2- and COOH-terminal alignment were made upon the addition of more opsins.

The assessment of pairwise grouping was made with bootstrap resampling to generate $n = 100$ sets of randomly sampled data as described above. Hamster β-adrenergic receptor (hamβADR1) was used as an outgroup to display the pairings. The consensus tree produced is given in Fig. 3.

Some opsin pairings are resolved with assignment of a significant level of confidence upon the evaluation of the larger opsin database and additional sequence

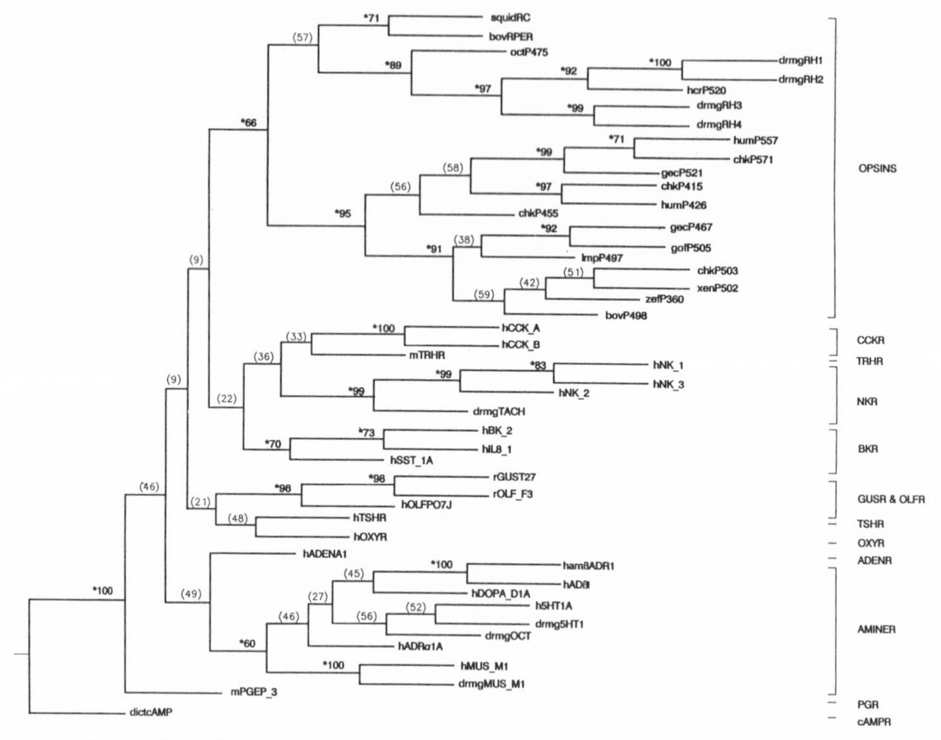

Figure 2. Relationships among selected G-protein coupled receptors. The source of amino acid sequences and notations are given in Tables I and II. The inferences are generated using the fixed alignment given in Fig. 1, scoring with the PAM-250 (001) matrix of Dayhoff et al. (1978) and the neighbor-joining method of Saitou and Nei (1987). To provide a statistical value of confidence in pairing, the bootstrap resampling method (Felsenfeld, 1985) was applied for $n = 100$ data sets. Scores are given at the node of the pairings. Confidence in pairing is apparent for scores >60 which are marked with *. Values of <60 given in parentheses indicate the appropriate pairing remains unresolved. Subgroups of receptors that are well formed are indicated at the right. The tree is unrooted, but constructed with the dictcAMP receptor as an outgroup.

length used. The squid retinochrome and mammalian retinochromelike receptors appear to be more related to the invertebrate opsins and clearly form a pair of their own. The lamprey rod opsin (lamP497) is well paired with other vertebrate rod opsins and several pairings within the rod opsins are now more significant. However, the pairing of blue and violet cone pigments (gofP441, chkP455, humP426) with either the rod group or the long wavelength cone group remains unresolved.

Although the blue/violet pigments usually pair with the rods in single analyses, as has been shown by Okano et al. (1992) and Johnson et al. (1993), their pairing changes depending upon which region(s) of the sequence is sampled. Okano et al. (1992) has noted previously that the bootstrap scores are insufficient to resolve the relationship. It is disappointing that a level of confidence is lacking to indicate whether the blue/violet pigments might have been the ancestors of the rod pigments. No resolution is obtained for whether the chicken rod opsin (chkP503) might pair with the mammalian rod opsins or the amphibian opsins. The human (humP498) and canine (canP498) opsins vary in their pairing with other mammalian pairs. For the

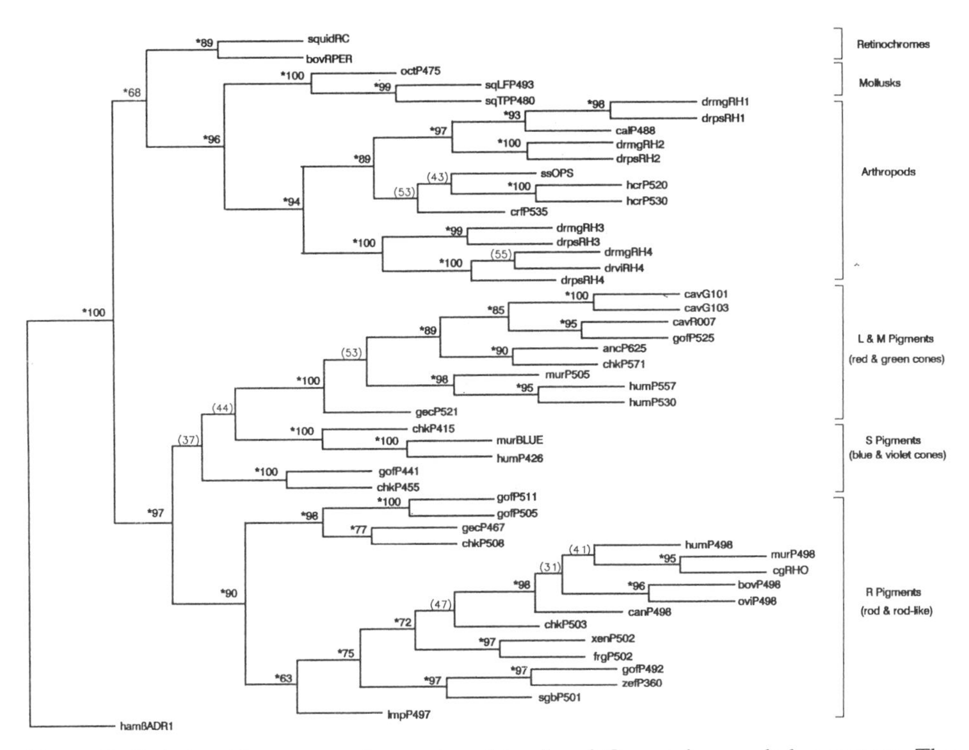

Figure 3. Relationships among the opsin subfamily of G-protein coupled receptors. The inferences were generated as described in the legend of Fig. 2 and the Methods section. hamβADR1 was included in the calculations for reference. The tree is unrooted, but constructed with hamβADR1 as an outgroup for presentation.

latter opsins, an analysis of the rod opsins alone was carried out (data not shown). The differences in Dayhoff matrix scoring within the mammalian rod opsin group is less than 1% (Dayhoff et al., 1978). A single amino acid change in alignment in the COOH termini can change the pairing. Thus, the pairing depends upon the alignment chosen and which areas of sequence are compared. While the PAM-250 matrix (Dayhoff et al., 1978) has been found appropriate for scoring distantly related sequences (20–30% similarity) (Schwartz and Dayhoff, 1978; Feng et al., 1985), alternative comparison matrices should be explored to evaluate the highly similar sequences.

The relationships generated for the current opsin database do suggest that the generation of absorption spectra of given wavelength maxima (produced upon opsin covalently binding a retinaldehyde to form a visual pigment) has been achieved multiple times. Pigments with like-absorption maxima are observed in different branches (particularly invertebrate and vertebrate branches) suggesting functional convergent evolution. It remains to be clarified if more than one structural arrangement can achieve the same end. Some caution in the interpretation about convergent evolution, however, needs to heeded. Convergent evolution has been suggested upon noting that the human P530 and P557 pigments are more related to one another than to their teleost green and red counterparts (Yokohama et al., 1990). The human pigments, however, are present as a tandem gene repeat in the q28 locus of the X chromosome. The P530 and P557 genes have undergone considerable unequal homologous recombination (Nathans, Plantanida, Eddy, Shows, and Hogness, 1986*a*; Nathans, Thomas, and Hogness, 1986*b*). Thus, the latter are probably more homogeneous than the teleost genes, even though each may have derived from distinctly different green (medium) and red (long) wavelength lines. Indeed, given the extensive size of the G-protein–coupled superfamily, it is possible that recombination events may have played a major role in generating and scrambling the receptors over the millenna, such that it may prove difficult to resolve their relationships and evolutionary history.

In summary, the above preliminary evaluation of the relationships among G-protein–coupled receptors, has raised several problems. Pairwise groupings vary depending upon alignment, length or region of the sequence compared, and the matrix used to score the pairings. Application of some method of scoring for levels of confidence is important in presentation of relationships generated (Felsenstein, 1988). Evaluation of pairwise relationships among the members of the G-protein–coupled receptors using the bootstrap resampling method of generating a statistical level of confidence shows that confidence levels for many pairings are low. Further evaluation of the receptor family will be necessary to resolve the several issues here raised. Some insight may be gained with yet larger databases, but the computation time becomes exorbitant and will call for larger computer capacity and better algorithms.

Acknowledgements

The author would like to thank Denis Kaznadzey for participating in the initial phase of this work and making an excellent set of programs available for analysis. The generous advice and insights of Bill Ballard are sincerely appreciated. The author is grateful for the ready availability of the PHYLIP package from J. Felsenstein, its documentation, and its ease of use.

References

Applebury, M. L. 1987*a*. An evolutionary comparison of visual pigments. *In* Retinal Proteins. Proceedings of an International Conference. Lake Baikal, USSR. Y. A. Ovchinnikov, editor. VNU Science Press, The Netherlands. 15–24.

Applebury, M. L. 1987*b*. Evolutionary relationships of visual pigments and other signal receptors. Predicted elements of structure-function. *In* Discussions in Neurosciences. Sensory

Transduction. A.J. Hudspeth, P.R. MacLeish, F.L. Margolis, and T.N. Wiesel, editors. Vol. IV (No 3). Foundation for the Study of the Nervous System, Geneva, Switzerland. 28–37.

Applebury, M. L., and P. A. Hargrave. 1986. Molecular biology of the visual pigments. *Vision Research.* 26:1881–1895.

Britt, S. G., R. Feiler, K. Kirschfeld, and C. S. Zuker. 1993. Spectral tuning of rhodopsin and metarhodopsin *in vivo. Neuron.* 11:29–39.

Dayhoff, M. O., R. M. Schwartz, and B. C. Orcutt. 1978. A model of evolutionary change in proteins. *In* Atlas of Protein Sequencing and Structure. M. O. Dayhoff, editor. Vol. 5 (Supplement 3) National Bioscience Research Foundation, Washington, DC. 345–352.

Doolittle, R. F. 1981. Similar amino acid sequences: chance or common ancestry? *Science.* 214:149–159.

Felsenstein, J. 1985. Confidence limits on phylogenies: An approach using the bootstrap. *Evolution.* 39:783–791.

Felsenstein, J. 1989. Phylip. Phylogeny inference package. *Cladistics.* 5:164–166.

Felsenstein, J. 1988. Phylogenies from molecular sequences: Inference and reliability. *Annual Review of Genetics.* 22:521–565.

Feng, D. F., M. S. Johnson, and R. F. Doolittle. 1985. Aligning amino acid sequences: Comparison of commonly used methods. *Journal of Molecular Evolution.* 21:112–125.

Goldsmith, T. H. 1990. Optimization, constraint, and history in the evolution of eyes. *Quarterly Review of Biology.* 65:281–321.

Gotoh, O. 1982. An improved algorithm for matching biological sequences. *Journal of Molecular Biology.* 162:705–708.

Hardie, R. C. 1985. Functional organization of the fly retina. *In* Progress in Sensory Physiology. D. Ottoson, editor. Vol. 5. Springer-Verlag, Berlin. 1–79.

Inglese, J., N. J. Freedman, W. J. Koch, and R. J. Lefkowitz. 1993. Structure and mechanism of the G protein-coupled receptor kinases. *Journal of Biological Chemistry.* 268:23735–23738.

Jiang, M., S. Pandey, and H. K. W. Fong. 1993. An opsin homologue in the retina and pigment epithelium. *Investigative Ophthalmology and Visual Sciences.* 34:3669–3678.

Johnson, R. L., K. B. Grant, T. C. Zankel, M. F. Boehm, S. L. Merbs, J. Nathans, and K. Nakanishi. 1993. Cloning and expression of goldfish opsin sequences. *Biochemistry.* 32:208–214.

Kawamura, S., and S. Yokoyama. 1993. Molecular characterization of the red visual pigment gene of the American chameleon. *(Anolis carolinensis). FEBS Letters.* 323:247–251.

Lythgoe, J. N. 1972. List of vertebrate visual pigments. *In* Handbook of Sensory Physiology. H.J.A. Dartnall, editor. Vol. VII/1. Springer-Verlag, Berlin. 604–624.

Myers, E. W., and W. Miller. 1988. Optimal alignments in linear space. *Computations in Applied Biosciences.* 4:11–17.

Nathans, J., T. P. Plantanida, R. L. Eddy, T. B. Shows, and D. S. Hogness. 1986*a*. Molecular genetics of inherited variation in human color vision. *Science.* 232:203–232.

Nathans, J., D. Thomas, and D. S. Hogness. 1986*b*. Molecular genetics of human color vision: The genes encoding blue, green, and red pigments. *Science.* 232:193–202.

Needleman, S. B., and C. D. Wunsch. 1970. A general method applicable to the search for similarities in the amino acid sequence of two proteins. *Journal of Molecular Biology.* 48:443–453.

Okano, T., D. Kojima, Y. Fukada, Y. Shichida, and T. Yoshizawa. 1992. Primary structures of chicken cone visual pigments: vertebrate rhodopsins have evolved out of cone visual pigments. *Proceedings of the National Academy of Sciences, USA.* 89:5932–5936.

Probst, W. C., L. A. Snyder, D. I. Schuster, J. Brosius, and S. C. Sealfon. 1992. Sequence alignment of the G-protein coupled receptor superfamily. *DNA and Cell Biology.* 11:1–20.

Saitou, N., and M. Nei. 1987. The neighbor-joining method: a new method for reconstructing phylogenetic trees. *Molecular Biology and Evolution.* 4:406–425.

Schuler, G. D., S. F. Altschul, and D. J. Lipman. 1991. A workbench for multiple alignment construction and analysis. *Proteins: Structure, Function, and Genetics.* 9:180–190.

Schwartz, R. M., and M. O. Dayhoff. 1978. Matrices for detecting distant relationships. *In* Atlas of Protein Sequence and Structure. M. O. Dayhoff, editor. Vol. 5 (Supplement 3) National Biomedical Research Foundation, Washington, DC. 353–358.

Smith, T. F., M. S. Waterman, W. M. Fitch. 1981. Comparative biosequence metrics. *Journal of Molecular Evolution.* 18:38–46.

Wilkie, T. M. 1994. G protein-mediated signal transduction: evolution of the mammalian G protein α subunit multigene family. *Journal of General Physiology. In* Molecular Evolution of Physiological Processes. Society of General Physiologists Series. Vol. 49. D. M. Fambrough, editor. 249–270.

Wilson, C. J., and M. L. Applebury. 1993. Arresting G-protein coupled receptor activity. *Current Biology.* 3:683–686.

Yokoyama, R., and S. Yokoyama. 1990. Convergent evolution of the red- and green-like visual pigment genes in fish, *Astyanax fasciatus,* and human. *Proceedings of the National Academy of Sciences, USA.* 87:9315–9318.

Evolution of the G Protein Alpha Subunit Multigene Family

Thomas M. Wilkie* and Shozo Yokoyama‡

**Pharmacology Department, UT Southwestern Medical Center, Dallas, Texas 75235-9041; and ‡Department of Biology, Syracuse University, Syracuse, New York 13244*

G protein-mediated signal transduction systems have been identified in a diverse group of eukaryotic organisms, including yeast, plants, *Dictyostelium* and animals. G protein signaling components have been identified in many of these organisms, from the seven transmembrane domain receptors to distinct α, β and γ subunits of the heterotrimeric G protein and the intracellular effectors which they regulate. Their broad distribution and sequence conservation implies that genes encoding the components of G protein signaling evolved with early eukaryotes. Their subsequent proliferation among eukaryotic organisms provides an opportunity to study the coevolution of these interacting multigene families. We have focused our interests on G protein α subunits, which bind and hydrolyze GTP and interact with receptors and effectors. Gene structure and nucleotide sequence comparisons provided a comprehensive picture of Gα evolution. Sequence comparisons identified three major groups of Gα genes, termed the GPA, the Gα-I and Gα-II Groups. Gα genes within the three Groups have evolved at different rates. The GPA Group is primarily composed of Gα genes from fungi, plants, and slime mold. Within the Gα-I and Gα-II Groups, four classes of genes have been identified based upon sequence comparisons and functional similarities; G_i , G_q, G_{12}, and G_S. Members of all four classes are expressed in invertebrates and vertebrates but not in other eukaryotes, suggesting that this quartet evolved with metazoan progenitors.

Introduction

G protein signaling constitutes a fundamental mechanism of intercellular communication used by a diverse group of eukaryotes. G protein signaling pathways have several components, each encoded by distinct multigene families, including the seven transmembrane domain receptors, the heterotrimeric G protein α, β, and γ subunits and various intracellular effector proteins (Gilman, 1987). Heterotrimeric G proteins fulfill an essential role in coupling the reception of extracellular signals to intracellular second messenger cascades generated by effector proteins. In lower eukaryotes, G proteins are required to generate cellular responses to environmental cues during development (Dietzel and Kurjan, 1987; Miyajima, Nakafuku, Nakayama, Brenner, Miyajima, Kaibuchi, Arai, Kaziro, and Matsumoto, 1987; Firtel, van Haastert, Kimmel, and Devreotes, 1989; Pitt, Gunderson, Lilly, Pupillo, Vaughan, and Devreotes, 1990). G proteins have also been shown to be important in *Drosophila*

Address correspondence to Thomas M. Wilke, University of Texas Southwestern Medical Center, Dallas, TX 75235-9041.

TABLE I
Different G Protein α Subunit Genes and Their Data Source

Gene	Organism	Number of codons	Notation of cDNA*	Data source
Gnas	Human	394	$s\alpha_{Hs}$	Kozasa et al. (1988)
	Bovine	394	$s\alpha_{Bt}$	Nukada et al. (1986)
	Rat	394	$s\alpha_{Rn}$	Jones and Reed (1987)
	Mouse	394	$s\alpha_{Mm}$	Rall and Harris (1987)
	Frog	379	$s\alpha_{Xl}$	Olate et al. (1990)
	Drosophila	379	$s\alpha_{Dm}$	Quan et al. (1989)
	Schistosome	379	$s\alpha_{Sm}$	Iltzsch et al. (1992)
	Mouse	377	αs_{Mm}	Sullivan et al. (1986)
dgf	*Drosophila*	398	dgf_{Dm}	Quan et al. (1993)
Gnai1	Human	354	$i\alpha 1_{Hs}$	Bray et al. (1987)
	Bovine	354	$i\alpha 1_{Bt}$	Nukada et al. (1986)
	Frog	354	$i\alpha 1_{Xl}$	Olate et al. (1990)
	Rat	354	$i\alpha 1_{Rn}$	Jones and Reed (1987)
dgαi1	*Drosophila*	354	$i\alpha 1_{Dm}$	Provost et al. (1988)
Gnai2	Human	355	$i\alpha 2_{Hs}$	Beals et al. (1987)
	Mouse	355	$i\alpha 2_{Mm}$	Sullivan et al. (1986)
	Rat	355	$i\alpha 2_{Rn}$	Jones and Reed (1987)
Gnai3	Human	354	$i\alpha 3_{Hs}$	Beals et al. (1987)
	Rat	354	$i\alpha 3_{Rn}$	Jones and Reed (1987)
Gnao	Human	354	$o\alpha_{Hs}$	Tsukamoto et al. (1991)
	Bovine	354	$o\alpha_{Bt}$	van Meurs et al. (1987)
	Hamster	354	$o\alpha_{Ma}$	Hsuet al. (1990)
	Frog	354	$o\alpha_{Xl}$	Olate et al. (1989)
	Rat	354	$o\alpha_{Rn}$	Jones and Reed (1987)
	Nematode	354	$o\alpha_{Ce}$	Lochrie et al (1991)
dgol	*Drosophila*	354	$o\alpha 1_{Dm}$	Yoon et al. (1989)
dgoll	*Drosophila*	354	$o\alpha 2_{Dm}$	Schmidt et al. (1989)
Gnaz	Human	355	$z\alpha_{Hs}$	Fong et al. (1988)
	Rat	355	$x\alpha_{Rn}$	Matsuoka et al. (1988)
	Human	355	$x\alpha_{Hs}$	Matsuoka et al. (1990)
Gnat1	Bovine	350	$t1_{Bt}$	Medynski et al. (1985)
	Mouse	350	$t1_{Mm}$	Rapport et al. (1989)
	Frog	350	$t1_{Xl}$	Knox et al. (1993)
Gnat2	Bovine	354	$t2_{Bt}$	Lochrie et al. (1985)
Gnag	Rat	354	$gust_{Rn}$	McLaughlin et al. (1992)
Gna11	Mouse	358	$\alpha 11_{Mm}$	Strathmann and Simon (1990)
	Bovine	358	$\alpha 11_{Bt}$	Nakamura et al. (1991)
Gna12	Mouse	379	$\alpha 12_{Mm}$	Strathmann and Simon (1991)
Gna13	Mouse	377	$\alpha 13_{Mm}$	Strathmann and Simon (1991)
Gna14	Mouse	355	$\alpha 14_{Mm}$	Wilkie et al. (1991)
	Bovine	355	$\alpha 14_{Bt}$	Nakamura et al. (1991)
Gna15	Mouse	374	$\alpha 15_{Mm}$	Wilkie et al. (1991)
	Human	374	$\alpha 16_{HS}$	Amatruda et al. (1991)
Gnaq	Mouse	359	αq_{Mm}	Strathmann and Simon (1990)
dgq	*Drosophilia*	360	αq_{Dm}	Lee et al. (1991)
cta	*Drosophila*	457	cta_{Dm}	Parks and Wieschaus (1991)

TABLE I
(*continued*)

Gene	Organism	Number of codons	Notation of cDNA*	Data source
gpa-1	Nematode	357	$gpa1_{Ce}$	Lochrie et al. (1991)
gpa-2	Nematode	356	$gpa2_{Ce}$	Silva and Plasterk (1990)
gpa-3	Nematode	354	$gpa3_{Ce}$	Lochrie et al. (1991)
GPA1	Tomato	384	$\alpha1_{Le}$	Ma et al. (1991)
GPA1	*Arabidopsis*	383	$\alpha1_{At}$	Ma et al. (1990)
GPA1	*Candida*	430	$\alpha1_{Ca}$	Sadhu et al. (1992)
GPA	*Pombe*	407	α_{Sp}	Obara et al. (1991)
GPA1	Yeast	360	$\alpha1_{Sc}$	Nakafuku et al. (1987)
GPA2	Yeast	360	$\alpha2_{Sc}$	Nakafuku et al. (1988)
gna-1	*Neurospora*	353	$gna1_{Nc}$	Turner and Borkovich (1993)
gna-2	*Neurospora*	355	$gna2_{Nc}$	Turner and Borkovich (1993)
GPA1	Slime mold	354	$\alpha1_{Dd}$	Pupillo et al. (1989)
GPA2	Slime mold	348	$\alpha2_{Dd}$	Pupillo et al. (1989)
GPA4	Slime mold	345	$\alpha4_{Dd}$	Hadwiger et al. (1991)

*At (*Arabidopsis thaliana*), Bt (*Bos taurus*), Ca (*Candida albicans*), Ce (*Caenorhabditis elegans*), Dd (*Dictyostelium discoideum*), Dm (*Drosophila melanogaster*), Le (*Lycopersicon esculentum*), Hs (*Homo sapiens*), Ma (*Mesocricetus auratus*), Mm (*Mus musculus*), Nc (*Neurospora crassa*), Rn (*Rattus norvegicus*), Sc (*Saccharomyces cerevisiae*), Sm (*Schistosoma mansoni*), Sp (*Schizosaccaromyces pombe*), Xl (*Xenopus laevis*).

gastrulation (Parks and Wieschaus, 1991). In vertebrates and invertebrates, G proteins are essential components of sensory transduction systems and regulate a variety of other specialized functions in terminally differentiated cells (Gilman, 1987; Birnbaumer, 1990; Moss and Vaughan, 1991; Simon, Strathmann, and Ganta, 1991).

Specificity in the signaling pathways is achieved, in part, by the α subunits as a result of interactions with distinct receptors and effectors (recently reviewed by Conklin and Bourne, 1993). βγ subunits are also required for receptor activation of G proteins and subsequent regulation of some effectors (Whiteway, Hougan, Dignard, Thomas, Bell, Saari, Grant, O'Hara, and MacKay, 1989; Tang and Gilman, 1992; Camps, Hou, Sidiropoulos, Stock, Jakobs, and Gierschik, 1992; Katz, Wu, and Simon, 1992; reviewed in Iniguez-Lluhi, Kleuss, and Gilman, 1993). Remarkable specificity in G protein coupled receptor regulation of Ca^{2+} channel flux in GH3 cells was demonstrated by antisense oligonucleotide ablation of targeted α, β, and γ subunits (Kleuss, Hescheler, Ewel, Rosenthal, Schultz, Wittig, Kunkel, Roberts, and Zabour, 1991; Kleuss, Scherubl, Hescheler, Schultz, and Wittig, 1992, 1993). However, it is not known in these studies whether βγ contributes directly to signaling specificity through receptor and/or effector interactions or if specificity is mediated indirectly through α interactions. Although βγ subunits are clearly important in G protein signaling specificity, more is known about the genetics, biochemistry and pharmacology of Gα subunit specificity in signaling pathways. Thus, our focus is currently on the evolution of the Gα subunits.

The G protein α subunits are members of a superfamily of GTPases which also include *ras,* other small GTP binding proteins related to *ras* and translation elongation factors EF-Tu and EF-1α (Halliday, 1983; Lochrie, Hurley, and Simon, 1985).

This superfamily is distinguished by five amino acid motifs, termed G1 through G5, that are essential components of the GTP binding pocket and are critical to the GTPase activity exhibited by these proteins (Bourne, Sanders, and McCormick, 1990, 1991). Activity of the GTPases is regulated by a cycle of GTP binding and hydrolysis. The crystal structure of EF-Tu and ras reveal similarity both in their overall structure and conformational changes between the GTP and GDP bound proteins that presumably regulate their activity (Jurnak, 1985; de Vos, Tong, Milburn, Maniatis, Jancarik, Noguchi, Nishimura, Muira, Ohtsuka, and Kim, 1988; Tong, Milburn, De Vos, and Kim, 1989; Pai, Kabsch, Krengel, Holmes, John, and Wittinchofer, 1989). It is anticipated that this similarity between EF-Tu and ras will extend to the Gα subunits (Berlot and Bourne, 1992; Conklin and Bourne, 1993). The GTPase supergene family may have a common ancestor that evolved in early prokaryotes. EF-Tu and EF-1α are found ubiquitously in bacteria (as well as in mitochondria and chloroplasts) and eukaryotes, respectively. The small GTP-binding proteins have also been identified in eukaryotes and prokaryotes.

While other members of the GTPase superfamily are ubiquitous, G protein α, β and γ subunits have only been identified in eukaryotic organisms. The Gα subunits have been cloned from slime mold, fungi, plants and animals (see Table I). Less emphasis has been placed on identifying βγ subunits, but they are assumed to coexist in those organisms that express α subunits. G protein-coupled receptors and/or effectors have also been cloned from the same organisms that express Gα subunits. Biochemical characterization indicates that Gα subunits, β adrenergic-like receptors and adenylyl cyclase also exist in trypanosomes (Eisenschlos, Paladini, Molina, Vedia, Torres, and Flawia, 1986; de Castro and Oliveira, 1987; Fraidenraich, Pena, Isola, Lammel, Coso, Anel, Pongor, Baralle, Torres, and Flawia, 1994). Bacterial rhodopsin, cloned from *Halobacteria* in the Kingdom Archaebateria, is structurally related to the seven transmembrane domain receptors, but this protein is a proton pump that apparently does not couple to G proteins (Henderson and Schertler, 1990). Thus, G protein signaling appears to have evolved with eukaryotic organisms.

At present, 60 Gα sequences have been characterized in eukaryotes. In this paper, phylogenetic trees for these genes have been constructed and evolutionary rates of nucleotide substitutions were evaluated. The genomic structures of these genes were also compared to independently test the evolutionary relationships inferred from nucleotide sequence analysis.

Results and Discussion

Phylogenetic Tree

To study molecular phylogeny and evolutionary rates of G proteins, a total of 60 alpha subunit genes were considered (Table I). The nucleotide sequences of these genes were initially aligned by using a multiple alignment program in CLUSTAL V (Higgins, Bleasby, and Fuchs, 1992) and then adjusted manually to increase sequence similarity.

For the pairwise comparisons of the 60 genes, the number of nucleotide substitutions per site (d_{12}) was estimated by considering nucleotide sequence at the first and second position within each codon. This quantity was estimated by a formula $d_{12} = -(3/4) \ln [1 - (4/3)p]$, where p is the proportion of different nucleotides per site between two sequences (Jukes and Cantor, 1969). Using the amino acid sequences deduced from the α subunit genes, the number of amino acid replacements per site (K) was also estimated by $K = -\ln(1 - p)$, where p is the proportion of different amino acids per site between two polypeptides. For the present data set, d_{12} and K values ranged between 0.002–0.88 and 0.005–1.29, respectively. Topology and branch lengths of phylogenetic trees were evaluated by applying the neighbor-joining (NJ) method (Saitou and Nei, 1987) to the d_{12} and K values. The levels of bootstrap support for branches of the NJ trees were estimated by bootstrap analysis with 1000 replications (CLUSTAL V, Higgins et al., 1992).

Phylogenetic trees based on the d_{12} and K values differed somewhat in detail. The least complicated tree topology based on the d_{12} values is shown in Fig. 1 (this analysis did not include $\alpha 11_{Bt}$, $\alpha 14_{Bt}$, αq_{Dm} and $G\alpha olf_{Rn}$). Some low bootstrap supports are associated with different clusters. Nevertheless, it seems useful to classify the α subunit genes into three major groups: (*a*) GPA Group with 14 genes; (*b*) Gα-I Group consisting of G_o, G_t, G_x, and G_i genes; and (*c*) Gα-II Group consisting of G_q, G_{12}, and G_s genes (Fig. 1).

Among the 14 genes of the GPA Group, only four clusterings are highly reliable: (*a*) gpa-1_{Ce}, gpa-2_{Ce}, and gpa-3_{Ce}; (*b*) $\alpha 1_{Ca}$ and $\alpha 1_{Sc}$; (*c*) $\alpha 1_{At}$ and $\alpha 1_{Le}$; and (*d*) gna-2_{Nc} and α_{Sp}, each having a support value of 1.0 (see Fig. 1 and Table II). Otherwise, the bootstrap support values are often very low and the tree topology within the GPA group is unreliable. For example, in Fig. 1, gpa-1_{Ce}, gpa-2_{Ce}, gpa-3_{Ce}, $\alpha 1_{Ca}$, $\alpha 1_{Sc}$, $\alpha 1_{At}$, $\alpha 1_{Le}$, $\alpha 2_{Sc}$, $\alpha 1_{Dd}$, $\alpha 2_{Dd}$ and $\alpha 4_{Dd}$ form one cluster and gna-1_{Nc}, gna-2_{Nc}, and α_{Sp} form another with support values of only 0.18 and 0.58, respectively. When the NJ method was applied to the d_{12} values of all 60 genes, the 14 genes of the GPA Group were again distinguished into the same two groups, but they did not form a monophyletic group. When the K values of the 60 amino acid sequences were considered, the GPA Group was grouped into three polyphyletic clusters: (*a*) gpa-1_{Ce}, gpa-2_{Ce}, gpa-3_{Ce}, $\alpha 1_{Dd}$, and $\alpha 2_{Dd}$; (*b*) $P1\alpha_{Ca}$, $P1\alpha_{Sc}$, $P2\alpha_{Sc}$, $\alpha 1_{At}$, $\alpha 1_{Le}$, gna-1_{Nc}, gna-2_{Nc}, and α_{Sp}; and (*c*) $\alpha 4_{Dd}$.

Compared to the GPA Group, the evolutionary relationships of the genes within the Gα-I and Gα-II Groups are much clearer. It should be noted that when $\alpha 11_{Bt}$, $\alpha 14_{Bt}$, and αq_{Dm} are included in the analysis, they belong to the G_q group ($\alpha 11_{Bt}$ and $\alpha 14_{Bt}$ are closely related to $\alpha 11_{Mm}$ and $\alpha 14_{Mm}$, respectively, whereas αq_{Dm} is the most distantly related among the five genes in the G_q group) and $G\alpha olf_{Rn}$ belongs to the G_s group (Jones and Reed, 1989). The tree topology of Fig. 1 and those based on the d_{12} and K values for the 60 genes share important common features. That is, within the Gα-I Group, the clustering of four distinct groups are highly reliable: (*a*) G_o group; (*b*) G_t group (t1 and t2 genes and $Gust_{Rn}$); (*c*) G_x group (rat and human homologs); and (*d*) G_i group (α_{i1}, α_{i2}, and α_{i3} genes). The clustering of the Gα-I Group has a support value of 1.0 (Table II) and has a monophyletic origin (Fig. 1). Fig. 1 shows that G_i and G_x groups are most closely related, their common ancestor diverged from the ancestor of G_t group, and their common ancestor diverged from the ancestor of the G_o group. Because of rather low levels of bootstrap support (see Table II), the evolutionary relationships of the α subunit genes in the Gα-I Group cannot be firmly established. In fact, the NJ tree based on the K values for 60 amino acid sequences shows that the common ancestor of G_i and G_x genes diverged from the ancestor of G_o genes, and their common ancestor diverged from the ancestor of the G_t genes.

Among the Gα-II Group, three clusters can be distinguished: (*a*) G_q group (αq, α11, α14, α15, and α16 genes); (*b*) G_{12} group (α12 and α13 genes and cta_{Dm}); and (*c*)

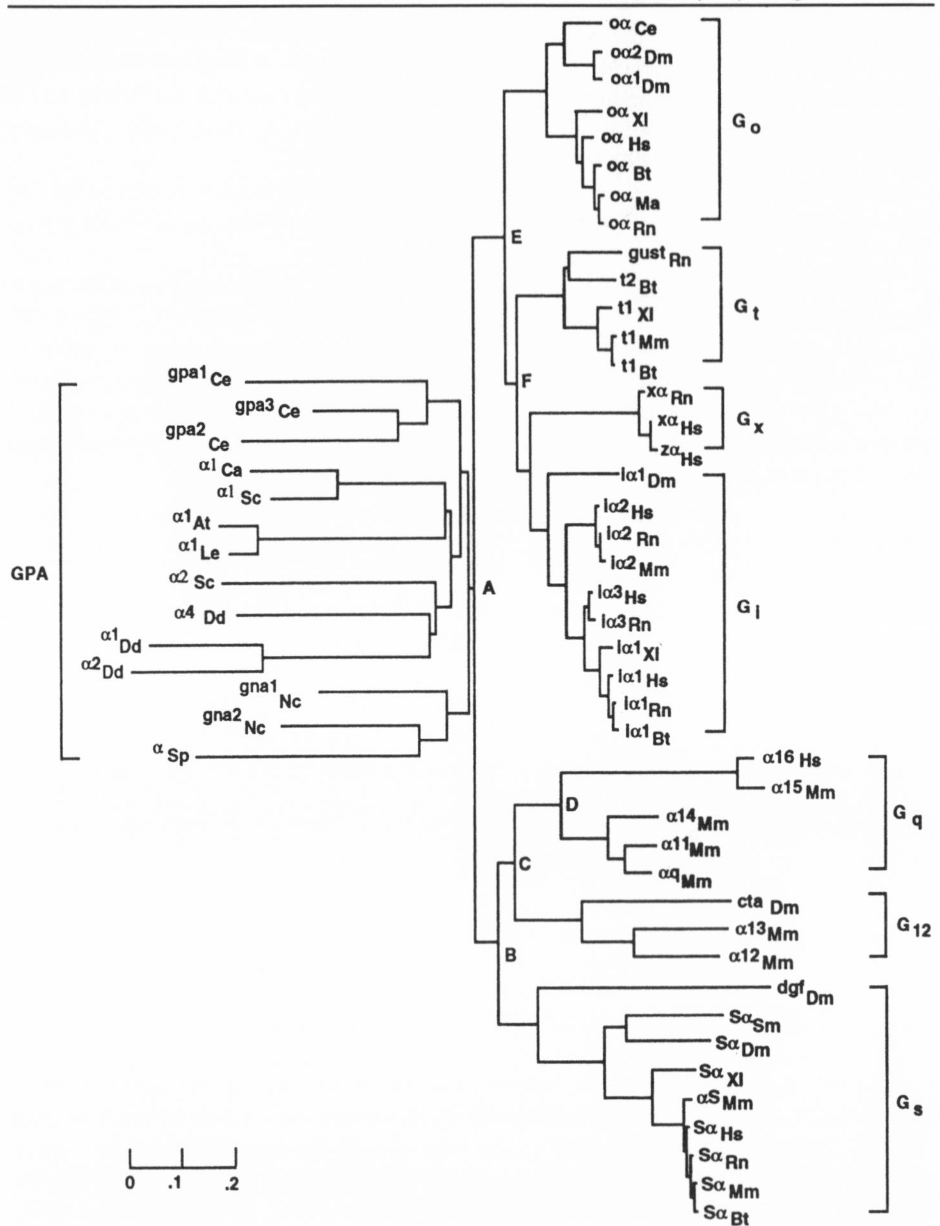

Figure 1. Phylogenetic tree of G protein α subunit genes ($\alpha 11_{Bt}$, $\alpha 14_{Bt}$, αq_{Dm} and αolf_{Rn} were not included). Branch lengths were estimated by the d_{12} values by means of the NJ method of Saitou and Nei (1987).

G_s group (sα and α_s genes and $Gf\alpha_{Dm}$). Fig. 1 shows that the G_{12} and G_q groups are most closely related and their common ancestor diverged from the ancestor of the G_s genes. The same tree topology is also obtained when the NJ method is applied to the K values for the 60 amino acid sequences. In Fig. 1, clustering of the G_q, G_{12}, and G_s groups have support values of 0.63, 1.0, and 0.56, respectively (Table II). When all 60 α subunit genes are considered, the respective support values improve dramatically (Table II). Thus, the inclusion of the $\alpha 11_{Bt}$, $\alpha 14_{Bt}$, and αq_{Dm} in the analysis makes the clustering of G_q genes and of G_s genes more reliable. However, the support value for the clustering of the Gα-II Group (which includes G_s, G_{12}, and G_q) is only ~0.6 (see Table II) and their clustering has not been well established.

This evolutionary relationship of the Gα-I and Gα-II groups is very similar to that obtained by Yokoyama and Starmer (1992) by considering 34 α subunit genes. However, the tree topology of Fig. 1 is drastically different from those in Simon et al. (1991), Strathmann and Simon (1991) and Wilkie, Gilbert, Olsen, Chen, Amatruda, Korenberg, Trask, Dejong, Reed, Simon, Jenkins, and Copeland (1992), where the G_i and G_q groups were most closely related, and their common ancestor and those of the G_{12} and G_s groups diverged at about the same time. The difference seems to have occurred because a constant rate of amino acid replacement was assumed in their analyses, whereas the present analysis finds that the Gα subunits evolved at different rates (for more details, see Yokoyama and Starmer 1992).

TABLE II
Bootstrap Support Levels for Major Gα Subunit Clusters

Cluster	57 sequences	60 sequences	60 sequences
G_o	1.0	1.0	1.0
G_t	1.0	1.0	1.0
G_x	1.0	1.0	1.0
G_i	0.89	0.99	0.99
G_x and G_i	0.77	0.76	0.60
G_x, G_i and G_t	0.72	0.76	N.A.*
Gα-I	1.0	1.0	0.99
G_q	0.63	1.0	1.0
G_{12}	1.0	1.0	1.0
G_s	0.56	0.92	0.90
G_q and G_{12}	0.55	0.93	0.49
Gα-II	0.66	0.59	0.64

*G_t group is most distantly related within the Gα-I group.

There is agreement that vertebrates and invertebrates express four classes of genes, G_s, G_i, G_q and G_{12}, based upon comparisons in cDNA sequence, gene structure and functional properties. The function of the α subunits within a given class are often similar. For example, α subunits of the G_s class (*Drosophila* and rat Gαs and rat Gαolf) activate adenylyl cyclase (Gilman, 1987; Jones and Reed, 1989; Quan et al., 1989) while α subunits of the G_q class mediate pertussis toxin-resistant activation of phospholipase Cβ (Taylor et al., 1991; Smrcka et al., 1991; Wu et al., 1992). G_i class α subunits regulate a more diverse set of effectors but their functions require fatty acid acylation (Taussig et al., 1993) and, with the exception of Gαz, each is inhibited by pertussis toxin-catalyzed ADP-ribosylation (Gilman, 1987; Birnbaumer et al., 1990). The signaling properties of the G12 class of α subunits have not been determined. Members of each of these four classes are probably exclusively found within metazoan organisms. Lower eukaryotes also express Gα subunits that regulate adenylyl cyclase or phospholipase C (Firtel, 1991), but these sequences appear to be distinct from any of the mammalian Gα class genes.

Branch lengths. From Fig. 1, it appears that the branch lengths from the common ancestor to the Gα-I and Gα-II Groups are different. Accordingly, branch lengths are calculated by taking a series of averages of lengths for different groups (Table III). In Table III, standard errors were computed from $[9p(1-p)/\{3-4p)^2 \times 704\}]^{1/2}$ and $p = (3/4) \times [1 - \exp\{-(4/3)\,d_{12}\}]$ (see Nei, 1987). The average number of nucleotide substitutions per site for branches A-Gα-I Group and A-Gα-II Group is 0.23 and 0.44, respectively, showing that significantly fewer nucleotide changes have accumulated in the former group than in the latter group at the first and second positions of codons. Similar tests show that differences between other comparable branches are not statistically significant (Table III). Quantitatively the same conclusion was reached when the d_{12} and K values for the 60 genes were considered.

Fig. 1 also shows that the length for D-αq_{Mm}, $\alpha 11_{Mm}$, and $\alpha 14_{Mm}$ (denoted as the G_q-I group) appears to be much shorter than that for D-$\alpha 15_{Mm}$ and $\alpha 16_{Hs}$ (G_q-II group). Branch lengths between the G_q-I and other groups have been similarly compared (Table IV). In Table IV, we can see that G_q-I genes have accumulated a

TABLE III
Branch Lengths Measured by the Number of Nucleotide Substitutions per Site (d_{12}) at the First and Second Positions of Codons

Lower branch	$d_{12} \times 100$	Upper branch	$d_{12} \times 100$
A–$G_q/G_{12}/G_s$	43.8 ± 3.18*	*A*–$G_o/G_t/G_x/G_i$	22.7 ± 2.02*
B–G_s	42.3 ± 3.10	*B*–G_q/G_{12}	36.0 ± 2.75
C–G_{12}	34.1 ± 2.65	*C*–G_q	32.0 ± 2.53
E–$G_t/G_x/G_i$	18.8 ± 1.80	*E*–G_o	16.5 ± 1.67
F–G_x/G_i	18.3 ± 1.77	*F*–G_t	15.6 ± 1.61

Note: see Fig. 1 for branch points *A*, *B*, *C*, *E*, and *F*. Standard errors were computed from $[9p(1-p)/\{(3-4p)^2 \times 704\}]^{1/2}$ and $p = (3/4) \times [1 - \exp(-3d/4)]$ (see Nei, 1987). $\alpha 11_{Bt}$, $\alpha 14_{Bt}$, αq_{Dm}, and $G\alpha olf_{Rn}$ were not included in the analysis.
*Difference in branch lengths is significant at the 1% level.

significantly larger number of nucleotide changes than Gα-I Group genes. In contrast, significantly fewer changes have accumulated in G_q-I genes than in G_q-II, G_{12}, and G_s genes. The longer length of branch D-G_q-II compared to the D-G_q-I branch may be explained by relaxed selective constraints for the G_q-II group relative to the G_q-I group after the gene duplication event at branch point D. In this regard, it is interesting that Gα15 and Gα16 are expressed specifically in hematopoietic cells, whereas Gαq, Gα11 and Gα14 are widely or ubiquitously expressed (Wilkie et al., 1991). Gα15 and Gα16 may interact with a more restricted set of receptors and effectors and function in a more permissive cell lineage than the other members of the G_q class, thus providing the relaxed selective constraints that would allow their faster rate of evolution. Another possibility is that *Gna15* (the gene that encodes Gα15 and Gα16 in mice and humans, respectively) has been recruited for novel functions in mouse and human hematopoietic cells that are driving evolutionary change of this gene at an accelerated rate relative to other mammalian G_q class genes.

Evolutionary rate. Yokoyama and Starmer (1992) studied the evolutionary rates of nucleotide substitution for the G protein α subunit genes during the last 75 million years of mammalian evolution. Because the α subunit genes from a larger number of organisms have now been characterized (see Table I), the evolutionary rates of the α subunit genes can be evaluated since the divergence between arthropods and mammals.

To estimate the rate of amino acid replacements per site per year (r), average branch lengths were calculated which were subsequently divided by the appropriate divergence times among groups (Table V). Divergence times between *Drosophila* and mammals and between frog and mammals were taken as 700 and 300 million years ago, respectively (Dayhoff, 1978; see also Figure 2.2 in Nei, 1987). Table V shows that the evolutionary rates for the Gα-I Group (G_o, G_t, G_x, and G_i groups) tend to be lower than those for the Gα-II Group (G_q, G_{12}, and G_s groups). That is, the r values are $0.04–0.35 \times 10^{-9}$/site/year for the Gα-I Group, whereas those for the Gα-II Group are $0.21–0.45 \times 10^{-9}$/site/year. Among the Gα-II Group, G_q has the slowest evolutionary rate. For comparison, the most highly conserved Gα genes have evolved somewhat faster than histones and triosephosphate isomerase but more

TABLE IV
Comparison of Branch Lengths Between the G_q and Other Groups

Gq-I branch	$d_{12} \times 100$	Out group branch	$d_{12} \times 100$
$\alpha q_{Mm}/\alpha 11_{Mm}/\alpha 14_{Mm}$	29.8 ± 2.41*	$G_o/G_t/G_x/G_i$	22.7 ± 2.02*
$\alpha q_{Mm}/\alpha 11_{Mm}/\alpha 14_{Mm}$	26.1 ± 2.21‡	G_s	42.3 ± 3.10‡
$\alpha q_{Mm}/\alpha 11_{Mm}/\alpha 14_{Mm}$	23.2 ± 2.05‡	G_{12}	34.1 ± 2.65‡
$\alpha q_{Mm}/\alpha 11_{Mm}/\alpha 14_{Mm}$	15.5 ± 1.61‡	$\alpha 15_{Mm} \pm /\alpha 16_{Mm}$	33.1 ± 2.59‡

Note: standard errors were computed from $[9p(1-p)/\{(3-4p)^2 \times 704\}]^{1/2}$ and $p = (3/4) \times [1 - \exp(-3d/4)]$ (see Nei, 1987). $\alpha 11_{Bt}$, $\alpha 14_{Bt}$, and αq_{Dm} were excluded from the analysis.
*Difference in branch lengths is significant at the 5% level.
‡Difference in branch lengths is significant at the 1% level.

slowly than creatine kinase (R. Doolittle, personal communication). Substitution rates in the most rapidly evolving mammalian Gα genes fall between the hemoglobin α and β genes.

The r values cited above for Gα sequences in the Gα-I and Gα-II Groups are much higher than the number of nonsynonymous substitutions per site per year in Yokoyama and Starmer (1992). This is probably because the evolutionary rates for the G protein α subunits (and α subunit genes) were much faster before the mammalian radiation. However, in order to establish the differences in evolutionary tempo of G protein α subunits, sequence data from additional other species are needed.

Comparative gene structure. The gene structures of 20 Gα subunits isolated from plants, fungi, a slime mold and a variety of animals have been determined (Fig. 2). Most of these genes are in the G_i class (Gα-I Group in Fig. 1), including two *Drosophila* and seven mammalian genes. The G_q class is represented by the *Drosophila* dgq gene, the G_s class by human *Gnas* and, based on similarities in nucleotide sequence and gene structure, *Drosophila* dgf (Figs. 1 and 2). The effectors regulated by dgf have not been identified, but based on its resemblance to Gαs, these proteins may interact with a similar set of effectors. The only G_{12} class gene characterized to date is *Drosophila concertina* (*cta*). Among the GPA Group, the *Arabidopsis Gpa1* gene structure is distinct whereas the three *C. elegans gpa* genes appear to have related structures, as do the three *Dictyostelium* genes.

By virtue of the high amino acid sequence conservation between Gα subunits, intron/exon boundaries can be identified relative to the consensus Gα sequence. The intron positions within each Gα gene will be referred to by their position within the consensus sequence shown in Fig. 3. Comparison of the 20 Gα genes indicates that three intron boundaries occur following exactly the same codon in the sequence alignment of almost all Gα genes, despite the evolutionary distance between these organisms. These highly conserved intron positions occur near amino acid 55, 120, and 175 in the consensus sequence (introns *a, d,* and *f,* respectively, in Fig. 2). The intron near position 225 (intron *g*) is nearly as well conserved and is present in most G_i and G_q class genes, human *Gnas* and two *C. elegans* genes. The introns near positions 270 and 330 (introns *h* and *i*) are found only in the G_i, G_q and *C. elegans*

TABLE V
Evolutionary Rates of Amino Acid Replacement Per Site Per Year ($\times 10^{-9}$)

Group	Amino acid sequences compared	r
G_{t1}	$t1_{Xl} - t1_{Mm}, t1_{Bt}$	0.35 ± 0.059
G_i	$i\alpha1_{Dm} - i\alpha2_{Rn}, i\alpha2_{Mm}, i\alpha2_{Hs}, i\alpha3_{Rn}, i\alpha3_{Hs}, i\alpha1_{Xl}, i\alpha1_{Hs}, i\alpha1_{Rn}, i\alpha1_{Bt}$	0.18 ± 0.028
G_{i1}	$i\alpha1_{Xl} - i\alpha1_{Hs}, i\alpha1_{Rn}, i\alpha1_{Bt}$	0.04 ± 0.020
G_o	(1) $o\alpha_{Xl} - o\alpha_{Bt}, o\alpha_{Rn}, o\alpha_{Ma}, o\alpha_{Hs}$	0.15 ± 0.038
	(2) $o\alpha1_{Dm}, o\alpha2_{Dm} - o\alpha_{Bt}, o\alpha_{Rn}, o\alpha_{Ma}, o\alpha_{Xl}, o\alpha_{Hs}$	0.15 ± 0.025
G_s	(1) $S\alpha_{Dm} - S\alpha_{Xl}, \alpha S_{Mm}, S\alpha_{Hs}, S\alpha_{Rn}, S\alpha_{Mm}, S\alpha_{Bt}$	0.24 ± 0.033
	(2) $S\alpha_{Xl} - \alpha S_{Mm}, S\alpha_{Hs}, S\alpha_{Rn}, S\alpha_{Mm}, S\alpha_{Bt}$	0.31 ± 0.056
G_q	$\alpha q_{Dm} - \alpha14_{Bt}, \alpha14_{Mm}, \alpha q_{Mm}, \alpha11_{Bt}, \alpha11_{Mm}$	0.21 ± 0.030
G_{12}	$cta_{Dm} - \alpha12_{Mm}, \alpha13_{Mm}$	0.45 ± 0.046

genes. Other intron boundaries are characteristic of one class of Gα genes, such as positions 75 and 150 (introns *c* and *e*) in the G_s class genes, or are particular to a specific gene. Among the three *Dictyostelium* genes, the position of the two introns in *α4,* near positions 120 and 175 (introns *d* and *f*), is conserved. *Dictyostelium α1* and *α2* have one intron each near position 120 that interrupt the coding sequence at a different nucleotide within this conserved codon. The *Saccharomyces cerevisiae GPA1* and *GPA2* and *S. pombe GPA1* genes lack introns (Nakafuku et al., 1987; 1988; Obara et al., 1991), as do most genes in these organisms, and are not included in Fig. 2.

The evolutionary relationships between the Gα cDNA sequences depicted in Fig. 1 is reflected in the structure of the Gα genes. For example, the *Dictyostelium, Arabidopsis, Drosophila cta* and mammalian G_i and G_s classes have clearly distinct gene structures. However, this generalization has several interesting exceptions; in particular, the similarity between the *Drosophila dgq* gene, the *C. elegans* genes, and the G_i class genes. The three *C. elegans gpa* genes have two introns near positions 270

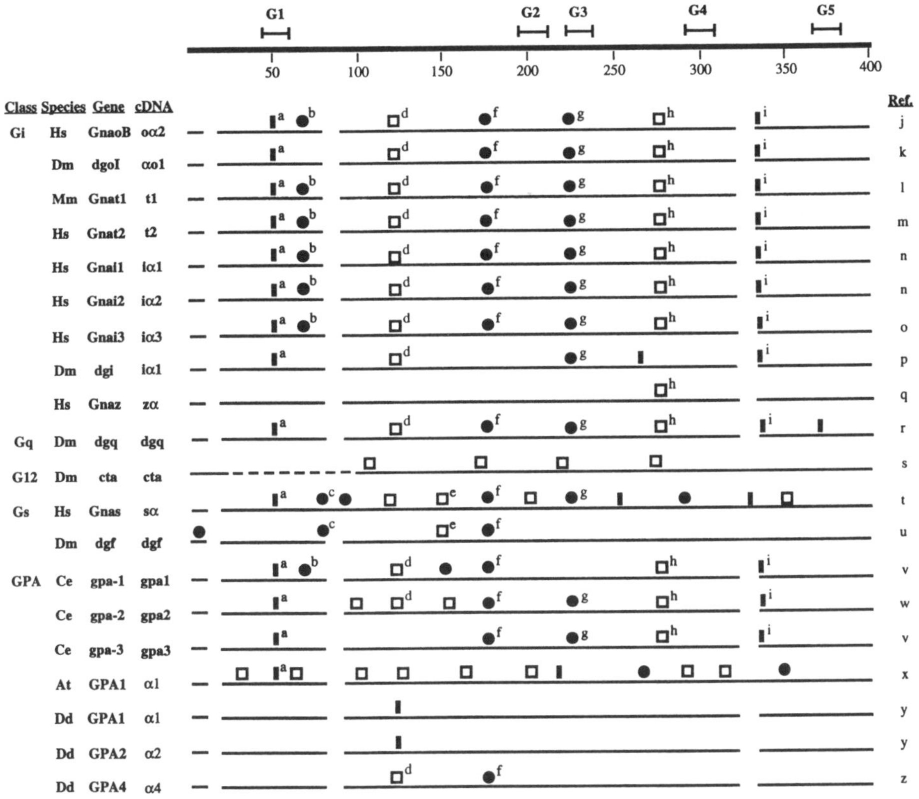

G1 G2 G3 G4 G5

50 100 150 200 250 300 350 400

Class	Species	Gene	cDNA	Ref.
Gi	Hs	GnaoB	oα2	j
	Dm	dgoI	αo1	k
	Mm	Gnat1	t1	l
	Hs	Gnat2	t2	m
	Hs	Gnai1	iα1	n
	Hs	Gnai2	iα2	n
	Hs	Gnai3	iα3	o
	Dm	dgi	iα1	p
	Hs	Gnaz	zα	q
Gq	Dm	dgq	dgq	r
G12	Dm	cta	cta	s
Gs	Hs	Gnas	sα	t
	Dm	dgf	dgf	u
GPA	Ce	gpa-1	gpa1	v
	Ce	gpa-2	gpa2	w
	Ce	gpa-3	gpa3	v
	At	GPA1	α1	x
	Dd	GPA1	α1	y
	Dd	GPA2	α2	y
	Dd	GPA4	α4	z

Figure 2. *Gna* gene structure. The consensus sequence of twenty G protein α subunits was obtained from an amino acid sequence alignment. The consensus sequence is depicted by a thick solid line above the count of amino acid residues within the alignment. Relative position of the amino acids within the GTP binding domains G1 through G5 are shown above. Each gene is identified by class, species, gene, and name of cDNA according to Fig. 1 and Table I. Thin solid lines depict the individual genes; gaps in each line represent gaps in the sequence alignments. Intron positions are indicated above each gene. Introns that are positioned after the first, second, or third nucleotide within a codon are identified by a thick vertical line, solid circle, or open square, respectively. The superscript letters *a* through *i* identify those introns that occupy exactly the same position in the alignment in more than one gene. Introns of unique position within the alignment are not lettered. The dashed line indicates the unanalyzed portion of the *Drosophila concertina* gene. References are: *j,* Tsukamoto et al., 1991; *k,* de Sousa et al., 1989, Yoon et al., 1989, Thambi et al., 1989; *l,* Rapport et al., 1989; *m,* Kubo et al., 1991; *n,* Itoh et al., 1988; *o,* Kaziro et al., 1991; *p,* Provost et al., 1988; *q,* Matsuoka et al., 1988; *r,* Lee et al., 1990; *s,* M. Wayne, personal communication; *t,* Kozasa et al., 1988; *u,* Quan et al., 1993; *v,* Lochrie et al., 1991; *w,* Silva and Plasterk, 1990; *x,* Ma et al., 1990; *y,* M. Brandon, personal communication; *z,* Hadwiger et al., 1991.

and 330 (introns *h* and *i,* and intron *b* in *gpa-1*) which are otherwise found only in the G_i class genes, *Drosophila dgq* and the mammalian G_q class genes (*Gna11* and *Gna15;* T. M. Wilke, unpublished observation). Furthermore, the *Drosophila dgq* gene is even more similar to the mammalian G_i class genes than is the *Drosophila Gnai* homolog; six intron positions are identical although *Drosophila dgq* has a unique intron near position 270 and is missing intron *b* near position 65, as are the

Drosophila dgo and *dgi* genes. The similar structure of the G_i class genes, Drosophila *dgq* and the *C. elegans* genes could indicate that they share a common ancestor in a monophyletic grouping, in contrast to the relationship depicted in Fig. 1 that splits the G_i, G_q and *C. elegans gpa* genes into separate groups. For example, *Drosophila dgq* gene structure is clearly more related to mammalian G_i class genes than to *Gnas*. In

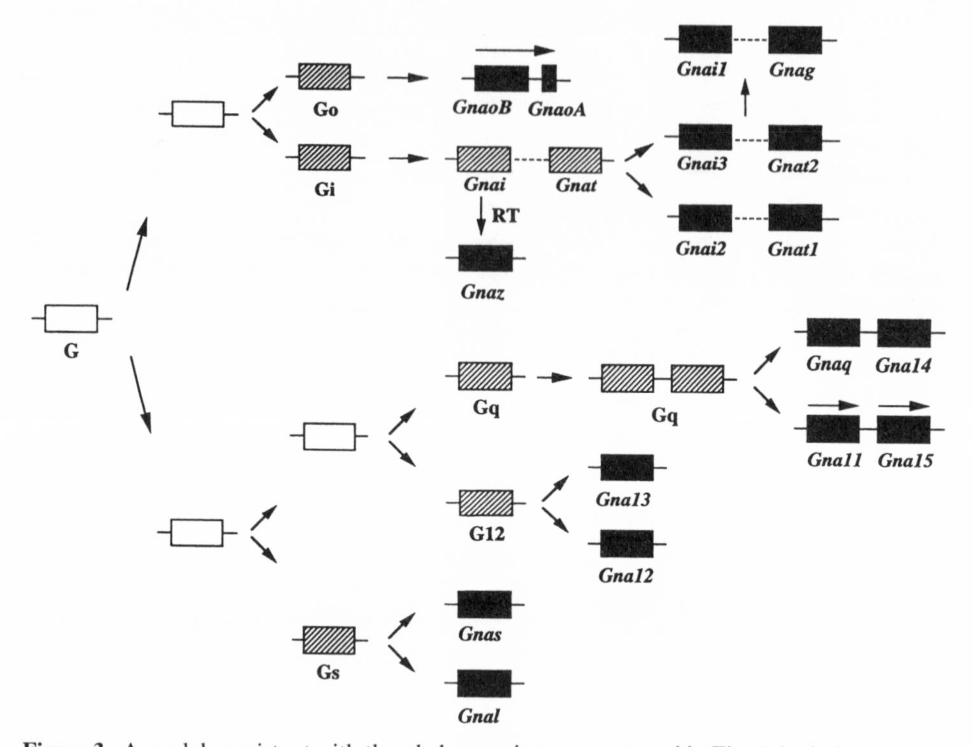

Figure 3. A model consistent with the phylogenetic tree presented in Fig. 1 depicting some of the possible evolutionary events that generated the vertebrate Gα subunit multigene family. The G_q and G_i class genes are predicted to have independently undergone tandem duplication, although *Gnao, Gnai* and *Gnat* genes may be descendent from a common tandemly duplicated progenitor. Solid boxes represent the genes. The dotted line between *Gnai* and *Gnat* genes indicate that the distance between the genes is unknown and may contain intervening genes. Arrows above the genes point towards the 3′ end of the transcription unit where their relative orientation is known. The short filled box above *Gnao*A depicts alternatively spliced exons. RT indicates that *Gnaz* may be derived from a *Gnai* progenitor through reverse transcription. Open and hatched boxes represent earlier and later progenitor genes, respectively. G_i, G_q, G_{12} and G_s represent progenitors of the four classes of G protein α subunits that are expressed in vertebrates and invertebrates. G symbolizes an ancient progenitor gene.

addition, the cDNA sequence of the three *C. elegans gpa* genes exhibit two features that are characteristic of G_i class genes; firstly, the consensus sequence for myristoylation at the amino terminus (MGXXXS) and secondly, with the exception of *gpa-2,* the cysteine four amino acids from the carboxy terminus that is ADP-ribosylated by pertussis toxin in G_i class genes. An alternative explanation for the similarity between the G_i, G_q and *C. elegans gpa* genes, consistent with the phylogenetic tree in Fig. 1, is that these genes evolved more slowly than the G_s class genes, at least until relatively recently, and therefore retained more of an ancestral sequence and gene structure.

The structure of *Drosophila cta* is unique among Gα genes (Fig. 2) although the positions of the last three introns are only slightly removed from highly conserved introns. The *Drosophila melanogaster cta* gene resides in a region of heterochromatic DNA. It is subject to reduced recombination (including gene conversion between related genes) and is apparently under relaxed selection (M. Wayne, personal communication). Therefore, *cta* may not be characteristic of the G_{12} class gene structure. It will be interesting to compare *cta* in *Drosophila melanogaster* to other *Drosophila* species and to mammalian *Gna12* and *Gna13* to determine if the drift in intron positions in *cta* is a recent evolutionary event.

In summary, the evolution of the Gα genes is complex and may exhibit the remnants of several mechanisms that generate genetic diversity. The position of several Gα introns are conserved among genes of different classes and in divergent eukaryotic organisms. However, none of the Gα intron positions are related to introns among other genes within the GTPase superfamily that are expressed in eukaryotes and prokaryotes. Conservation of intron position in other genes of ancient origin that have been characterized in plant, fungi, and animals, most notably triosephosphate isomerase and globin (Marchionni and Gilbert, 1986; Tittiger, Whyard, and Walker, 1993; Go, 1981), has been interpreted to support the notion that introns must have arisen early in evolution, perhaps in the RNA precursors of primordial genes (Gilbert, Marchionni, and McKnight, 1986). Alternative interpretations have been presented to explain both the origin of spliceosomal introns (Cavalier-Smith, 1991) and the apparent conservation of intron position within genes (Dibb and Newman, 1989; Wistow, 1993). Unlike intron positions in triosephosphate isomerase and globin genes, most intron positions are not conserved among the Gα genes from plants, invertebrates and vertebrates (for example, compare *Arabidopsis GPA1,* human *Gnas* and the G_i class genes from *Drosophila* and mammals, Fig. 3). The simplest explanation for the dissimilarity in intron position in many Gα genes is that introns were both acquired and lost during eukaryotic evolution. A similar conclusion was reached following comparison of serine protease and α and β actin and tubulin genes (Rogers, 1985; Dibb and Newman, 1989). Introns that are widely conserved within a multigene family may have been aquired early in eukaryotic evolution while the less well conserved introns may have appeared more recently. In addition, some sequences may favor intron aquisition during evolution of homologous genes in separate eukaryotic lineages. For example, within the Gα multigene family, three genes from three different phylogenetic groups (human *Gnas, C. elegans gpa2* and *Arabidopsis GPA1*) have introns near residue 80 that are each in a slightly different position. Alternatively, a mechanism that allows intron slippage may have operated on an intron position that was originally conserved in a common progenitor. *Drosophila concertina* is particularly interesting in this regard because each of its four known introns are near to introns found in either the G_i and G_s class genes.

Tandem gene duplication. The chromosomal location of all known murine Gα genes has been identified, and most Gα genes have also been mapped in humans (Wilkie et al., 1992). The mapping results revealed that members of the G_s and G_{12}

classes segregated as unlinked genes in mice and humans (i.e., they were not closely linked to any other Gα subunit gene). In contrast, most G_i and G_q class members segregated as pairs of closely linked genes. For example, mouse *Gnai3* and *Gnat2* cosegregated on chromosome 3, *Gnai2* and *Gnat1* cosegregated on chromosome 9, *Gnai1* and *Gnag* cosegregated on chromosome 5 (N. Copeland, N. Jenkins, and R. Margolskee, personal communication), *Gna11* and *Gna15* cosegregated on chromosome 10, and *Gnaq* and *Gna14* cosegregated on chromosome 19 (Wilkie et al., 1992). Analysis of the human and mouse *Gna11* and *Gna15* genes cloned from cosmid and lambda genomic libraries, respectively, showed that these genes reside within 100 kb in a tandem head-to-tail array (Wilkie et al., 1992). Since *Gnaq* and *Gna14* are closely related to *Gna11* and *Gna15* and are also closely linked in mouse, we presume that they are also tandemly duplicated.

Is it possible that G_q and G_i class genes were derived from common, tandemly duplicated progenitor genes? *Gnaz* and *Gnao* are apparently the only G_i or G_q class genes that are not paired with another gene in the same class. *Gnaz* is the most divergent member of the G_i class (Fig. 1). The intron organization in *Gnaz* is unique, with one intron in the 5′ untranslated region and a second intron within the coding region in place of the eight introns that are conserved in all other mammalian G_i class genes (Kaziro et al., 1991). The structure of *Gnaz* resembles an incompletely processed pseudogene suggesting that it may have been derived by reverse transcription. *Gnaz* was presumably functional before divergence of rodents and primates because its amino acid sequence is highly conserved in rats and humans (Fong et al., 1988; Matsuoka, 1988).

Gnao presents the most provocative exception to the model of tandemly duplicated G_i and G_q class genes. The complete gene structure of human *Gnao* resides within approximately 100 kb (Tsukamoto et al., 1991). Interestingly, in at least three mammalian species, *Gnao* encodes two functionally distinct proteins, GαoA and GαoB (Tsukamoto et al., 1991; Strathmann et al., 1991; Hsu et al., 1990; Kleuss et al., 1991). The amino acids encoded in the first six exons are common to both Gαo proteins but the remainder of the carboxy termini are encoded by two different pairs of exons that are selected through alternative splicing (Tsukamoto et al., 1991). This unusual gene organization may have derived from a pair of tandemly duplicated genes. Deletion of the transcription termination signal of the upstream gene, along with the amino terminal two-thirds of the downstream gene, could have generated the present *Gnao* gene structure. The *Drosophila Gnao* homologue also undergoes alternative splicing, but in this case two transcripts, containing individual translation start sites in their respective first coding exons, initiate from distinct promoters and then splice to a common acceptor site at the beginning of exon 2 (Yoon et al., 1989). The structure of *Drosophila Gnao* could also have resulted from internal deletion of a tandem gene pair, in this case removing nearly the entire coding region as well as the transcription termination signals of the upstream gene. Alternatively, duplication of these different *Gnao* exons could have occurred independently in the *Drosophila* and mammalian lineages.

If *Gnao* and G_q class genes are descendent from a common tandemly duplicated progenitor, then other G_i class genes must share this ancestry. However, tandem duplication of the Gi class genes may not have been rigorously conserved in invertebrates. For example, *Drosophila* express a *Gnai* homolog, but not transducin, the gene that is closely linked to *Gnai* in mammals. It is unlikely that a functional invertebrate transducin awaits discovery, as invertebrate vision is thought to be mediated by Gαq coupling to phospholipase C in contrast to transducin regulation of cGMP phosphodiesterase (Fain and Matthews, 1990; Ranganathan, Harris, and Zucker, 1991). The possibility remains that a transducin homolog with a different physiological function, such as gustducin, awaits discovery in *Drosophila.*

The Gnai/transducin gene pairs that cosegregated in mice were shown to colocalize on human chromosomes by fluorescence *in situ* hybridization (Wilkie et al., 1992). These results are consistent with the notion that the three *Gnai* and transducin gene pairs are tandemly duplicated and were derived from a common progenitor gene pair. However, the physical distance between the *Gnai* and transducin gene pairs is unknown and may be as great as 2.5 Mb in mouse and 4 Mb in human. Thus, it is possible that duplication of the G_i class genes occurred in the vertebrate lineage independently from tandem duplication in the G_q class genes or *GnaoA* and *GnaoB.* Linkage of *Gnai* and transducin genes might even result from a fortuitous chromosomal transposition event in a common ancestor to rodents and primates. Identification and characterization of YAC or cosmid contigs containing at least one of the *Gnai* and transducin gene pairs should help to resolve whether these genes are also tandemly duplicated.

Evolution of mammalian Gna genes. Multigene families are thought to be generated by a number of different mechanisms including (*a*) genome duplication or tetraploidization (for a review see Nadeau, 1991) (*b*) tandem gene duplication and (*c*) reverse transcription, a process that is likely responsible for the formation of intronless genes and processed pseudogenes. There is considerable evidence that the eukaryotic genome has undergone multiple genome duplication events with the most recent duplication event occurring ~300 million years ago, long before the divergence of the lineages leading to mouse and man. A simple model to explain the chromosomal mapping results is that the vertebrate Gα multigene family was created by at least one gene duplication event, followed by one tandem duplication that occurred in the progenitor of the G_i and G_q genes after its divergence from the G_s and G_{12} progenitors and subsequent duplications of the genome and smaller chromosomal segments (Wilkie et al., 1992). A more complex model is shown in Fig. 3 which postulates not one, but three independent tandem gene duplication events in the *Gnao, Gnai* and G_q lineages. This is a working model that is consistent with the evolutionary relationships depicted in Fig. 1.

Conclusions. Evolution of the G protein α subunit genes is complex. The Gα multigene family may have evolved in the common ancestor of fungi, plants, animals, slime molds and ciliates, after their divergence from simpler eukaryotes. Three groups of Gα sequences, the GPA, Gα-I and Gα-II Groups, were identified in a phylogenetic tree (Fig. 1). Gα subunits are at least 35% identical in amino acid sequence between Groups, suggesting that the invariant residues are critical for functions common to all Gα proteins. The Gα-I and Gα-II Groups are composed of four classes of Gα proteins that are expressed in metazoan organisms, termed the G_s, G_i, G_q, and G_{12} classes, based on sequence comparisons and effector interactions. Sequence comparison between these four classes of Gα proteins have identified residues that contribute to signaling specificity (Itoh and Gilman, 1991; Berlot and Bourne, 1992). Thus, phylogenetic trees can contribute to understanding both protein function and the evolution of multigene families.

Alignment availability. The alignment of the amino acid sequences of 60 G protein α subunits is available from the EMBL file server by sending the email message "get align: DS15369.dat" to NetServ@EMBL-Heidelberg.DE or by anonymous ftp to ftp.embl-heidelberg.de in pub/databases/embl/align/DS15369.dat.

Acknowledgments

T. M. Wilkie thanks Mel Simon, Nancy Jenkins, Neal Copeland and their colleagues for contributions to this work, parts of which have appeared in Wilkie et al. (1992). We thank Kathy Borkovich, Maureen Brandon, Neal Copeland, Nancy Jenkins, Robert Margolskee, Paul Sternberg, and Marta Wayne for sharing unpublished data and James Jaynes, Isabelle Davignon, Christiane Kleuss, and Jorge Iniguez-Lluhi for comments on the manuscript.

S. Yokoyama was supported by NIH grant GM-42379 and T. M. Wilkie, in part, by an NIH Postdoctoral Fellowship.

References

Amatruda, III, T. T., D. A. Steele, V. Z. Slepak, and M. I. Simon. 1991. Gα16, a G protein α subunit specifically expressed in hematopoietic cells. *Proceedings of the National Academy of Sciences, USA*. 88:5587–5591.

Beals, C. R., Wilson, C. B., Perlmutter, R. M. 1987. A small multigene family encodes G_i signal-transduction proteins. *Proceedings of the National Academy of Sciences, USA*. 84:7886–7890.

Berlot, C. H., and H. R. Bourne. 1992. Identification of effector-activating residues of $G_s\alpha$. *Cell.* 68:911–922.

Birnbaumer, L. 1990. G proteins in signal transduction. *Annual Reviews in Pharmacology and Toxicology*. 30:675–705.

Bourne, H. R., D. A. Sanders, and F. McCormick. 1990. The GTPase superfamily: A conserved switch for diverse cell functions. *Nature.* 348:125–132.

Bourne, H. R., D. A. Sanders, and F. McCormick. 1991. The GTPase superfamily: conserved structure and molecular mechanism. *Nature.* 349:117–127.

Bray, P., A. Carter, V. Guo, C. Puckett, J. Kamholz, A. Spiegel, M. Nirenberg. 1987. Human cDNA clones for an α subunit of G_i signal-transduction protein. *Proceedings of the National Academy of Sciences, USA*. 84:5115–5119.

Cavalier-Smith, T. 1991. Intron phylogeny: a new hypothesis. *Trends in Genetics.* 7:145–148.

Camps, M., C. Hou, D. Sidiropoulos, J. B. Stock, K. H. Jakobs, and P. Gierschik. 1992. Stimulation of phospholipase C by G-protein βγ-subunits. *European Journal of Biochemistry.* 206:821–831.

Conklin, B. R., and H. R. Bourne. 1993. Structural elements of Gα subunits that interact with Gβγ, receptors, and effectors. *Cell.* 73:631–641.

Dayhoff, M. O. 1978. Survey of new data and computer methods of analysis. Atlas of Protein and Structure. M. O. Dayhoff, editor. National Biomedical Research Foundation, Silver Springs, MD. 2–8.

de Castro, S. L., and M. M. Oliveira. 1987. Radioligand binding characterization of beta-adrenergic receptors in the protozoa *Trypanosoma cruzi. Comparative Biochemistry and Physiology.* 87C:5–8.

de Sousa, S. M., L. L. Hoveland, S. Yarfitz, and J. B. Hurley. 1989. The *Drosophila* $G_0\alpha$-like G protein gene produces multiple transcripts and is expressed in the nervous system and in ovaries. *Journal of Biological Chemistry.* 264:18544–18551.

de Vos, A. M., L. Tong, M. V. Milburn, P. M. Maniatis, J. Jancarik, S. Noguchi, S. Nishimura, K. Muira, E. Ohtsuka, and S. Kim. 1988. Three-dimensional structure of an oncogene protein: catalytic domain of human c-H-*ras* p21. *Science.* 239:888–893.

Dibb, N. J., and A. J. Newman. 1989. Evidence that introns arose at proto-splice sites. *EMBO Journal.* 8:2015–2021.

Dietzel, C., and J. Kurjan. 1987. The yeast SCG1 gene: a G alpha-like protein implicated in the a- and alpha-factor response pathway. *Cell.* 50:1001–1010.

Eisenschlos, C. D., A. A. Paladini, L. Molina, Y. Vedia, H. N. Torres, and M. M. Flawia. 1986. Evidence for the existence of an N_s-type regulatory protein in *Trypanosoma cruzi* membranes. *Biochemistry Journal.* 237:913–917.

Fain, G. L., and G. R. Matthews. 1990. Calcium and the mechanism of light adaptation in vertebrate photoreceptors. *Trends in Neuroscience.* 13:378–384.

Firtel, R. A. 1991. Signal transduction pathways controlling multicellular development in *Dictyostelium. Trends in Genetics.* 7:381–388.

Firtel, R. A., P. J. van Haastert, A. R. Kimmel, and P. N. Devreotes. 1989. G protein linked signal transduction pathways in development: *Dictyostelium* as an experimental system. *Cell.* 58:235–239.

Fong, H. K. W., K. K. Yoshimoto, P. Eversole-Cire, and M. I. Simon. 1988. Identification of a GTP-binding protein alpha subunit that lacks an apparent ADP-ribosylation site for pertussis toxin. *Proceedings of the National Academy of Sciences, USA.* 85:3066–3070.

Fraidenraich, D., C. Pena, E. L. Isola, E. M. Lammel, O. Coso, A. D. Anel, S. Pongor, F. Baralle, H. N. Torres, and M. M. Flawia. 1994. Stimulation of Trypanosoma cruzi adenylyl cyclase by an αD-globin fragment from Triatomoa hindgut. Effect on differentiation of epimastigote to trypanomastigote forms. *Proceedings of the National Academy of Sciences, USA.* In press.

Gilbert, W., M. Marchionni, and G. McKnight. 1986. On the antiquity of introns. *Cell.* 46:151–154.

Gilman, A. G. 1987. G proteins: transducers of receptor-generated signals. *Annual Reviews in Biochemistry.* 56:615–649.

Go, M. 1981. Correlation of DNA exonic regions with protein structural units in haemoglobin. *Nature.* 291:90–92.

Hadwiger, J. A., T. M. Wilkie, M. P. Strathmann, and R. A. Firtel. 1991. Identification of Dictyostelium Gα genes expressed during multicellular development. *Proceedings of the National Academy of Sciences, USA.* 88:8213–8217.

Halliday, K. 1983. Regional homology in GTP-binding proto-oncogene products and elongation factors. *Journal of Cyclic Nucleotides and Protein Phosphoylation Research.* 9:435–448.

Henderson, R., and G. F. Schertler. 1990. The structure of bacteriorhodopsin and its relevance to the visual opsins and other seven-helix G-protein coupled receptors. *Philosophical Transactions of the Royal Society of London B Biological Sciences.* 326:379–389.

Higgins, D. G., A. J. Bleasby, and R. Fuchs. 1992. Identification of a GTP-binding protein α subunit that lacks an apparent ADP-ribosylation site for pertussis toxin. *Computer Applied Biosciences.* 8:189–191.

Hsu, W. H., U. Rudolph, J. Sanford, P. Betrand, J. Olate, C. Nelson, L. G. Moss, A. E. Boyd, J. Codina, and L. Birnbaumer. 1990. Molecular cloning of a novel splice variant of the α subunit of the mammalian Go protein. *Journal of Biological Chemistry.* 265:11220–11226.

Iltzsch, M. H., D. Bieber, S. Vijayasarathy, P. Webster, M. Zurita, J. Ding, and T. E. Mansour. 1992. Cloning and characterization of a cDNA coding for the α-subunit of a stimulatory G protein from *schistosoma mansoni. Journal of Biological Chemistry.* 267:14504–14506.

Iniguez-Lluhi, J., C. Kleuss, and A. G. Gilman. 1993. The importance of G-protein βγ subunits. *Trends in Cell Biology.* 3:230–236.

Itoh, H., and A. G. Gilman. 1991. Expression and analysis of Gsα mutants with decreased ability to activate adenylylcyclase. *Journal of Biological Chemistry.* 266:16226–16231.

Itoh, H., R. Toyama, T. Kozasa, T. Tsukomoto, M. Matsuoka, and Y. Kaziro. 1988. Presence of three distinct molecular species of Gi protein alpha subunit. *Journal of Biological Chemistry.* 263:6656–6664.

Jones, D. T., and R. R. Reed. 1987. Molecular cloning of five GTP-binding protein cDNA species from rat olfactory neuroepithelium. *Journal of Biological Chemistry.* 262:14241–14249.

Jones, D. T., and R. R. Reed. 1989. Golf, an olfactor neuron specific G protein involved in odorant signal transduction. *Science.* 244:790–795.

Jukes, T. H., and C. R. Cantor. 1969. Evolution of protein molecules. Mammalian protein metabolism. H. N. Munro, editor. Academic Press, New York. 21–132.

Jurnak, F. 1985. Structure of the GDP domain of EF-Tu and location of the amino acids homologous to *ras* oncogene proteins. *Science.* 230:32–36.

Katz, A., D. Wu, and M. I. Simon. 1992. Subunits beta gamma of heterotrimeric G protein activate beta 2 isoform of phospholipase C. *Nature.* 360:686–689.

Kaziro, Y., H. Itoh, T. Kozasa, M. Nakafuku, and T. Satoh. 1991. Structure and function of signal-transducing GTP-binding proteins. *Annual Reviews in Biochemistry.* 60:349–400.

Kleuss, C., J. Hescheler, C. Ewel, W. Rosenthal, G. Schultz, B. Wittig, T. A. Kunkel, J. D. Roberts, and R. A. Zabour. 1991. Assignment of G-protein subtypes to specific receptors inducing inhibition of calcium currents. *Nature.* 353:43–48.

Kleuss, C., H. Scherubl, J. Hescheler, G. Schultz, and B. Wittig. 1993. Selectivity in signal transduction determined by gamma subunits of heterotrimeric G proteins. *Science.* 259:832–894.

Kleuss, C., H. Scherubl, J. Heschler, G. Schultz, and B. Wittig. 1992. Different β-subunits determine G-protein interaction with transmembrane receptors. *Nature.* 358:424–426.

Knox, B. E., L. Scalzetti, J.-Q. Wang, and S. Batni. 1994. Molecular cloning of *Xenopus* rod opsin and transducin alpha subunit. *Experimental Eye Research.* In press.

Kozasa, T., H. Itoh, T. Tsukamoto, and Y. Kaziro. 1988. Isolation and characterization of the human $G_s\alpha$ gene. *Proceedings of the National Academy of Sciences, USA.* 85:2081–2085.

Kubo, M., T. Hirano, and M. Kakinuma. 1991. Molecular cloning and sequence analysis of cDNA and genomic DNA for the human cone transducin α subunit. *Federation of European Biochemical Societies.* 291:245–248.

Lee, Y. J., M. B. Dobbs, M. L. Verardi, and D. R. Hyde. 1990. dgq: a *drosophila* gene encoding a visual system-specific G α molecule. *Neuron.* 5:889–898.

Lochrie, M. A., J. B. Hurley, and M. I. Simon. 1985. Sequence of the alpha subunit of photoreceptor G protein: homologies between transducin, *ras,* and elongation factors. *Science.* 228:96–99.

Lochrie, M. A., J. E. Mendel, P. W. Sternberg, and M. I. Simon. 1991. Homologous and unique G protein alpha subunits in the nematode *Caenorhabditis elegans. Cell Regulation.* 2:135–154.

Ma, H., M. F. Yanofsky, and H. Huang. 1991. Isolation and sequence analysis of *TGA1* cDNAs encoding a tomato G protein α subunit. *Gene.* 107:189–195.

Ma, H., M. F. Yanofsky, and E. Meyerowitz. 1990. Molecular cloning and characterization of GPA1, a G protein alpha subunit gene from *Arabidopsis thaliana. Proceedings of the National Academy of Sciences, USA.* 87:3821–3825.

Marchionni, M., and W. Gilbert. 1986. The triosephosphate isomerase gene from maize: introns antedate the plant-animal divergence. *Cell.* 46:133–141.

Matsuoka, M., H. Itoh, and Y. Kaziro. 1990. Characterization of the human gene for Gxα, a pertussis toxin-insensitive regulatory GTP-binding protein. *Journal of Biological Chemistry.* 265:13215–13220.

Matsuoka, M., H. Itoh, T. Kozasa, and Y. Kaziro. 1988. Sequence analysis of cDNA and genomic DNA for a putative pertussis toxin-insensitive guanine nucleotide-binding regulatory protein alpha subunit. *Proceedings of the National Academy of Sciences, USA.* 85:5384–5388.

McLaughlin, S. K., P. J. McKinnon, and R. F. Margolskee. 1992. Gustducin is a taste-cell-specific G protein closely related to the transducins. *Nature.* 357:563–569.

Medynski, D. C., K. Sullivan, D. Smith, C. Van Dop, F. Chang, B. K. Fung, P. H. Seeburg, and H. R. Bourne. 1985. Amino acid sequence of the α subunit of transducin deduced from the cDNA sequence. *Proceedings of the National Academy of Sciences, USA.* 82:4311–4315.

Miyajima, I., M. Nakafuku, N. Nakayama, C. Brenner, A. Miyajima, K. Kaibuchi, K. Arai, Y. Kaziro, and K. Matsumoto. 1987. *GPA1,* a haploid-specific essential gene, encodes a yeast homolog of mammalian G protein which may be involved in mating factor signal transduction. *Cell.* 50:1011–1019.

Moss, J., and M. Vaughan, editors. 1990. ADP-ribosylating toxins and G Proteins. American Society of Microbiology, Washington, DC.

Nadeau, J. H. 1991. Advanced techniques in chromosome research. K. W. Adolph, editor. Marcell Dekker, New York. 269–296.

Nakafuku, M., H. Itoh, S. Nakamura, and Y. Kaziro. 1987. Occurrence in *Saccharomyces cerevisiae* of a gene homologous to the cDNA coding for the α subunit of mammalian G proteins. *Proceedings National Academy Sciences, USA.* 84:2140–2144.

Nakafuku, M., T. Obara, K. Kaibuchi, I. Miyajima, A. Miyajima, H. Itoh, S. Nakamura, K. Arai, K. Matsumoto, and Y. Kaziro. 1988. Isolation of a second yeast *Saccharomyces cerevisiae* gene (*GPA2*) coding for guanine nucleotide-binding regulatory protein: studies on its structure and possible functions. *Proceedings of the National Academy of Sciences, USA.* 85:1374–1378.

Nakamura, F., K. Ogata, K. Shiozaki, K. Kameyama, K. Ohara, T. Haga, and T. Nukada. 1991. Identification of two novel GTP-binding protein α-subunits that lack apparent ADP-ribosylation sites for pertussis toxin. *Journal of Biological Chemistry.* 266:12676–12681.

Nei, M. 1987. Molecular Evolutionary Genetics. Columbia University Press, New York.

Nukada, T., T. Tanabe, H. Takahashi, M. Noda, T. Hirose, S. Inayama, and S. Numa. 1986. Primary structure of the α-subunit of bovine adenylate cyclase-stimulating G-protein deduced from the cDNA sequence. *Federation of Biochemical Societies Letters.* 195:220–224.

Obara, T., M. Nakafuku, M. Yamamoto, and Y. Kaziro. 1991. Isolation and characterization of a gene encoding a G-protein α subunit from *Schizosaccharomyces pombe:* involvement in mating and sporulation pathways. *Proceedings of the National Academy of Sciences, USA.* 88:5877–5881.

Olate, J., H. Jorquera, P. Purcell, J. Codina, L. Birnbaumer, and J. E. Allende. 1989. Molecular cloning and sequence determination of a cDNA coding for the α-subunit of a Go-type protein of *Xenopus laevis* oocytes. 244:188–192.

Olate, J., S. Martinez, P. Purcell, H. Jorquera, J. Codina, L. Birnbaumer, and J. Allende. 1990. Molecular cloning and sequence determination of four different cDNA species coding for alpha-subunits of G proteins from *Xenopus laevis* oocytes. *Federation of Biochemical Societies Letters.* 268:27–31.

Pai, E. F., W. Kabsch, U. Krengel, K. C. Holmes, J. John, and A. Wittinghofer. 1989. Structure of the guanine-nucleotide-binding domain of the Ha-ras oncogene product p21 in the triphosphate conformation. *Nature.* 341:209–214.

Parks, S., and E. Wieschaus. 1991. The drosophila gastrulation gene *concertina* encodes a Gα-like protein. *Cell.* 64:447–458.

Pitt, G. S., R. E. Gundersen, P. J. Lilly, M. B. Pupillo, R. A. Vaughan, and P. N. Devreotes. 1990. G protein-linked signal transduction in aggregating *Dictyostelium. Society of General Physiologists Series.* 45:125–131.

Provost, N. M., D. E. Somers, and J. B. Hurley. 1988. A *Drosophila melanogaster* G protein α subunit gene is expressed primarily in embryos and pupae. *Journal of Biological Chemistry.* 263:12070–12076.

Pupillo, M., A. Kumagai, G. S. Pitt, R. A. Firtel, and P. N. Devreotes. 1989. Multiple α subunits of guanine nucleotide-binding proteins in *Dictyostelium. Proceedings of the National Academy of Sciences, USA.* 86:4892–4896.

Quan, F., W. J. Wolfgang, and M. Forte. 1993. A *Drosophila* G-protein α subunits, Gfα, expressed in a spatially and temporally restricted pattern during *Drosophila* development. *Proceedings of the National Academy of Sciences, USA.* 90:4236–4240.

Quan, F., W. J. Wolfgang, and M. A. Forte. 1989. The *Drosophila* gene coding for the α subunit of a stimulatory G protein is preferentially expressed in the nervous system. *Proceedings of the National Academy of Sciences, USA.* 86:4321–4325.

Rall, T., and B. A. Harris. 1987. Identification of the lesion in the stimulatory GTP-binding protein of the uncoupled S49 lymphoma. *Federation of Biochemical Societies Letters.* 224:365–371.

Ranganathan, R., W. A. Harris, and C. S. Zucker. 1991. The molecular genetics of invertebrate phototransduction. *Trends in Neuroscience.* 14:486–493.

Rapport, C. J., B. Dere, and J. B. Hurley. 1989. Characterization of the mouse rod transducin α subunit gene. *Journal of Biological Chemistry.* 264:7122–7128.

Rogers, J. 1985. Exon shuffling and intron insertion in serine protease genes. *Nature.* 315:458–459.

Sadhu, C., D. Hoekstra, M. J. McEachern, S. I. Reed, and J. B. Hicks. 1992. A G-protein α subunit from asexual *Candida albicans* functions in the mating signal transduction pathway of *Saccaromyces cerevisiae* and is regulated by the a1-α2 repressor. *Molecular and Cellular Biology.* 12:1977–1985.

Saitour, N., and M. Nei. 1987. The neighbor-joining method: a new method for reconstruction of phylogenetic trees. *Molecular Biological Evolution.* 4:406–425.

Schmidt, C. J., S. Garen-Fazio, Y-K. Chow, E. J. Neer. 1989. Neuronal expression of a newly identified *Drosophila melanogaster* G protein α_o subunit. *Cell Regulation.* 1:125–134.

Silva, I. F., and R. H. A. Plasterk. 1990. Characterization of a G-protein α-subunit gene from the nematode *Caenorhabditis elegans. Journal of Molecular Biology.* 215:483–487.

Simon, M. I., M. P. Strathmann, and N. Gautam. 1991. Diversity of G proteins in signal transduction. *Science.* 252:802–808.

Smrcka, A. V., J. R. Hepler, K. O. Brown, and P. C. Sternweis. 1991. Regulation of polyphosphoinositide-specific phospholipase C activity by purified G_q. *Science.* 251:804–807.

Strathmann, M., and M. I. Simon. 1990. G proteins diversity: a distinct class of α subunits is present in vertebrates and invertebrates. *Proceedings of the National Academy of Sciences, USA.* 87:9113–9117.

Strathmann, M. P., and M. I. Simon. 1991. Gα12 and Gα13 subunits define a fourth class of G protein α subunits. *Proceedings of the National Academy of Sciences, USA.* 88:5582–5586.

Strathmann, M. P., T. M. Wilkie, and M. I. Simon. 1991. Alternative splicing produces transcripts encoding two forms of the alpha subunit of GTP-binding protein G_0. *Proceedings of the National Academy of Sciences, USA.* 87:6477–6481.

Sullivan, K. A., Y. Liao, A. Alborzi, B. Beiderman, G. Chang, S. B. Masters, A. D. Levinson, and H. R. Bourne. 1986. Inhibitory and stimulatory G proteins of adenylate cyclase: cDNA and amino acid sequences of the α chains. *Proceedings of the National Academy of Sciences, USA.* 83:6687–6691.

Tang, W., and A. G. Gilman. 1992. Type-specific regulation of adenylyl cyclase by G protein βγ subunits. *Science.* 254:1500–1503.

Taussig, R., J. A. Iniguez-Lluhi, and A. G. Gilman. 1993. Inhibition of adenylyl cyclase by $G_{i\alpha}$. *Science.* 261:218–221.

Taylor, S. J., H. Z. Chae, S. G. Rhee, and J. H. Exton. 1991. Activation of the β1 isozyme of phospholipase C by α subunits of the G_q class of G proteins. *Nature.* 350:516–518.

Thambi, N. C., F. Quan, W. J. Wolfgang, A. Spiegel, and M. Forte. 1989. Immunological and molecular characterization of Goα-like proteins in the *Drosophila* central nervous system. *Journal of Biological Chemistry.* 264:18552–18560.

Tittiger, C., S. Whyard, and V. K. Walker. 1993. A novel intron site in the triosephosphate isomerase gene from the mosquito *Culex tarsalis. Nature.* 461:470–472.

Tong, L., M. V. Milburn, A. M. De Vos, and S. H. Kim. 1989. Structure of *ras* proteins. *Science.* 245:244.

Tsukamoto, T., R. Toyama, H. Itoh, T. Kozasa, M. Matsuoda, and Y. Kaziro. 1991. Structure of the human gene and two rat cDNAs encoding the α chain of GTP-binding regulatory protein G_0: two different mRNAs are generated by alternative splicing. *Proceedings of the National Academy of Sciences, USA.* 88:2974–2978.

Turner, G., and K. A. Borkovich. 1993. Identification of a G protein alpha subunit from *Neurospora crassa* that is a member of the Gi family. *Journal of Biological Chemistry.* 268:14805–14811.

van Meurs, K. P., C. W. Angus, W. Lavu, H. Kung, S. K. Czarnecki, J. Moss, and M. Vaughn. 1987. Deduced amino acid sequence of bovine retinal $G_0\alpha$: similarities to other guanine nucleotide-binding proteins. *Proceedings of the National Academy of Sciences, USA.* 84:3107–3111.

Whiteway, M., L. Hougan, D. Dignard, D. Y. Thomas, L. Bell, G. C. Saari, F. J. Grant, P. O'Hara, and V. L. MacKay. 1989. The STE4 and STE18 genes of yeast encode potential β and γ subunits of the mating factor receptor-coupled G protein. *Cell.* 56:467–477.

Wilkie, T. M., D. J. Gilbert, A. S. Olsen, X. N. Chen, T. T. Amatruda, J. R. Korenberg, B. J. Trask, P. Dejong, R. R. Reed, M. I. Simon, N. A. Jenkins, and N. G. Copeland. 1992. Evolution of the mammalian G protein α subunit multigene family. *Nature Genetics.* 1:85–91.

Wilkie, T. M., P. A. Scherle, M. P. Strathmann, V. Z. Slepak, and M. I. Simon. 1991. Characterization of G-protein α subunits in the G_q class: expression in murine tissues and in stromal and hematopoietic cell lines. *Proceedings of the National Academy of Sciences, USA.* 88:10049–10053.

Wistow, G. 1993. Protein structure and introns. *Nature.* 364:107–108.

Wu, D., C. H. Lee, S. G. Rhee, and M. I. Simon. 1992. Activation of phospholipase C by the α subunits of the G_q and G_{11} proteins in transfected Cos-7 cells. *Journal of Biological Chemistry.* 267:1811–1817.

Yokoyama, S., and W. T. Starmer. 1992. Phylogeny and evolutionary rates of G protein α subunit genes. *Journal of Molecular Evolution.* 35:230–238.

Yoon, J., R. D. Shortridge, B. T. Bloomquist, S. Schneuwly, M. H. Perdew, and W. L. Pak. 1989. Molecular characterization of *Drosophila* gene encoding $G_o\alpha$ subunit homolog. *Journal of Biological Chemistry.* 264:18536–18543.

Molecular Evolution of the Calcium-Transporting ATPases Analyzed by the Maximum Parsimony Method

Yan Song and Douglas Fambrough

Department of Biology, The Johns Hopkins University, Baltimore, Maryland 21218

The calcium-transporting adenosine triphosphatases (Ca^{2+}-ATPases, or calcium ion pumps) are a group of membrane proteins that actively transport calcium ions across cellular membranes. Because of the role of calcium ions as a universal secondary signal, the concentration of free Ca^{2+} in the cytosol must be tightly controlled. In higher vertebrates, cytosolic Ca^{2+} concentration is maintained at ~ 0.1 μM, four orders of magnitude lower than the extracellular level (Campbell, 1983). A comparable level of Ca^{2+} is observed in the cytoplasm of plants, which is at least a thousand times lower than in the extracellular space, the apoplast (Evans, Briars, and Williams, 1991).

Two types of calcium pumps are involved in the regulation of cellular calcium ion levels: the plasma membrane Ca^{2+}-ATPase and an organellar Ca^{2+}-ATPase, which remove Ca^{2+} from the cytosol by, respectively, extruding the ion from the cell and sequestering it into intracellular reservoirs. In recent years, genes and transcripts for calcium pump proteins have been identified and characterized. Many molecular biological studies have provided insights into the structural and functional features of these ion pumps. The currently known calcium pumps belong to the P-type cation transporting ATPases, which are a distinct subgroup of ion-motive ATPases (Pedersen and Carafoli, 1987*a,b;* Green and MacLennan, 1989; Läuger, 1991). Many of these P-type ATPases, including the calcium pumps, have been demonstrated to consist of a single subunit that spans the membrane many times (a generic model of P-type ATPases is depicted in Fig. 1). During ion transport, a phosphorylated intermediate is formed, where phosphate is bound to an aspartyl residue located on a large cytosolic loop. These ion pumps are characteristically inhibited by vanadate, which acts as a transition state analog of phosphate. P-type ATPases appear to be ubiquitous, found in both prokaryotic and eukaryotic cells.

The sequences considered in this paper include those of the two main types of calcium pumps, namely the plasma membrane Ca^{2+}-ATPase, conventionally referred to as the PMCA, and the Ca^{2+}-ATPase of the sarcoplasmic and endoplasmic reticulum, or the SERCA (See Burk, Lytton, MacLennan, and Shull, 1989). The two pumps are distinguished by their separate locations in the cell, by differential sensitivities to pump inhibitors, and by differing amino acid sequences. Also included are several P-type ATPase sequences that have been proposed to encode calcium pumps in two fungi and the protozoan *Leishmania.*[1]

[1] Shull, Clarke, and Gunteski-Hamblin (1992) more recently cloned a new calcium pump associated with the Golgi apparatus, which they consider to be related to the fungal secretive pathway calcium pump. It is not included here.

The present study is focused on the molecular evolution of these proteins, a subject that has received little discussion, despite the fact that the identification of calcium pump genes has been almost solely based on nucleotide sequence similarities rather than on functional studies of gene products. There has been no systematic study devoted to understanding the evolution of calcium pump genes, and the time is ripe to begin addressing this issue, thanks to the good number of genes cloned from a reasonably diverse variety of organisms. For general information as well as for discussions on particular functional and structural aspects of these calcium ATPases, see Sachs and Munson (1991), Carafoli, Kessler, Falchetto, Helm, Quadroni, Krebs, Strehler, and Vorherr, (1992), Inesi and Kirtley (1992), Inesi, Cantilina, Yu, Nikic, Sagara, and Kirtley, (1992), Jencks (1992), MacLennan, Toyofuku, and Lytton (1992), Wuytack and Raeymaekers (1992), Wuytack et al. (1992).

Several molecular evolutionary questions are addressed in this study. First, how are the pump-encoding genes related to one another? Intriguingly, at least three different genes have been discovered in higher vertebrates to encode separate forms of the SERCA-type calcium pumps. The isoforms, as these genetically distinct pumps are commonly referred to, are structurally and functionally similar, but their

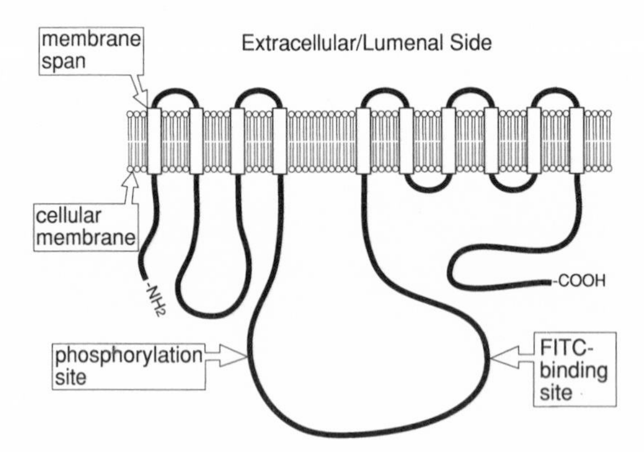

Figure 1. A diagrammatic structural model of a P-type ATPase. The number of transmembrane domains is still being debated, and some P-type ATPases have an additional transmembrane domain at the carboxyl terminus. A small subunit, found in certain P-type pumps such as the Na/K-ATPase, is not shown here.

biological differences are not quite clear. The same situation holds for the PMCAs. Characteristically, the isoforms have been found to have differential distributions in various tissues (e.g., Kaprielian and Fambrough, 1987; Kaprielian, Bandman, and Fambrough 1991), and such tissue-specific distributional patterns appear to be parallel in different organisms that have multiple isoforms. For instance, SERCA1 is typically abundant in fast twitch muscle cells in both mammals and birds. Pairwise comparisons have suggested a general tendency for synonymous isoforms (i.e., corresponding isoforms by the same numerical name) found in different organisms to have higher sequence similarities between each other than between different isoforms found in the same species. However, using sequence similarity alone to determine a transcript's identity can be misleading, as similarities may not necessarily impart the true relationships. Indeed, an inference of the evolutionary relationships of a gene family must be based on validated homology or common origin, just like the deduction of organismal phylogeny. A robust systematic framework for the calcium ATPases is currently lacking, and such a framework can only be obtained by examining the overall relationships among the pump-coding genes.

A second question addressed here is: how did the calcium pump isoforms arise? Pump isoforms have been found in higher vertebrates (birds and mammals). Though the paucity of pump isoform sequence data in invertebrates and other groups may well be due to limited cloning efforts in those organisms to date, the relationship between pump-encoding genes in organisms with multiple isoforms and those without isoforms should help clarify the patterns of isoform diversification and thereby provide instructive insights into pathways of isoform evolution. Finally, this study addresses the questions: are PMCA and SERCA indeed two distinct pumps? And, if so, are they sister groups or is there an ancestral-descendent connection?

Pump-encoding Sequences and Their Analyses

Sequences and Alignment

A total of 25 amino acid sequences of calcium-motive ATPases were selected from those retrieved from the data banks on the computer system Genmenu (the data banks, including Genebank, EMBL, and Swissprot, were updated to September 1992). Protein sequences were used to avoid artifacts caused by possible codon biases between organismal groups. Table I lists the sequences used in this study, arranged by the organisms in which the genes were identified. The numbers in column 3

TABLE I
Calcium-ATPase Sequences Used in the Current Study (September, 1992)

Organismal species	Gene	Acc. No.	Reference
Homo sapiens (human)	PMCA1	a30802	Verma et al., 1988
Homo sapiens (human)	PMCA3	m25874	Strehler et al., 1990
Homo sapiens (human)	SERCA2	m23115	Lytton and MacLennon, 1988
Oryctolagus cuniculus (rabbit)	PMCA	s17179	Khan and Grover, 1991
Oryctolagus cuniculus (rabbit)	SERCA1	m12898	Brandl et al., 1986
Oryctolagus cuniculus (rabbit)	SERCA2	j04703	Lytton et al., 1989
Rattus norvegicus (rat)	PMCA1	j03753	Shull and Greeb, 1988
Rattus norvegicus (rat)	PMCA2	j03754	Shull and Greeb, 1988
Rattus norvegicus (rat)	PMCA3	a34308	Greeb and Shull, 1989
Rattus norvegicus (rat)	SERCA1	m99223	Wu and Lytton, 1993
Rattus norvegicus (rat)	SERCA2	a31982	Gunteski-Hamblin et al., 1988
Rattus norvegicus (rat)	SERCA3	a34307	Burk et al., 1989
Sus scrofa (pig)	PMCA 1	p23220	De Jaegere et al., 1990
Sus scrofa (pig)	SERCA2	p11606	Eggermont et al., 1991
Felis catus (cat)	SERCA	z11500	Gambel et al., 1992
Gallus gallus (chicken)	SERCA1	m26064	Karin et al., 1989
Gallus gallus (chicken)	SERCA2	a40812	Campbell et al., 1991
Rana esculenta (frog)	SERCA	s18884	Vilsen and Andersen, 1991
Artemia sp. (brine shrimp)	SERCA	s07526	Palmero and Sastre, 1989
Drosophila melanogaster (fruit fly)	SERCA	a36691	Magyar and Varadi, 1990
Lycopersicon esculentum (tomato)	SERCA	m96324	Wimmers et al., 1992
Leishmania donovani (protozoa)	CA	j04004	Meade et al., 1989
Saccharomyces cerevisiae (baker's yeast)	CA1	p13586	Rudolph et al., 1989
Saccharomyces cerevisiae (baker's yeast)	CA2	p13587	Rudolph et al., 1989
Schizosaccharomyces pombe (fission yeast)	CA	j05634	Ghislain et al., 1990

denote the accession numbers by which the sequences were retrieved from the data banks. Only full-length (completed) sequences were used. In each case where there are multiple "pseudo-isoforms," or messenger RNAs derived by alternative processing of the primary transcripts (the term pseudo-isoform is proposed for this context, to indicate that such transcripts are not encoded by separate genes), only a single primary sequence was used.

A multiple alignment of the selected protein sequences was conducted using the program PILEUP of GCG (a software package from the University of Wisconsin). Sometimes computer-aided aligning such as the one used in this study has been faulted for producing less-than-perfect results, but this approach has the advantage of applying consistent criteria in objective search for homology. The alignments (partially presented in Fig. 2) appear to give satisfactory residue matching pattern,

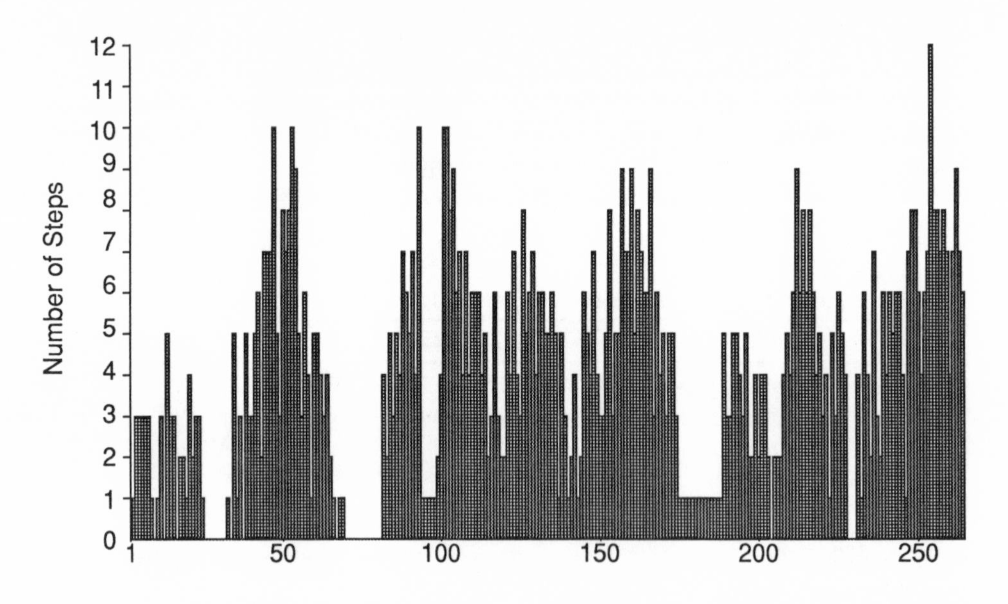

Figure 2. A portion of the computer-aided multiple alignment of calcium-ATPase sequences used in the cladistic analyses. Clearly, the two definitive functional domains, i.e., the phosphorylation site and the FITC-binding site, are well aligned, indicating that the two sites are among the most conserved motifs in the pumps. The residues are numbered arbitrarily.

featuring perfectly aligned domains such as the phosphorylation site and the ATP-binding site (Figs. 2 and 3), both of which are considered definitively homologous among P-type ATPases.

Phylogenetic Analyses Using PAUP

Four sets of analyses were conducted using PAUP (Phylogenetic Analysis Using Parsimony, edition 3.1; Swofford, 1993) designed for use on MacIntosh computers. In all analyses, the most parsimonious cladistic trees (cladograms) were searched out by the "branch-and-bound" method. Each analysis was repeated after the sequences were reshuffled to alter their order of entry.

Deterministic searches (e.g., exhaustive and branch-and-bound) for most parsimonious trees using PAUP involves very large amounts of computation time, which can be sometimes inhibitingly long. There are two ways to circumvent this problem. One is by limiting the number of sequences to the minimal, the other by using a minimal yet sufficiently informative part of the sequence. The first analysis utilized only the region encompassing the phosphorylation and the FITC (fluorescein isothiocyanate)-binding sites, the latter site believed to be involved in ATP-binding. These two sites are characteristically conserved throughout the gene family, but there is considerable sequence variation in between these two sites (Fig. 3).

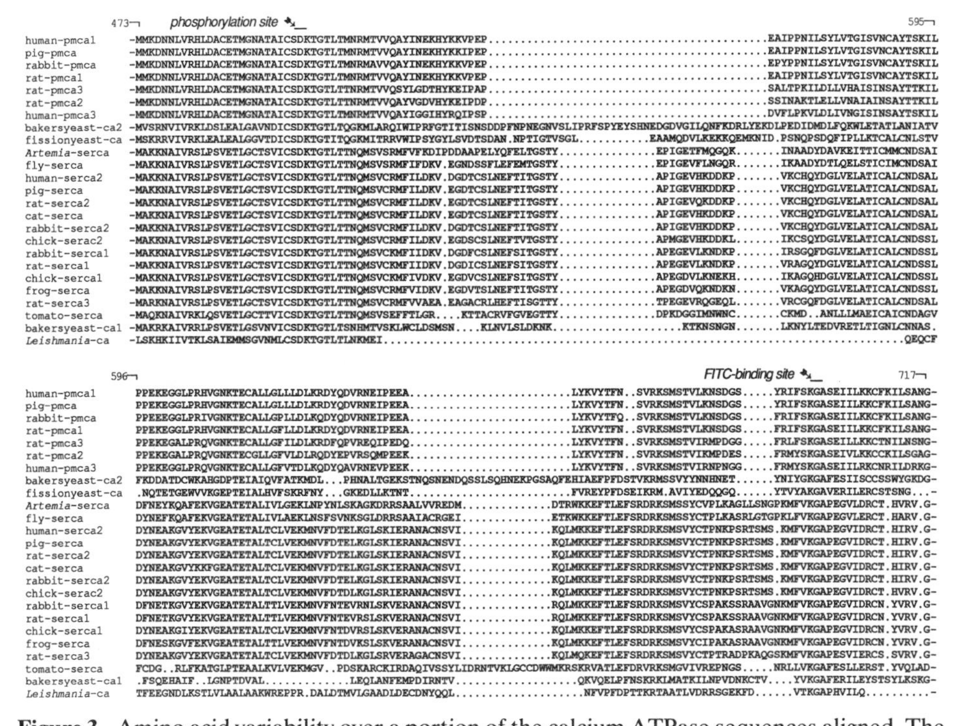

Figure 3. Amino acid variability over a portion of the calcium ATPase sequences aligned. The variability is measured by the number of residue substitutions at each position, or number of "steps" (see Maddison and Maddison, 1992), which was calculated, using the program MacClade 3, based on a most parsimonious cladogram reconstructed from the PAUP analysis of the 25 partial calcium pump sequences (see Fig. 4). Note that the phosphorylation site (around residue 498 in Fig. 2) and the FITC-binding site (around residue 700 in Fig. 2) are the best conserved positions. The apparent low variability around positions 75 and 180 on the histogram were caused by gaps generated in the alignment.

Entire protein sequences were used in the other analyses in order to improve the resolution. In the second analysis, the *Drosophila* Na/K-ATPase sequence (Lebovitz, Takeyasa, and Fambrough, 1989) was included as outgroup in order to place the root of a tree encompassing a subset of sequences used in analysis 1. Adding this outgroup further confirmed the relationship between the PMCAs and the SERCAs. The PMCAs and the SERCAs were examined separately in analyses 3 and 4.

What Do the Cladograms Say about the Molecular Evolution of PMCAS and SERCAS?

Cladograms (phylogeny-inferring tree figures) from the PAUP analyses are presented in Figs. 4–7. Their main features and evolutionary implications are discussed below.

The PMCA and SERCA Are Two Evolutionarily Distinct Calcium-pumping ATPases

Fig. 4 shows one of the nine most parsimonious trees resultant from the branch-and-bound search. The nine trees have minor topological differences, but all have the same polytomy, i.e., that of mammalian SERCA2 (a polytomous node has more than two descendant nodes, which can result from uncertainty in relationships due to lack of resolution in the data, or from a simultaneous multiple branching event, see

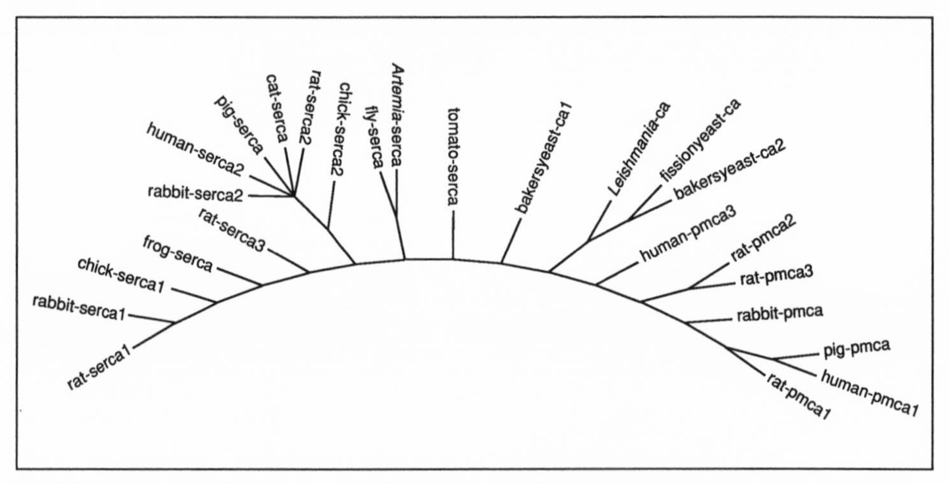

Figure 4. A redrawn rootless tree from the branch-and-bound analysis using the twenty-five partial calcium pump sequences. A dichotomy between the plasma membrane calcium ATPase (PMCA) and the organellar ATPase (SERCA) is suggested by the separation of the two pumps.

Maddison, 1989; and Maddison and Maddison, 1992). The most striking feature of this rootless tree is the parity of the PMCAs and the SERCAs. They are manifestly the two most distant clades, though it is unclear how the two groups are related to each other in terms of ancestral-descendent polarity because the tree is rootless.

By including the fly Na/K-ATPase as the outgroup, the root could be placed between the PMCAs and the SERCAs (Fig. 5). Interestingly, in this analysis the fungal and the protozoan calcium pumps were grouped with the PMCA clade, though they have been presented as functionally more SERCA-like and though sequence similarity data also indicate otherwise. A cladistic analysis of the entire P-type ion pump gene group can provide insights into the evolutionary relationships among the different ion-specific pumps. One issue that we hope can be better addressed by such an analysis is how the two calcium pumps are related to each other as well as to other ion pumps.

Isoformity Is an Essential Feature of the Molecular Evolution in the Two Calcium Pump Gene Families

Multiple genes encoding different calcium-pumping ATPases have been found in the PMCAs, in the SERCAs and in the fungal calcium pump genes. The isoforms have been numerically named based on sequence similarity data, and synonymous isoforms have been found in different species. There is yet not enough experimental data to provide a good explanation of isoformity, but one can speculate that it must have involved gene multiplication and must have been subjected to selective processes, perhaps mostly of a developmental nature.

Six equally most parsimonious trees were found by PAUP for the SERCAs (Fig. 6). Using the nonvertebrate genes as outgroups, the vertebrate genes clearly show a tripartite classification. The SERCA1 clade displays a hierarchical branching pattern that is congruent with the systematic ranking order of the four vertebrate taxa represented here (two mammalians, one bird, and one amphibian). Among the

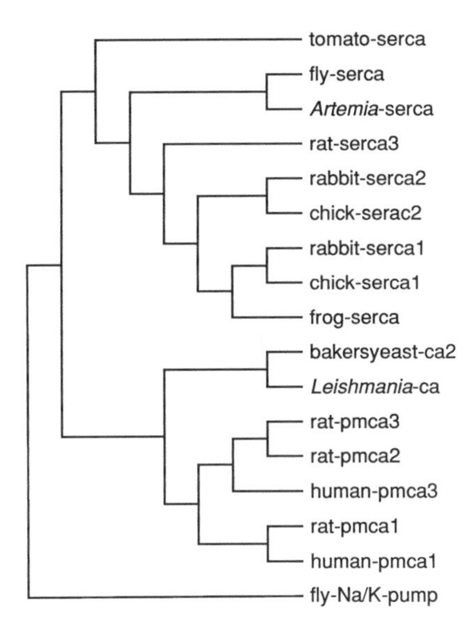

Figure 5. The rootless tree from the branch-and-bound analysis using the full-length sequences of a subset of the calcium pumps. The *Drosophila* sodium/potassium pump is included in order to locate the rooting of the cladogram. The PMCAs and the SERCAs are separated, and the fungal and the *Leishmania* pumps are suggested to be more closely related to the PMCAs than to ther SERCAs.

SERCA2s, the five mammalian genes are so closely related to each other that there is no definite resolution to their polytomy in the partial sequence cladogram, except for the apparently stable clade between the human and the rabbit SERCA2 genes. There is only one SERCA3 included in this study, and we expect new SERCA3 genes will help strengthen this isoform's position.

A similar delineation of the PMCA isoforms can be observed in the single most parsimonious cladogram for the PMCAs and the fungal and the protozoan calcium ATPases, with the two clades being outgroups to each other (Fig. 7). A puzzling result from this analysis, however, is the designation of the rat PMCA2 gene to the otherwise PMCA3 clade. Pairwise distances between these genes are considerably larger than those in the PMCA1 group. This incongruency may have been caused by a designation problem. The human PMCA in this clade was originally published as isoform 3 (Strehler, James, Fischer, Heim, Vorherr, Filoteo, Penniston, and Carafoli, 1990). It has since been referred to as isoform 4 (e.g., Brandt, Neve, Kanmesheidt, Rhoads, and Vanaman, 1992), in which case each of the three members of this clade represents a different isoform. Alternatively, the various PMCAs may have resulted from a diversifying process following different and perhaps more complex pathways than the SERCAs and the Na/K pumps in that isoforms originated in a nested fashion. In any case, the number of PMCA genes available is too few to provide a satisfactory resolution. The current PMCA data consist solely of mammalian genes, unlike the SERCAs where a wider range of organisms are represented, and unlike the Na/K-ATPases (Shull, Greeb, and Lingrel, 1986; Takeyasu, Lemas, and Fambrough, 1990; Lingrel, 1992) where there seems to be consistent isoform groupings between mammals and birds.

The Modern Molecular Diversity of the Calcium Pump Genes May Have Arisen Via Diverse Evolutionary Processes

An important task for molecular evolutionary studies is to explore the mechanisms by which genes and genomes evolve. Despite the relatively small and skewed data set,

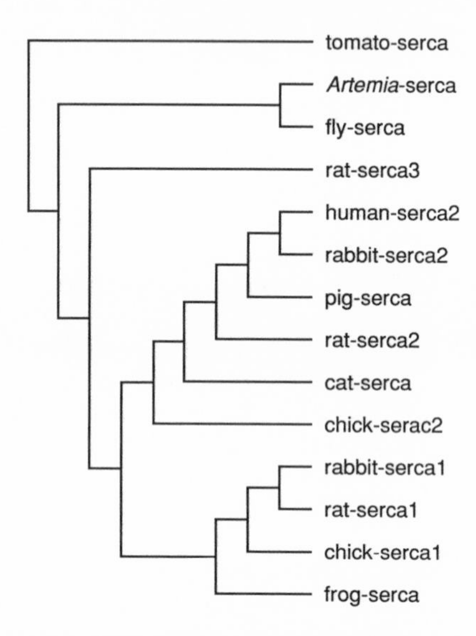

Figure 6. One of the six rootless trees from the branch-and-bound analysis using the full length sequences of the SERCA-type calcium pumps. With the nonvertebrate genes as outgroup, there is a tripartite delineation of the vertebrate genes. The SERCA2 clade is unstable in terms of the ranking order of the mammalian members.

results from the current study have provided several insights into how the molecular biological diversity of the calcium-motive ATPases came into being.

Our first observation is that the gene multiplication that gave rise to the currently observed isoforms seems to have involved an evolutionarily event after the vertebrates separated from the rest of the biota on earth in the late Cambrian and the early Ordovician periods, more than 500 million years before the present time (Valentine, 1985; Carroll, 1988). This is best demonstrated by the case of SERCAs. Mammalian, avian and amphibian SERCAs fit very well into distinct isoform clades and are clearly separated from nonvertebrate genes. Searching for new genes in the lower vertebrates (fishes, more amphibians, reptiles) should help determine (*a*) exactly when, on the evolutionary scale, the isoform-generating event occurred, and (*b*) which of the isoforms has been the "house-keeping" or "stem" SERCA. Because of the tissue specificity of the isoforms, the latter question is of particular evolutionary interest as its answer will shed light on the historical development of organismal complexity in the vertebrates.

Furthermore, the development of isoforms in other organismal groups does not seem to be related to that in the vertebrates. It is not yet known whether there are multigenic isoforms of calcium ATPases in the three nonvertebrate taxa, though the *Artemia* SERCA has been shown to have two pseudo-isoforms resulting from alternative splicing of the primary transcript (Escalante and Sastre, 1993). Nonetheless, no overlap was found between any of the nonvertebrate genes and any of the vertebrate SERCAs, the latter showing a definite pattern of isoform subdivision. The two fungal gene isoforms, on the other hand, have already demonstrated that isoformity is not confined to the vertebrate calcium pumps. Given that the vertebrates represent but a small fraction of the biological diversity on Earth, the molecular evolution for their calcium pump isoforms should be interpreted neither as unique nor necessarily as a general model. It would not be surprising should new

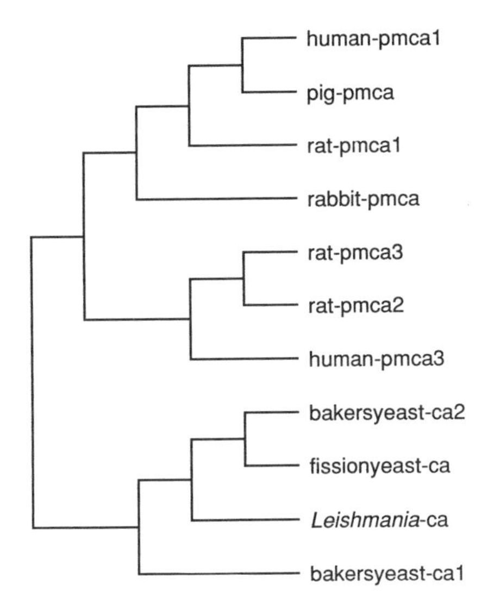

Figure 7. The rootless tree from the branch-and-bound analysis using the full-length sequences of the PMCA-type calcium pumps. The incongruency in mammalian PMCA2-3 clade can perhaps be resolved if the human-PMCA3 is redesignated as a PMCA4 (see text).

patterns of isoform development be discovered in other groups. Much more data are needed to explore the possibility of multigenic pump isoforms, in addition to the multiple transcript-products, in higher invertebrates such as insects and crustaceans, as these organisms display developmental and physiological sophistications that to some degree are comparable to those observed in the vertebrates.

The final observation from this study is that the history of the calcium-transporting ATPases appears to be one with intertwined paralogous and orthologous evolutionary events. Orthology and paralogy are two critical concepts related to homology (Fitch, 1970; Patterson, 1988). The former has been defined as "common ancestry by speciation," and the latter "ancestry by gene duplication" (Miyamoto and Cracraft, 1991). The vertebrate SERCA isoformity appears to be a classic example of paralogy. Orthologous gene evolution, which by definition must be congruent with organismal systematics, can be observed on more than one level. First, the SERCA1 clade immediately reminds one of the phylogeny of those species.

Despite the uncertainties in the ranking order among the rat, the cat and the pig genes, the human and the chick SERCA2 genes appear to be placed in positions that also properly reflect the phylogenetic positions of the species. Orthology is also evidenced by the obvious lack of overlap between genes from very different organismal groups. As noted above, there seems to be no connection of the arthropod, the higher plant, and the fungal genes with any of the vertebrate isoforms, and nonvertebrate isoformity, if any, is thus predicted to have evolved completely independently from that of the vertebrates. Finally, the counter-intuitive designation of rat PMCA2 to the supposed PMCA3 clade suggests that new isoforms can evolve independently within a particular genome during its evolutionary history, although the limited data to date do not allow for an assertive conclusion to be made.

Acknowledgements

We thank Vikas Nanda for assistance with data processing on GCG, and David Swofford for advice with PAUP.

This study has been possible thanks to grants from the National Institutions of Health (NS-23241 to D. Fambrough and PO1 HL27867 to D. Fambrough and colleagues).

References

Brandl, C., N. Green, B. Korczak, and D. MacLennen. 1986. Two Ca^{2+}ATPase genes: homologies and mechanistic implications of deduced amino acid sequences. *Cell.* 44:597–607.

Brandt, P., R. L. Neve, A. Kammesheidt, R. E. Rhoads, and T. C. Vanaman. 1992. Analysis of the tissue-specific distribution of mRNAs encoding the plasma membrane calcium-pumping ATPases and characterization of an alternately spliced form of PMCA4 at the cDNA and genomic levels. *Journal of Biological Chemistry.* 267:4376–4385.

Burk, S., J. Lytton, D. MacLennan, and G. Shull. 1989. cDNA cloning, functional expression, and mRNA tissue distribution of a third organellar Ca^{2+} pump. *Journal of Biological Chemistry* 2364:18561–18568.

Campbell, A. 1983. Intracellular Calcium, Its Universal Role as Regulator. John Wiley & Sons, Inc., New York.

Campbell, M., P. D. Kessler, Y. Sagara, G. Inesi, and D. Fambrough. 1991. Nucleotide sequences of avian cardiac and brain SR/ER Ca^{2+}-ATPases and functional comparisons with fast twitch Ca^{2+}-ATPase. Calcium affinities and inhibitor effects. *Journal of Biological Chemistry.* 266:16050–16055.

Carafoli, E., F. Kessler, R. Falchetto, R. Heim, M. Quadroni, J. Krebs, E. Strehler, and T. Vorherr. 1992. The molecular basis of the modulation of the plasma membrane calcium pump by calmodulin. *Annals of the New York Academy of Sciences.* 671:58–69.

Carroll, R. L. 1988. Vertebrate Paleontology and Evolution. W. H. Freeman, New York. 698 pp.

De Jaegere, S., F. Wuytack, J. A. Eggermont, H. Verboomen, and R. Casteels. 1990. Molecular cloning and sequencing of the plasma-membrane Ca^{2+} pump of pig smooth muscle. *Biochemical Journal.* 271:655–660.

Eggermont, J. A., F. Wuytack, and R. Casteels. 1991. Characterization of the 3′ end of the pig sarcoplasmic/endoplasmic-reticulum Ca^{2+} pump gene 2. *Biochemica et Biophysica Acta.* 1088:448–451.

Escalante, R., and L. Sastre. 1993. Similar alternative splicing events generate two sarcoplasmic or endplasmic reticulum Ca-ATPase isoforms in the crustacean *Artemia franciscana* and in vertebrates. *Journal of Biological Chemistry*. 268:14090–14095.

Evans, D. E., S. Briars, and L. E. Williams. 1991. Active calcium transport by plant cell membranes. *Journal of Experimental Botany*. 42:285–303.

Fambrough, D., and E. Bayne. 1983. Multiple forms of (Na^+ + K^+)-ATPase in the chicken. *Journal of Biological Chemistry*. 258:3926–3935.

Fitch, W. 1970. Distinguishing homologous from analogous proteins. *Systematic Zoology*. 19:99–113.

Ghislain, M., A. Goffeau, D. Halachmi, and Y. J. Eilam. 1990. Calcium homeostasis and transport are affected by disruption of cta3, a novel gene encoding Ca^{2+}-ATPase in *Schizosaccharomyces pombe*. *Biological Chemistry*. 265:18400–18407.

Greeb, J., and G. E. Shull. 1989. Molecular cloning of a third isoform of the calmodulin-sensitive plasma membrane Ca^{2+}-transporting ATPase that is expressed predominantly in brain and skeletal muscle. *Journal of Biological Chemistry*. 264:18569–18576.

Green, M., and D. MacLennan. 1989. ATP drive ion pumps: an evolutionary mosaic. *Biochemical Society Transactions*. 17:819–822.

Gunteski-Hamblin, A. M., J. Greeb, and G. E. Shull. 1988. A novel Ca^{2+} pump expressed in brain, kidney, and stomach is encoded by an alternative transcript of the slow-twitch muscle sarcoplasmic reticulum Ca-ATPase gene. *Journal of Biological Chemistry*. 263:15032–15040.

Inesi, G., T. Cantilina, X. Yu, D. Nikic, Y. Sagara, and M. E. Kirtley. 1992. Long-range intramolecular linked functions in activation and inhibition of SERCA ATPases. *Annals of the New York Academy of Sciences*. 671:32–48.

Inesi, G. T., and M. E. Kirtley. 1992. Structural features of cation transport ATPases. *Journal of Bioenergetics and Biomembranes*. 24:271–283.

Jencks, W. P. 1992. On the mechanism of ATP-driven Ca^{2+} transport by the calcium ATPase of sarcoplasmic reticulum. *Annals of the New York Academy of Sciences*. 671:49–57.

Kaprielian, Z., and D. Fambrough. 1987. Expression of fast and slow isoforms of the Ca^{2+}-ATPase in developing chicken skeletal muscle. *Developmental Biology*. 124:490–503.

Kaprielian, Z., E. Bandman, and D. Fambrough. 1991. Expression of Ca^{2+}-ATPase isoforms in denervated, regenerating, and dystrophic chicken skeletal muscle. *Developmental Biology*. 144:199–211.

Karin, N. J., Z. Kaprielian, and D. Fambrough. 1989. Expression of avian Ca^{2+}-ATPase in cultured mouse myogenic cells. *Molecular Cell Biology*. 9:1978–1986.

Khan, I., and A. K. Grover. 1991. Expression of cyclic-nucleotide-sensitive and -insensitive isoforms of the plasma membrane Ca^{2+} pump in smooth muscle and other tissues. *Biochemical Journal*. 277:345–349.

Läuger, P. 1991. Electrogenic Ion Pumps. Sinauer Associates, Inc., Sunderland, MA. 313 pp.

Lebovitz, R. M., K. Takeyasu, and D. M. Fambrough. 1989. Molecular characterization and expression of the (Na^+ + K^+)-ATPase alpha-subunit in *Drosophila melanogaster*. *EMBO Journal*. 8:193–202.

Lingrel, J. B. 1992. Na, K-ATPase: isoform structure, function, and expression. *Journal of Bioenergetics and Biomembranes*. 24:263–270.

Lytton, J., and D. MacLennan. 1988. Molecular cloning of cDNAs from human kidney coding for two alternative spliced products of the cardiac Ca^{2+}-ATPase gene. *Journal of Biological Chemistry.* 263:15024–15031.

Lytton, J., A. Zarain-Herzberg, M. Periasamy, and D. H. MacLennan. 1989. Molecular cloning of the mammalian smooth muscle sarco(endo) plasmic reticulum Ca^{2+}-ATPase. *Journal of Biological Chemistry.* 264:7059–7065.

MacLennan, D. H., C. J. Brandl, B. Korczak, and N. M. Green. 1985. Amino-acid sequence of a Ca^{2+} + Mg^{2+}-dependent ATPase from rabbit muscle sarcoplasmic reticulum, deduced from its complementary DNA sequence. *Nature.* 316:696–700.

MacLennan, D. H., T. Toyofuku, and J. Lytton. 1992. Structure-function relationships in sarcoplasmic or endoplasmic reticulum type Ca^{2+} pumps. *Annals of the New York Academy of Sciences.* 671:1–10.

Maddison, W. P. 1989. Reconstruction character evolution on polytomous cladograms. *Cladistics.* 5:365–377.

Maddison, W. P., and D. R. Maddison. 1992. MacClade: Analysis of Phylogeny and Character Evolution (3.0). Sinauer Associates, Inc., Sunderland, MA. 398 pp.

Magyar, A., and A. Váradi. 1990. Molecular cloning and chromosomal localization of a sarco/endoplasmic reticulum-type Ca^{2+}-ATPase of *Drosophila melanogaster. Biochemical and Biophysical Research Communications.* 173:872–877.

Meade, J. C., K. M. Hudson, S. L. Stringer, and J. R. Stringer. 1989. A tandem pair of *Leishmania donovani* cation transporting ATPase genes encode isoforms that are differentially expressed. *Molecular Biochemical Parasitolology.* 33:81–91.

Miyamoto, M. M., and J. Crcraft, editors. 1991. *Phylogenetic Analysis of DNA Sequences.* Oxford University Press, NY. 358 pp.

Palmero, I., and L. J. Sastre. 1989. Complementary DNA cloning of a gene highly homologous to mammalian sarcoplasmic reticulum Ca-ATPase from the crustacean *Artemia. Molecular Biology.* 210:737–748.

Patterson, C. 1988. Homology in classical and molecular biology. *Molecular Biology and Evolution.* 5:603–625.

Pedersen, P., and E. Carafoli 1987*a.* Ion motive ATPases. I. Ubiquity, properties, and significance to cell function. *Trends in Biochemical Sciences.* 12:146–150.

Pedersen, P., and E. Carafoli 1987*b.* Ion motive ATPases. II. Energy coupling and work output. *Trends in Biochemical Sciences.* 12:186–189.

Rudolph, H. K., A. Antebi, G. R. Fink, C. M. Buckley, T. E. Dorman, J. Levitre, L. S. Davidow, J. I. Mao, and D. T. Moir. 1989. The yeast secretory pathway is perturbed by mutations in PMR1, a member of a Ca^{2+} ATPase family. *Cell.* 58:133–145.

Sachs, G., and K. Munson. 1991. Mammalian phosphorylating ion-motive ATPases. *Current Opinions in Cell Biology.* 3:685–694.

Shull, G. E., and J. J. Greeb. 1988. Molecular cloning of two isoforms of the plasma membrane Ca^{2+}-transporting ATPase from rat brain. *Biological Chemistry.* 263:8646–8657.

Shull, G. E., J. Greeb, and J. B. Lingrel. 1986. Molecular cloning of the three distinct forms of the Na^+, K^+-ATPase alpha-subunit from rat brain. *Biochemistry.* 25:8125–8132.

Shull, G. E., D. M. Clarke, and A. M. Gunteski-Hamblin. 1992. cDNA cloning of possible mammalian homologs of the yeast secretory pathway Ca^{2+}-transporting ATPase. *Annals New York Academy of Sciences.* 671:70–80.

Strehler, E. E., P. James, R. Fischer, R. Heim, T. Vorherr, A. G. Filoteo, J. T. Penniston, and E. Carafoli. 1990. Peptide sequence analysis and molecular cloning reveal two calcium pump isoforms in the human erythrocyte membrane. *Journal of Biological Chemistry.* 265:2835–2842.

Swofford, D. 1993. PAUP: Phylogenetic Analysis Using Parsimony, 3.1. Illinois Natural Hisory Survey, Champaign, IL. 263 pp.

Takeyasu, K., V. Lemas, and D. Fambrough. 1990. Stability of Na^+-K^+-ATPase alpha-subunit isoforms in evolution. *American Journal of Physiology.* 259:c619–c630.

Valentine, J. W., editor. 1985. Phanerozoic Profiles in Macroevolution. Princeton University Press, Princeton, NJ. 441 pp.

Våradi, A., M. Gilmore-Heber, and E. Benz, Jr. 1989. Amplification of the phosphorylation site-ATP-binding site cDNA fragment of the Na^+, K^+-ATPase and the Ca^{2+}-ATPase of *Drosophila melanogaster* by polymerase chain reaction. *FEBS Letters.* 258:203–207.

Verma, A. K., A. G. Filoteo, D. R. Stanford, E. D. Wieben, J. T. Penniston, E. E. Strehler, R. Fischer, R. Heim, G. Vogel, S. Mathews, M. A. Strehler-Page, P. James, T. Vorherr, J. Krebs, and E. Carafoli. 1988. Complete primary structure of a human plasma membrane Ca^{2+} pump. *Journal of Biological Chemistry.* 263:14152–14159.

Wimmers, L., N. Ewing, and A. Bennett. 1992. Higher plant Ca^{2+}-ATPase; primary structure and regulation of mRNA abundance by salt. *Proceedings of the National Academy of Sciences, USA.* 89:9205–9209.

Wu, K.-D., and J. Lytton. 1993. Molecular cloning and quantification of sarcoplasmic reticulum Ca^{2+}-ATPase isoforms in rat muscles. *American Journal of Physiolology.* 264:333–341.

Wuytack, F., and L. Raeymaekers. 1992. The Ca^{2+}-transport ATPases from the plasma membrane. *Journal of Bioenergetics and Biomembranes.* 24:285–300.

Wuytack, F., L. Raeymaekers, H. De Smedt, J. A. Eggermont, L. Missiaen, L. Van Den Bosch, S. De Jaegere, H. Verboomen, L. Plessers, and R. Casteels. 1992. Ca^{2+}-transport ATPases and their regulation in muscle and brain. *Annals of the New York Academy of Sciences.* 671:82–91.

Appendix

Computer Programs for Phylogenetic Analysis

MacClade 3: Exploring character evolution using phylogeny. Wayne Maddison (Ecology and Evolutionary Biology, University of Arizona, Tucson, AR) and David Maddison (Entomology, University of Arizona, Tucson, AR).

Just as evolutionary change follows the branches of the phylogenetic tree, so should our explanations of similarities and differences among organisms. We need to think vertically, in terms of the tree and evolutionary changes along its branches. MacClade 3 was designed to help our mind's eye see the phylogenetic tree and explore hypotheses of character evolution. It consists of a computer program and a book, which contains an overview of phylogenetic theory and of the need for a phylogenetic perspective in biology, as well as instructions for use of the program. The program includes many components, such as a spreadsheet data editor, a tree window, charting windows, and so on, but perhaps its central function is to reconstruct and graphically display the history of character evolution on the phylogenetic tree. As the user explores alternative phylogenetic trees by moving branches and other manipulations, MacClade gives feedback via tree graphics, statistics, and charts. Characters can be discrete- or continuous-valued, and various assumptions are available in determining the reconstruction of ancestral states. Charts and statistics summarize the reconstructions and can be used to test hypotheses such as those of correlated character evolution. The program shares a common file format with the phylogenetic programs PAUP and COMPONENT, allowing complementary use of the programs. MacClade functions on Apple Macintosh computers. It is available from Sinauer Associates, 23 Plumtree Road, Sunderland, MA 01375-0407, (413) 549-4300.

PAUP: A computer program for phylogenetic inference using maximum parsimony. David Swofford (Laboratory of Molecular Systematics MRC 534, National Museum of Natural History, Smithsonian Institution, Washington, DC 20560).

One of the impediments to the use of phylogenetic analysis by biologists who are not primarily systematists has been the complexity of available computer software for estimating evolutionary trees. The PAUP (Phylogenetic Analysis Using Parsimony) program addresses this difficulty by combining powerful exact and approximate tree-searching algorithms with a friendly user interface and clear, easily interpretable output. Input can be in the form of traditional qualitative characters (e.g., morphological attributes, restriction site presence/absence) or macromolecular sequences. Although the program is primarily intended for phylogenetic estimation under the maximum parsimony criterion, a wide range of assumptions can be accommodated (e.g., ordered versus unordered character states; free reversibility versus irreversibility; character weighting, including differential weighting for first, second, and third positions; transformation weighting where the user specifies the "cost" of a change from each state to each other state). PAUP also provides a variety of tools for comparing, manipulating, and drawing trees (e.g., filtering of trees according to topological and goodness-of-fit criteria; comparing sets of trees using operations on files of saved trees).

PAUP shares a common file format (NEXUS) with other programs including MacClade and Component, and can import and export data files in several other formats (e.g., PHYLIP, GCG/MSF). A graphical user interface is available for the Apple Macintosh. Other supported platforms (as of July, 1994) include DOS, Unix, and VMS microcomputers, workstations, and mainframes. The non-Macintosh platforms use the "portable" (command-line) interface. For information on how to obtain the program, contact the author at the address above or send e-mail to swofford@onyx.si.edu.

List of Contributors

Meredithe L. Applebury, Visual Sciences Center, The University of Chicago, Chicago, Illinois 60637

Maria Jesús Arrizubieta, Department of Food Service and Technology, University of California at Davis, Davis, California 95616

Francisco J. Ayala, Department of Ecology and Evolutionary Biology, University of California at Irvine, Irvine, California 92717

Everett Bandman, Department of Food Science Technology, University of California at Davis, Davis, California 95616

Michael V. L. Bennett, Albert Einstein College of Medicine, Bronx, New York 10461

Paul Blount, Laboratory of Molecular Biology and Department of Genetics, University of Wisconsin, Madison, Wisconsin 53706

W. Ford Doolittle, Department of Biochemistry, Canadian Institute for Advanced Research, Dalhousie University, Halifax, Nova Scotia, B3H 4H7, Canada

Stewart R. Durell, Laboratory of Mathematical Biology, National Cancer Institute, National Institutes of Health, Bethesda, Maryland 20891

Douglas M. Fambrough, Department of Biology, The Johns Hopkins University, Baltimore, Maryland 21218

Walter M. Fitch, Department of Ecology and Evolutionary Biology, University of California at Irvine, Irvine, California 92717

Christine Fyrberg, Department of Biology, The Johns Hopkins University, Baltimore, Maryland 21218

Eric Fyrberg, Department of Biology, The Johns Hopkins University, Baltimore, Maryland 21218

Holly V. Goodson, Department of Biochemistry, Beckman Center for Molecular and Genetic Medicine, Stanford University Medical School, Stanford, California 94305

H. Robert Guy, Laboratory of Mathematical Biology, National Cancer Institute, National Institutes of Health, Bethesda, Maryland 20891

David M. Helfman, Department of Molecular Genetics, Cold Spring Harbor Laboratory, Cold Spring Harbor, New York 11724

Lauri Herman, Department of Food Service and Technology, University of California at Davis, Davis, California 95616

David M. Hillis, Department of Zoology, The University of Texas at Austin, Austin, Texas 78712

John P. Huelsenbeck, Department of Zoology, The University of Texas at Austin, Austin, Texas 78712

Timothy Jegla, Department of Anatomy and Neurobiology, Washington University School of Medicine, St. Louis, Missouri 63110

Patrick J. Keeling, Department of Biochemistry, Canadian Institute for Advanced Research, Dalhousie University, Halifax, Nova Scotia, B3H 4H7, Canada

Maura Kenton, Department of Biology, The Johns Hopkins University, Baltimore, Maryland 21218

Ching Kung, Laboratory of Molecular Biology and Department of Genetics, University of Wisconsin, Madison, Wisconsin 53706

Leslie Leinwand, Department of Microbiology and Immunology, Albert Einstein College of Medicine, Bronx, New York 10461

Boris Martinac, Department of Pharmacology, The University of Western Australia, Nedlands, Western Australia 6009

D. C. Ghislaine Mayer, Department of Microbiology and Immunology, Albert Einstein College of Medicine, Bronx, New York 10461

Lisa McNally, Department of Biology, The Johns Hopkins University, Baltimore, Maryland 21218

Laurie A. Moore, Department of Food Science and Technology, University of California at Davis, Davis, California 95616

Peter Parham, Departments of Cell Biology and Microbiology and Immunology, Stanford University, Stanford, California 94305-5400

David J. Patterson, School of Biological Sciences, The University of Sydney, Sydney, New South Wales 2006, Australia

Thomas D. Pollard, Departments of Cell Biology and Anatomy, The Johns Hopkins Medical School, Baltimore, Maryland 21205

Robin R. Preston, Department of Physiology, Medical College of Pennsylvania, Philadelphia, Pennsylvania 19129

Stephen Quirk, Departments of Biophysics and Biophysical Chemistry, The Johns Hopkins Medical School, Baltimore, Maryland 21205

Andrew J. Roger, Department of Biochemistry, Canadian Institute for Advanced Research, Dalhousie University, Halifax, Nova Scotia, B3H 4H7, Canada

Liza Ryans, Department of Biology, The Johns Hopkins University, Baltimore, Maryland 21218

Yoshiro Saimi, Laboratory of Molecular Biology and Department of Genetics, University of Wisconsin, Madison, Wisconsin 53706

Lawrence Salkoff, Departments of Anatomy and Neurobiology, and Genetics, Washington University School of Medicine, St. Louis, Missouri 63110

Michael J. Sanderson, Department of Biology, University of Nevada at Reno, Nevada 89557

John C. Schimenti, The Jackson Laboratory, Bar Harbor, Maine 04609

Mitchell L. Sogin, Marine Biological Laboratory, Woods Hole, Massachusetts 02543

Yan Song, Department of Biology, The Johns Hopkins University, Baltimore, Maryland 21218

Allan Spradling, Department of Embryology, Howard Hughes Medical Institute Research Laboratories, Carnegie Institution of Washington, Baltimore, Maryland 21210

Sergei Sukharev, Laboratory of Molecular Biology and Department of Genetics, University of Wisconsin, Madison, Wisconsin 53706

William E. Tidyman, Department of Food Service and Technology, University of California at Davis, Davis, California 95616

Patricia O. Wainright, Institute of Marine and Coastal Sciences, Cook Campus, Rutgers University, New Brunswick, New Jersey 08903-0231

Alison Weiss, Department of Microbiology and Immunology, Albert Einstein College of Medicine, Bronx, New York 10461

Macdonald Wick, Department of Food Service and Technology, University of California at Davis, Davis, California 95616

Thomas M. Wilke, Pharmacology Department, Southwestern Medical Center, Dallas, Texas 75235-9041

Shozo Yokoyama, Department of Biology, Syracuse University, Syracuse, New York 13244

Xin Zheng, Marine Biological Laboratory, Woods Hole, Massachusetts 02543

Xin-Liang Zhou, Laboratory of Molecular Biology and Department of Genetics, University of Wisconsin, Madison, Wisconsin 53706

Subject Index

Acanthamoeba
myosin types, 149–150
profilin structure, 118, 123
RNA analysis, 42
Actin
binding
tropomyosin system, 108–110
See also Profilins.
protein superfamily, 173–178
Adenosine triphosphatases, calcium-transporting (Ca-ATPases), 271–283
cladogram analysis, 276–280
evolutionary processes, 278–280
isoformity, 277–278
PMCA and SERCA, 276
pump-encoding sequences, 273–275
phylogenetic analysis, 274–275
sequences and alignment, 273–274
Alcohol dehydrogenase (ADH)
exon-shuffling, 28
Algae
multicellularity, 43
Algorithms
clustering, 57
phylogenetic, 13, 17
ALIGN program, 120, 121
Alignment
Ca-ATPases, 273–274
G protein alpha subunit, 263–264
Alleles
of class I MHC genes, 93, 96–99
Alveolates
RNA analysis, 40–41
Amerindians
polymorphism of HLA genes, 97–101
Amino acid(s)
of G proteins. *See* G protein.
replacements, and molecular clocks, 3–4
sequence analysis
Ca-ATPases, 273–274
G protein alpha subunit genes, 252–257, 263–264
G-protein-coupled receptors, 236–238
myosin superfamily, 141–157
See also Sequence analysis.
and voltage-gated ion channels, 200–201
Amoeba
profilin structure, 118–120
Animals
actin protein superfamily, 173–178
G protein alpha subunit genes, 257–261
monophyletic origin, 39–53
biochemical analyses, 40–43
choanoflagellate and sponge similarities, 43
extracellular matrix, 44–46
fossil history, 40
and fungi, 46–47
historical theories, 39–40
metazoa, 43
multicellularity, 43–44
Annelids
voltage-gated potassium channels, 214
Anopheles gambiae
chromosome inversions, 79–80
Antigens. *See* Major histocompatibility complex genes.
Arabidopsis
G protein alpha subunit genes, 258–261
ATPases
calcium-transporting. *See* Adenosine triphosphatases.
muscle fiber types, 159–161

Bacillus subtilis
ion channels, 188
Bacteria
spliceosomal origin, 33
Bacteriophage T7 phylogenetic analysis, 62–64
Basement membrane
development of, 44–46
Bias
in maximum likelihood, 14
BN21, HLA-B lineage, 98–99
Bootstrap methods
G protein alpha subunit genes, 253–255
G-protein-coupled receptors, 235–248
myosin superfamily, 143
monophyletic origin of animals, 41–42
Brain
myosin isoforms, 164
Brush border myosin proteins, 150

Caenorhabditis elegans
actin protein superfamily, 174
G protein alpha subunit genes, 258–261
voltage-gated potassium channels, 214
Calcium
ATPases. *See* Adenosine triphosphatases, calcium-transporting.
ion channels. *See* Ion channels.
Calmodulin
in tropomyosin system, 107

Cardiac genes
myosin-based motility, 161, 164
Cell junctions, 45–46
Centractin, 174
Charcot-Marie-Tooth disease, 229
Chicken
myosin types, 149
See also Myosin, heavy chain multigene family, sarcomeric.
Chitin
in fungi and animals, 46
Chloride
ion channels. *See* Ion channels.
Choanoflagellates
extracellular matrix, 44
See also Animals, monophyletic origin.
Chromosomes
G protein alpha subunit genes, 261–263
myosin-based motility, 161
myosin heavy chain genes, 131
Y, heterochromatin insertion, 73–74
Cladograms, 15
Ca-ATPases, 276–280
Classification
myosin superfamily, 141
CLUSTAL V program, 120, 252–253
Cnidarians
voltage-gated potassium channels, 213–222
See also Animals, monophyletic origin.
Codons
G protein alpha subunit genes, 258–261
myosin heavy chain gene, 129–130
variability, and molecular clocks, 5–10
Cold viruses
human immune response, 95
Collagens
in extracellular matrix, 44
RNA splicing, 106
Combined immunodeficiency disease, 95
Computer programs
phylogenetic analysis, 55–56
profilin sequence analysis, 120–121
See also specific programs.
Connexins, 223–233
phylogenetic trees, 229–231
sequence comparisons, 224–229
CONSENSE program, 121, 240
Contractile apparatus
tropomyosin system, 107–112
Covarions, 9–10
Crystallography
actin protein superfamily, 175–176
HLA proteins, 95–96
profilin structure, 118
Ctenophora
See Animals, monophyletic origin.
Cyclase associated protein
profilin effects, 120
Cysteine
in connexins, 226
Cytoskeleton
muscle and nonmuscle proteins, 163–164
tropomyosin system, 111
See also Profilins.

Definitions
molecular clock, 3
Dendrograms, 15
Desmosomes, 45–46
Diabetes
human immune response, 95
Dictyostelium
G protein alpha subunit genes, 258–261
ion channels, 183
myosin types, 149
profilin, 123–124
Diploblasts, potassium channels, 217–219
Distance matrix methods
G-protein-coupled receptors, 235–248
myosin superfamily, 143
profilin sequences, 121, 124
DNA
gene conversion, 85–89
repetitive sequences, in heterochromatin, 75–77
evolution of, 78–79
transposons, 33–35
Dolichos biflorus
spliceosomal introns, 34
Dp1187 minichromosome, 74, 78
DRAWTREE program, 239
Drosophila
actin protein superfamily, 174, 175
codon variability, 6–7
G protein alpha subunit genes, 257–261
myosin heavy chain gene, 129
P elements, and heterochromatin evolution. *See* Heterochromatin.
profilin, 124–125
RNA splicing, 106–108
voltage-gated potassium channels, 214–219
Dyneins, tropomyosin system, 111
Dystrophin gene
RNA splicing, 106

Elongation factors, 46
Endoplasmic reticulum
Ca-ATPase (SERCA). *See* Adenosine triphosphatases, calcium-transporting.
Entamoeba hystolytica
intron positions, 29
ENTREZ database, 236
Error rates
maximum likelihood, 14

Escherichia coli
ion channels, 185–188
Exon-shuffling. *See* Introns.
Exons
connexins, 223
myosin heavy chain gene, 129–130
See also Proteins, isoform diversity by alternative RNA splicing.
Extracellular matrix
development of, 44–46

Felsenstein zone, 61
Ferritin
in fungi and animals, 46
Fibroblasts
tropomyosin system, 107–112
Fibronectin
in extracellular matrix, 44
FITCH program, 121
Fossils, Precambrian and Cambrian, 40
Fungi
G protein alpha subunit genes, 257–261
gene conversion, 85–86
ion channels, 184–185
See also Animals, monophyletic origin.

G protein
alpha subunit multigene family, 249–270
phylogenetic tree, 252–264
alignment availability, 263–264
branch lengths, 256
comparative gene structure, 257–261
evolutionary rate, 257
mammalian Gna genes, 263
tandem gene duplication, 261–263
receptors, opsin subfamily, 235–248
Galton-Watson model of gene duplication, 18
Gap junctions, 45–46
See also Connexins
Genes
Ca-ATPases. *See* Adenosine triphosphatases, calcium-transporting.
conversion, and gene families in mammals, 85–91
duplication
G protein alpha subunit genes, 261–263
in mammals, 85
maximum likelihood, 18–21
eukaryotic
and prokaryotic genes, 30
spliceosomal introns, 28–29
myosin. *See* Motility, myosin-based; Myosin, heavy chain multigene family.
Giardia lamblia
intron positions, 29
GIRK1, 216
Globin
gene evolution, maximum likelihood, 21–22
Glyceraldehyde-3-phosphate dehydrogenase
exon-shuffling, 28

Heterochromatin, and transposable elements, 69–83
DNA, tandemly repetitive sequences, 75–77, 78–79
DNA blots, 71
genetic strains, 71
hybridization, in situ, 71
P element repression, 74–77
selection or side effects, 79–80
subtelomeric, 78–79
Human immunodeficiency virus, 95
Human leukocyte antigen (HLA)
See Major histocompatibility complex genes.
Hydroxyproline
in fungi and animals, 46

Immune response, MHC genes, 94–95
Immunosuppressive drugs
and polymorphism of HLA genes, 97
Influenza virus, 95
Insects
profilin, 125
Introns, 27–37
connexins, 223
G protein alpha subunit genes, 258–261
insertional models, 30–35
phylogenetic distribution of Group II and spliceosomal introns, 31–33
prokaryotic origin, 33
spliceosome evolution from retroposing Group II introns, 30–31
transposition of Group II introns, 33–35
introns-early theory, 28–30
exon-shuffling, 28
positions, 28
protein structure, 28
spliceosomal
phylogenetic distribution in eukaryotic genes, 28–29
in prokaryotic and eukaryotic genes, 30
myosin heavy chain gene, 129–130
types, 27
See also Proteins, isoform diversity by alternative RNA splicing.
Invertebrates
RNA splicing, 107–108
Ion channels
of microbes, 179–195
Bacillus subtilis, 188
Dictyostelium, 183
Escherichia coli, 185–188
paramecium, 179–183
Saccharomyces cerevisiae, 183–184
Schizosaccharomyces pombe, 184–185
yeast, 183–184

Ion channels *(continued)*
voltage-gated proteins, 197–212
models of transmembrane topology and functional mechanisms, 197–202
three-dimensional molecular models, 202–209
IRK1, 216

Japanese, HLA-B lineage, 98–99
Jellyfish, potassium channels, 217–219

Kimura model of nucleotide substitutions, 61
Kinesins, tropomyosin system, 111

Lactate dehydrogenase
exon-shuffling, 28
lacZ gene conversion, 86–89
Laminin
in extracellular matrix, 44
Lipoprotein
and ion channels, 187
Liver
myofibroblasts, 164–165
MYH expression in Ito cells, 165–167

Magnesium
ion channels. *See* Ion channels.
Major histocompatibility complex (MHC) genes, functional polymorphism, 93–103
disease specfic selection, 100
diversification, 97–98
evolution of HLA-B lineage, 98–99
frequency dependent selection, 101
human class I HLA proteins, 95–96
gene polymorphism, 96–97
gene structure, 96
immune response, 94–95
immunological functions of antigens, 94
other polymorphisms, 99
overdominant selection, 100–101
selection role, 99–100
Malaria, and HLA alleles, 100
Mammals
G protein alpha subunit genes, 263
gene conversion, 85–91
myosin heavy chain genes, 134–136
Maximum likelihood
in phylogenetic analysis, 13–26
algorithms, 13
application problems, 14–17
error rates, 14
gene duplication, 18–21
globin gene evolution, 21–22
hybrid approach, 17–18
in phylogenetic analysis, 64–66
See also Animals, monophyletic origin.
McCAW program, 239
Meiosis
and gene conversion, 85–89
Membranes
ion channels. *See* Ion channels.
plasma membrane Ca-ATPase (PMCA). *See* Adenosine triphosphatases.
Metazoa
voltage-gated potassium channels, 213–222
See also Animals, monophyletic origin.
Microfilaments
tropomyosin system, 108–111
Mitochondria
origin theory, 31
Molecular clocks, 3–12
accuracy, 3
amino acid replacements, 3–4
codon variability, 5–10
definition, 3
maximum likelihood, 16–18
myosin superfamily, 143–144, 149, 152
nucleotide substitutions, 4–5
superoxide dismutase (SOD) as example, 5, 8–11
Motility, myosin-based, 159–171
coexpression of muscle and nonmuscle cytoskeletal proteins, 163–164
evolutionary comparisons, 161–163
human skeletal gene family, 161
muscle fiber types, 159–161
MYH gene expression and isoform distribution, 161
myofibroblasts in liver, 164–165
sarcomeric MYH expression in liver Ito cells, 165–167
Multicellularity, evolution of, 43–44
Mus
voltage-gated potassium channels, 214–217
Muscle contraction
tropomyosin system, 107–112
See also Myosin, heavy chain multigene family.
Mycoplasma capricolum
spliceosomal origin, 33
Myofibroblasts
liver, 164–165
MYH expression in Ito cells, 165–167
Myosin
heavy chain multigene family, sarcomeric, 129–139
conserved and diverging domains in myosin rod, 133–134
encoding regulation, 129–131
evolutionary relationship of chicken and mammalian, 134–136
gene conversion, 132–133
mammalian and avian organization, 131–132
molecular evolution, phylogenetic techniques, 141–157
classes in each organism, 148–149

molecular clock, 152–153
myosin I, 149–151
myosin II, 151–152
motility, multiple genes and functions, 159–171
coexpression and muscle and nonmuscle cytoskeletal proteins, 163–164
evolutionary comparisons, 161–163
gene expression and isoform distribution patterns, 161
human skeletal gene family, 161
muscle fiber types, 159–161
myofibroblasts in liver, 164–165
sarcomeric expression in ITO cells, 165–167
tropomyosin system, 111

Neighbor-joining methods
G protein alpha subunit genes, 253–255
G-protein coupled receptors, 239–240
in numerical simulations, 58, 61
Nuclear ribonuclear particles, small, 106
Nucleotides
G protein alpha subunit genes, 252–257
metazoan RNA, 41
substitutions
class I MHC genes, 93
Kimura model, 61
and molecular clocks, 4–5
myosin heavy chain genes, 133

Opsin
G-protein-coupled receptors, 235–248
Orthology, 279–280

P elements
and heterochromatin evolution. *See* Heterochromatin.
three-dimensional models, 204–208
and voltage-gated channels, 199–201
Paralogy, 279–280
Paramecium
ion channels, 179–183
voltage-gated potassium channels, 214, 217, 219
Parsimony analysis
Ca-ATPases, 271–283
and maximum likelihood, 18
monophyletic origin of animals, 43
myosin superfamily, 143
profilin sequences, 121, 123–124
PAUP computer program, 43, 274–275
Peptidylglycan
and ion channels, 187
Phenylalanine
in connexins, 226
Photoreceptors
opsin, G-protein-coupled, 235–248
PHYLIP program, 121, 143, 239, 243
Phylogenetic analysis
accuracy of methods, 55–67
criteria, 57–58
computational speed, 57
consistency, 57
discrimination, 57–58
power, 57
robustness, 57
versatility, 58
experimental phylogenies, 62–64
limits of numerical simulations, 58–62
numerical vs. biological simulations, 55–57
rating of major methods, 64–66
Ca-ATPases, 271–283
connexins, 223–233
G protein alpha subunit genes, 252–264
alignment availability, 263–264
branch lengths, 256
comparative gene structure, 257–261
evolutionary rate, 257
mammalian Gna genes, 263
tandem gene duplication, 261–263
intron distribution, 31–33
maximum likelihood, 13–26
algorithms, 13
application problems, 14–17
error rates, 14
gene duplication, 18–21
globin gene evolution, 21–22
hybrid approach, 17–18
myosin superfamily, 141–157
potassium channel evolution, 217
Physarum profilin, 123–124
PILEUP program, 274
Piman Indians, HLA-B lineage, 98–99
Plants
actin protein superfamily, 173
G protein alpha subunit genes, 257–261
profilin, 125
voltage-gated potassium channels, 214, 216, 220
PMCA (plasma membrane Ca-ATPase). *See* Adenosine triphosphatases.
Podospora
group II introns, 31
Polymorphism, in MHC genes, 93–103
disease specific selection, 100
diversification, 97–98
evolution of HLA-B lineage, 98–99
frequency dependent selection, 101
human class I HLA proteins, 95–96
gene polymorphism, 96–97
gene structure, 96
immune response, 94–95
immunological functions of antigens, 94
other polymorphisms, 99
overdominant selection, 100–101
selection role, 99–100

Polyphosphoinositides
binding to profilins, 119
Potassium
channels
molecular evolution, 213–222
in cnidarians and ciliate protozoans, 217
in diploblasts and triploblasts, 217–219
in high metazoa, 214–215
ion channel gene family, 215–216
other channel genes, 216–217
in paramecium, 219
signaling in metazoans, 220
See also Ion channels.
PREALIGN program, 120
PRODIST program, 239, 240, 243
Profilins, 117–128
sequence analysis, 120–121
and evolutionary relationships, 121–126
structure and function, 117–120
Prokaryotes
spliceosomes, 33
Proline
binding to profilins, 119
PROTDIST program, 121
Proteins
actin related, 173–178
isoform diversity by alternative RNA splicing, 105–115
evolutionary concerns, 112
overview, 105–106
tropomyosin system, 107–112
muscle and nonmuscle cytoskeletal, 163–164
structure, and intron positions, 28
voltage-gated ion channels, 197–212
models of transmembrane topology, 197–202
three-dimensional molecular models, 202–209
Proterospongia haeckeli, 39
Protozoans, ciliate
voltage-gated potassium channels, 213–222
PROTPARS program, 121, 143–146
Pseudogenes
HLA proteins, 96–97
Pyruvate kinase
exon-shuffling, 28

Receptors
G-protein-coupled, 235–248
transmembrane domain. *See* G protein.
Recombination
polymorphism of HLA genes, 97–98
somatic, in mammals, 87–88
Replacement substitutions, 5
Reverse transcription
spliceosomal introns, 31, 34
Ribonuclease
codon variability, 7–8
RNA
alternative splicing. *See* Proteins, isoform diversity
by alternative RNA splicing.
ribosomal sequences, metazoan, 40–43
ROMK1, 216

Saccharomyces
actin protein superfamily, 174
G protein alpha subunit genes, 258–261
group II introns, 31
ion channels, 183
Sarcomeres
myosin. *See* Myosin, heavy chain multigene family, sarcomeric.
Sarcoplasm
Ca-ATPase (SERCA). *See* Adenosine triphosphatases, calcium-transporting.
Saxitoxin binding sites, 199
Schizosaccharomyces pombe
actin protein superfamily, 173
ion channels, 184–185
Sea urchin profilin, 122
Selection, and HLA polymorphism, 99–100
disease specific, 100
frequency dependent, 101
overdominant, 100–101
SEQBOOT program, 121
Sequence analysis
connexins, 224–229
profilins, 120–121
evolutionary relationships, 121–126
voltage-gated ion channel proteins, 197–212
SERCA (sarcoplasmic and endoplasmic reticulum Ca-ATPase). *See* Adenosine triphosphatases, calcium-transporting.
Sex determination
in *Drosophila*, 106
Signal transduction, G protein-mediated. *See* G protein.
Skeletal muscle
fiber types, 159–161
Slime mold
G protein alpha subunit genes, 257–261
ion channels, 183
multicellularity, 43
Sodium
ion channels. *See* Ion channels.
Species differences
codon variability, 9
connexins, 231
HLA proteins, 96
myosin heavy chain genes, 162
Spliceosomes. *See* Introns.
Sponges
extracellular matrix, 44
See also Animals, monophyletic origin.

Stramenophiles
  RNA analysis, 40
Superoxide dismutase (SOD)
  as molecular clock, 3, 5, 8–11

Tetrahymena profilin, 121, 124
Tetrodotoxin binding sites, 199
Three-dimensional molecular models, 202–209
TPYR program, 121
Transplantation
  polymorphism of HLA genes, 97
Transposable elements
  and heterochromatin evolution. *See* Heterochromatin.
Transposons. *See* Introns.
Trichomonas vaginalis
  intron positions, 29
Trichoplax adhaerans, 39
Triose isomerase
  exon-shuffling, 28, 29
Triploblasts, potassium channels, 217–219
Tropomyosin
  See Proteins, isoform diversity by alternative RNA splicing.

Vaccinia
  profilin, 124–125
Voltage-gated ion channels. *See* Ion challens, voltage-gated proteins.

Wound healing
  muscle and nonmuscle cytoskeletal proteins, 164

Y chromosomes
  heterochromatin insertion, 73–74
Yeast
  actin protein superfamily, 174
  ion channels, 183
  myosin types, 149
  profilin effects, 120
Yule model of gene duplication, 18–21